Operations Research
Evolving Frontiers and Diverse Applications

Editors

Vilda Purutçuoğlu
Professor in Statistics, Department of Statistics
Middle East Technical University, Ankara University, Turkey
&
Chair of Biomedical Engineering Programme
Department of Biomedical Engineering
Middle East Technical University, Ankara, Turkey

Gerhard Wilhelm Weber
Professor of Applied Mathematics, Faculty of Engineering
Management & Chair of Marketing and Economic Engineering
Poznan University of Technology, Poland

Hajar Farnoudkia
Assistant Professor, Business Department
Başkent University, Ankara, Turkey

CRC Press
Taylor & Francis Group
Boca Raton London New York

CRC Press is an imprint of the
Taylor & Francis Group, an **informa** business
A SCIENCE PUBLISHERS BOOK

First edition published 2025
by CRC Press
2385 NW Executive Center Drive, Suite 320, Boca Raton FL 33431

and by CRC Press
4 Park Square, Milton Park, Abingdon, Oxon, OX14 4RN

CRC Press is an imprint of Taylor & Francis Group, LLC

Library of Congress Cataloging-in-Publication Data (applied for)

ISBN: 978-1-032-84304-9 (hbk)
ISBN: 978-1-032-88935-1 (pbk)
ISBN: 978-1-003-54043-4 (ebk)

DOI: 10.1201/9781003540434

Typeset in Times New Roman
by Prime Publishing Services

Preface

This original book on Operational Research-New Paradigms and Emerging Applications, Series 2 surveys and details the newest technology in analytics and intelligence computing, artificial intelligence and operational research which are reducing the dimensions of data coverage as well as variables worldwide. This compendium discusses the code of intelligent optimization which can be applied in various branches of optimization, social studies, data mining and clustering, economic/finance and medical applications. Involving modern and emerging techniques of OR and AI, and applying them together with strong and evolutionary algorithms to real-life problems for strategic, but also daily applications, this compendium elaborates on all areas of OR results, methods and applications. Through the rich diversity of this handbook, the state-of-the-art developments in quickly advancing key technologies are covered. In this way, with our reference work, we hope to be useful for students and emerging scientists of engineering and science, management and economics, social science and the art, researchers and scholars who are employed in OR-supported industries, for decision-makers and designers of tomorrow's World.

We editors hope that the chosen subjects and picked areas reflect a core sample of international OR research facing emerging, challenging, complex and even long-enduring problems of our environment and their field in supply chain management, production planning and inventory management, various types of signal procession from satellite images to medical instruments, economics and finance and social sciences, through the results and tools of OR-Analytics. We thank our respected authors for their diligent work and readiness to share their newest findings, insights and results with over international community. We hope that the research of our authors collected and edited by us will be an inspiration for cooperation and joint implementation, improvement and friendship at the global stage and premium level.

We convey special thanks to our families. Vilda Purutçuoğlu would like to thank deeply Prof. Dr. Gerhard Wilhelm Weber for his great support throughout her academic career, and Dr. Hajar Farnoudkia for her hardworking and willingness in this book. Finally, her utmost sincerest thanks go to her parents, Yurdagül, Ahmet, Eda Purutçuoğlu and her sweet daughter Vera for their love and support. Gerhard Wilhem Weber would like to cordially thank Prof. Dr. Vilda Purutçuoğlu for her vision, care and inspiration, and Dr. Hajar Farnoudkia for her dedicated hard-work. Hajar Farnoudkia would like to thank both Prof. Dr. Gerhard Wilhelm Weber, Prof. Dr. Vilda Purutçuoğlu and Assoc. Prof. Dr. Babek Erdebill who encouraged and helped her to be one of the academic world members.

Contents

Part II: Operation Research in Production Planning and Inventory Management

Part III: Operation Research in Signal Processing

Part IV: Operation Research in Social Sciences

Introduction

This book aims to provide Operation Research (OR) application in four different fields those are supply chain management, production planning and inventory management, signal processing and social sciences. The first part, as mentioned, is allocated to OR in supply chain management and composed of three chapters. The first chapter presents a bibliographic analysis for the vulnerable and resilient supply chain management by initially defining keywords and then, constructing network model of those selected keywords in order to investigate the main research topic in supply chain management. Then, in the second chapter, the logical network model is generated to represent the interconnection of gates and buildings. The third chapter in this part represents the time series modeling of consumer confidence index whose decision period is detected by the hidden Markov model.

The second part of the book is titled as the production planning and inventory management in three chapters. The first chapter of this part is based on a comprehensive review about data-driven approaches for decision-making. The second chapter of this portion is dedicated to introduce a new game model for aggregate production planning. The model is introduced with the estimation of model parameters and shown via a real life example. The last third chapter of this part discusses another game model under balancing cooperation and competition with a specific application in business.

The third portion of this book is based on different computational methods used in signal processing. The signals can come from various sources like climatologic studies, genomic experiments or medical instruments. The first chapter considers the signal analyses from satellite images to estimate the snow-cover area over complex mountain regions since the correct detection of these areas has the challenges of erroneous classification of shadows and clouds as snow, particularly over complex mountainous terrains. The second chapter discusses the signal processing in protein-protein interaction networks whose data includes directional information. The chapter proposes a circular graphical model whose model parameters are estimated via optimization methods. The third chapter of this part presents a specific problem in signal processing. It is based on the imputation of missing data from electromyography (EMG) instrument whose data is one of the common data types in many biomedical and biostatistical analyses. Here, the missing imputation is done via different approaches and their performances are compared via accuracy measures. The last chapter of the third part is about the application of copula and alternative regression models to describe Hodgkin-Huxley signal processing data which is studied in neuroscience research to investigate the generation and propagation of action potentials in neurons.

The last part of this book represents OR in social sciences in three chapters. The first chapter aims to find out how and why students make their decisions, in different applied cases that relate to machine learning experiences. The chapter proposes an optimization process for master students under different countries, universities and study programs and under this diversity it discusses the decision-making process, the representation and reflections. The second chapter suggests a model for multi-response optimization that can be typically faced with in social studies. The application of the model is shown via real dataset. Finally, the third chapter of the last part presents the comparison of various modeling approaches of a big survey, called the TAYA (Research on Family Structure in Turkey survey) that is repeated every 5-year, consisting of a detailed questionnaire to understand the complex structure of Turkish households and the individuals that are living in these households. The chapter specifically focuses on the factors that can affect the decision to be a foster family for Turkish people.

In conclusion, we think that the book can give some ideas about the variety of application of OR techniques in different problems. We already aimed to introduce some of these techniques in the first series of this book. This current series presents different aspects of OR techniques in distinct fields as the continue of the first series. We hope that the references used in each part can be useful for the new researchers while learning deeply the theoretical aspects of the selected methods. Hence, we wish the book can open new revenues for novel researches.

Contributors

Aghalar, Hamed
Isfahan University of Technology, Isfahan, Iran

Akdoğan, Kutay
Iskenderun Technical University, Iskenderun, Turkey

Aksu, Cansu
Middle East Technical University, Ankara, Turkey

Akyurek, Zuhal
Middle East Technical University, Ankara, Turkey

Alparslan Gök, Sırma Zeynep
Suleyman Demirel University, Isparta, Turkey

Atar, Gizem
Middle East Technical University, Ankara, Turkey

Can, Yavuz
T-Online, Germany and Haliç University, İstanbul, Turkey

Defterli, Özlem
Çankaya University, Ankara, Turkey

Doğan Dar, Elif
Heidelberg University, Heidelberg, Germany

Dominik Śledziński, Jacek
Poznan University of Technology, Poznan, Poland

Farnoudkia, Hajar
Baskent University, Ankara, Turkey

Gencağa, Orhan Deniz
Antalya Bilim University, Antalya, Turkey

Goli, Alireza
University of Isfahan, Isfahan, Iran

Gütmen, Selma
Poznan University of Technology, Poznan, Turkey

Hamidoğlu, Ali
Bahçeşehir University, Istanbul, Turkey

Jebreili, Sahel
Urmia University, Urmia, Iran

Kalayci, Betül
Middle East Technical University, Ankara, Turkey

Kuter, Semih
Cankiri Karatekin University, Cankiri, Turkey

Özcan, İsmail
Süleyman Demirel University, Isparta, Turkey

Özceylan, Eren
Gaziantep University, Gaziantep, Turkey

Ozturk, Basak Akteke
Middle East Technical University, Ankara, Turkey

Parikh, Jinal
Ahmedabad University, Ahmedabad, India

Purutçuoğlu, Vilda
Middle East Technical University, Ankara, Turkey

Rahmani, Sourena
Isfahan University of Technology, Isfahan, Iran

Reisach, Ulrike
Neu-Ulm University of Applied Sciences, Neu-Ulm, Germany

Sarasir, Fatemeh
Middle East Technical University, Ankara, Turkey

Şengül, Ayan
Antalya Bilim University, Antalya, Turkey

Uğur, Ömür
Middle East Technical University, Ankara, Turkey

Wilhelm Weber, Gerhard
Poznan University of Technology, Poznan, Poland

I

OPERATION RESEARCH IN SUPPLY CHAIN MANAGEMENT

Chapter 1

A Bibliometric and Descriptive Review of Mathematical Models for Vulnerable and Resilient Supply Chain Management

Kutay Akdoğan,[1,*] *Eren Özceylan*[2] and *Gerhard Wilhelm Weber*[3]

1.1 Introduction

Supply chain (SC) managers aim to ensure that all members of the chain work effectively and efficiently. During these processes, all members of the chain must work in harmony with each other. The success of a single point affects the suc-

[1]Iskenderun Technical University.

[2]Gaziantep University.

[3]Poznan University of Technology.

* Corresponding author: kutay.akdogan@iste.edu.tr

cess of the entire system. However, all members of the supply chain system face both internal and external risks [11]. Although globalization provides advantages, such as increasing the target customer base and accessing cost-effective raw materials and services, it brings risks and uncertainty on a global scale. Causes such as pandemics, economic crises, political conflicts, natural disasters, wars, and climate change can inevitably cause disruptions in supply chains as well as sector-related causes. However, while dealing with these disruptions requires a good analysis of vulnerabilities and risks, it is not always easy to accurately predict all risks and vulnerabilities in advance. In the event of risks and disruptions, various negative effects are encountered, some of which are very severe and irreversible. In order to overcome this handicap, researchers and managers have been working to create a resilient supply chain (RSC) in recent years. Resilience can be defined as returning to an original or better state after being disturbed [11]. Also, a resilient organization must be able to lessen the negative impacts caused by disruptions [29]. These definitions only outline recovery and adjustment elements of resilience. However, since supply chains are multi-actor structures that spread widely and are exposed to global effects, the concept of resilience should be looked at in a more complex way. Ponomarov and Holcomb (2009) [36] define supply chain resilience as "the adaptive capability of the SC to prepare for unexpected events, respond to disruptions, and recover from them by maintaining continuity of operations at the desired level of connectedness and control over structure and function." This definition brings more elements to consideration such as preparedness, responding, maintaining function and control. Preparedness requires proactive strategies before disruptive events occur and responding is achieved with reactive strategies. Safety stock-keeping, resilient supplier selection, multi-sourcing, effective design, risk management, and collaboration among chain members can be given as examples of strategies that can be used. Although the issue of resilience is important, it appears as a relatively new topic in the literature and there is no definition or criterion of resilience agreed upon by the academic and business world. Various definitions have been made by various researchers. However, the resiliency of a supply chain is often studied as a qualitative approach. Therefore, this review aims to evaluate the mathematical model-based studies in the related literature.

1.2 Methodology

The review method applied in this study is inspired by the methodologies proposed by Denyer and Tranfield (2018) [12] and Barbosa-Póvoa (2009) [7]. The review process consists of six steps. These steps are defining research questions, collecting relevant articles, defining inclusion and exclusion criteria and screening accordingly, descriptive, and bibliometric analysis, paper evaluation and presentation. The details of each step are given in the following subsections and the results of the analysis are given in the next section.

1.2.1 Defining Research Questions

In this study, the current state of the mathematical model-based resilient supply chain management literature is analyzed. In order to draw a good roadmap for this study and to facilitate the follow-up of important points, the associated questions of interest are determined as follows:

1. What is the publication year of mathematical model-based RSC studies?
2. Who is working on RSC and are there significant collaborations?
3. Which keywords are used the most by researchers?
4. Which country contributed the most to the mathematical model-based RSC literature?
5. Which documents are cited the most?
6. What kind of mathematical models are developed for RSC studies, linear or nonlinear models, integer or mixed integer models?
7. Which mathematical models are more commonly used, single objective models or multi-objective models and, what is the most evaluated objective?
8. How is the uncertainty handled in papers?
9. What changes in multi-period models are evaluated?
10. Are heuristics - metaheuristics developed for solution processes?

1.2.2 Collecting Relevant Papers

This part consists of two stages, namely, choosing the database where the documents will be collected and determining the query terms that will enable access to the relevant documents. Scopus is chosen as the database. The Scopus database is also an important resource in terms of providing datasets in a suitable format for bibliometric analysis. Query terms are searched for in the title, abstract and keywords sections and no year limitation is applied. Query terms are applied with the help of built-in Boolean operators and 95 papers are obtained. Query terms can be shown as below; (TITLE-ABS-KEY ("mathematical model") AND TITLE-ABS-KEY ("supply chain management" OR "supply chain") AND TITLE-ABS-KEY ("vulnerable" OR "resilient" OR "vulnerability" OR "resilience")).

1.2.3 Defining Inclusion and Exclusion Criteria

95 studies which were obtained from the previous step are evaluated using the following inclusion and exclusion criteria. As a result of this evaluation, 48 studies were found suitable for this study. Inclusion criteria follows below:

- Papers must be written in English.
- Resilience or vulnerability must be the target of the study or resilience criteria must be taken into consideration.
- A mathematical modeling approach must be used in the study.
- Papers written on individual operations that are considered within the scope of resilient supply chain management are included.

Accordingly, the exclusion criteria are presented below.

- Review papers and other papers that do not present a new mathematical model approach are excluded.
- Qualitative papers are excluded.
- Studies based on using surveys and statistical software are excluded.
- Papers written on individual operations that are not considered within the scope of resilient supply chain management are excluded.

1.3 Descriptive and Bibliometric Analysis

In this section, besides giving initial data statistics, it is aimed to answer the review questions from one to five. A total of 48 papers were reviewed in relation to the relevant topic. It is seen that 39 articles and 9 conference papers were published in 17 years. To answer Question 1 in Section 1.2, the distribution of the paper types and their years are presented in Figure 1.1 and there has been a remarkable increase in recent years. Since the year 2023 has not been completed, this increase is expected to continue. In addition, it is quite remarkable that this increasing trend in the number of publications occurred after the COVID-19 epidemic.

The second question in the same section is answered with the help of the information obtained from the Scopus database and the VOSviewer software co-authorship analysis [47]. According to analysis, out of 130 authors, only Dimitri

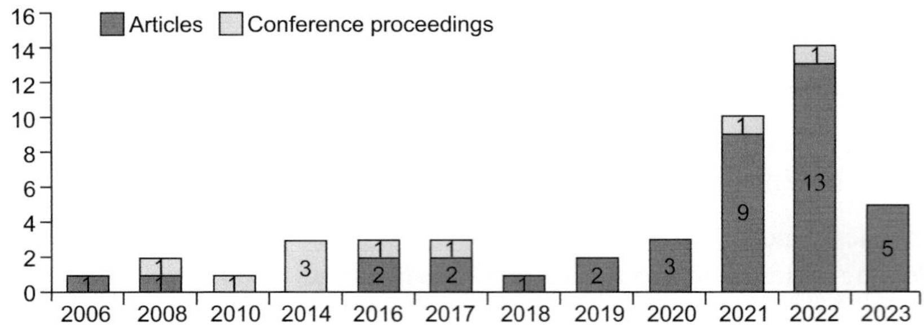

Figure 1.1: Distribution of the papers.

Table 1.1: Authors with more than one paper and their total link strength.

Author	Documents	Total Link Strength	Author	Documents	Total Link Strength
Ivanov D.	3	8	Nayeri S.	2	5
Gholami-Zanjani S.M.	2	6	Sazvar Z.	2	5
Jabalameli M.S.	2	6	Shen G.	2	5
Klibi W.	2	6	Aldrighetti R.	2	4
Najafi E.	2	6	Battini D.	2	4
Pishvaee M.S.	2	6	Dutta P.	2	4
Ekanayake E.M.A.C.	2	5	Shrivastava H.	2	4
Foroozesh N.	2	5	Zhao L.	2	3
Karimi B.	2	5	Roghanian E.	2	2
Kumaraswamy M.	2	5	Vali-Siar M.M.	2	2
Mousavi S.M.	2	5			

Ivanov stands out from the rest with three papers [2, 35, 39]. 109 authors contributed with only one paper and 20 authors published two papers each. This situation can be explained by the fact that it is a relatively new field of study, and the number of studies is not very high yet. Table 1.1 presents the authors with three and two papers and their total link strength. Total link strengths are calculated by VOS viewer software and indicate how many times an author a collaborated with other authors. Figure 1.2 shows the co-authorship network of the author with the highest total link strengths. However, no other remarkable collaboration network is found during co-authorship analysis and that can be tied to authors not producing many articles on the review subject yet.

A keyword bibliometric analysis is conducted on 48 papers to understand which keywords were chosen by the authors to reflect their RSC research. This analysis also gives an answer to Question 3 in Section 1.2. The top ten most frequent keywords and their total link strength are given in Table 1.2. The network

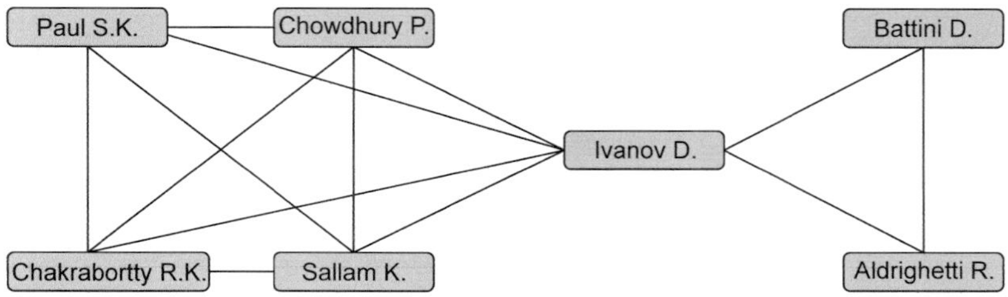

Figure 1.2: Co-authorship.

Table 1.2: Keyword occurrences and total link strength.

Keyword	Occurrences	Total Link Strength
Resilience	9	47
Supply chain resilience	6	29
Sustainability	6	27
Resiliency	6	25
Stochastic programming	5	26
Supply chain network design	5	22
Resilient supply chain	5	20
Covid-19	4	23
Supply chain management	4	21
Optimization	3	18

representation of most related keywords is also given in Figure 1.3. Keyword total link strength presents how many times a keyword occurred together with another keyword. As it can be seen from Table 1.2 and Figure 1.3, the most preferred author keywords reflect RSC clearly as four of the keywords directly mention resiliency, and three of them mention supply chain. The keyword occurrence network gives more insight into the nature of the reviewed papers as it supplies a summary of methods, industries, decisions, and such.

To answer Question 4 a country co-authorship analysis is performed. The threshold value is set to display the countries producing more than two documents. This analysis also shows the co-authorship relationship between countries. As a result of this analysis, it is seen that 11 countries out of 23 produced two or more documents. Unlike author analysis, country analysis presents a clear dominance of document production. Table 1.3 points out that Iran is the most productive country among 23 countries that participated in RSC research. Iran contributed literature with 21 papers and China and the United States followed with 9 papers each. Hence, it can be said that mathematical model-based RSC is a popular research topic in Iran. The most active countries, citation counts, and total link strength are given in Table 1.3.

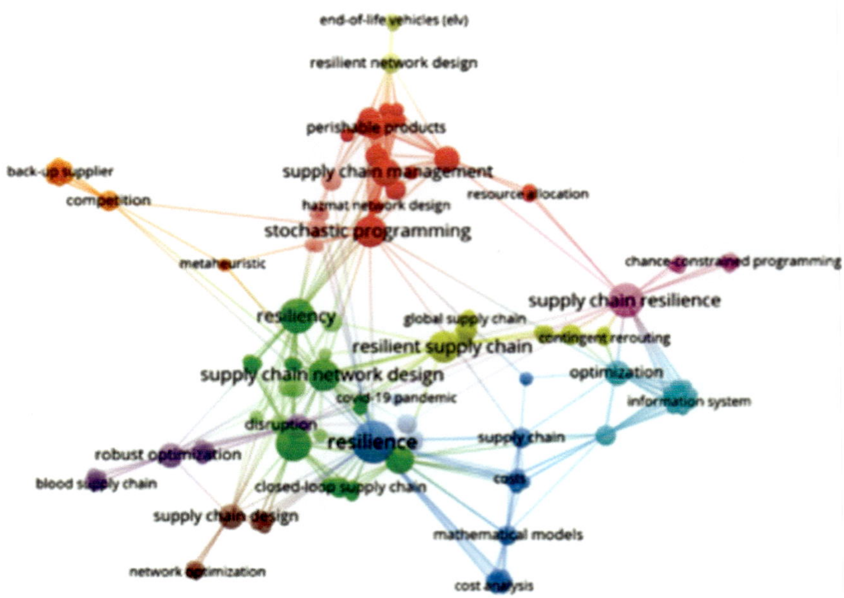

Figure 1.3: Keyword occurrence network.

Table 1.3: Most active countries.

Country	Documents	Citations	Total Link Strength
Iran	21	500	11
China	9	67	6
United States	9	299	4
Australia	5	219	6
Germany	4	46	4
India	4	27	3
Canada	3	47	5
France	2	82	3
Hong Kong	2	13	1
Mexico	2	23	2
Italy	2	6	1

Figure 1.4 shows the co-authorship network of countries that produced more than one paper. When Table 1.3 and Figure 1.4 are analyzed together, it can be seen that Iran made collaborations with six other countries 11 times which makes them also the most collaborative country. Furthermore, Question 5 intends to indicate the most influential articles on the review topic. The top ten documents with the highest citation counts are given in Table 1.4. The article "Marrying Supply Chain Sustainability and Resilience: A Match Made in Heaven" by Fahimnia

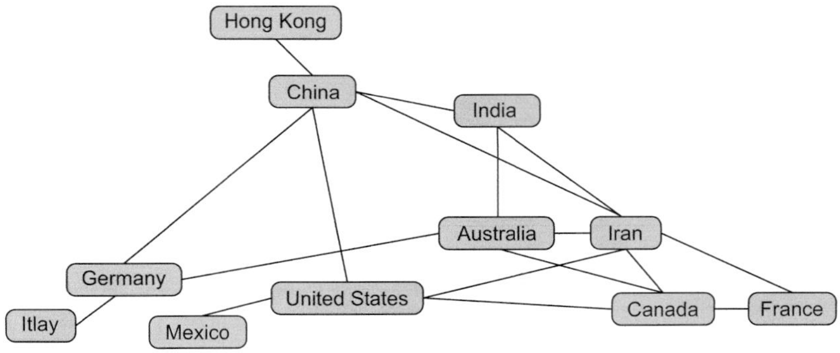

Figure 1.4: Co-authorship network of countries.

and Jabbarzadeh (2016) [16] has the highest citation count. When Table 1.4 is examined, it is seen that although the oldest document that is the subject of this study is in the top 10 list, it is not in the top three. In addition, it is seen that four documents were published after 2020, which is thought to be related to the increase in the number of studies in the field after 2020.

1.4 Paper Evaluation and Presentation

The paper evaluation and presentation process aims to answer research questions six to ten. Accordingly, six categories are determined to answer these questions and can be listed as follows: Linearity and integerness of the models, objective functions of the models, uncertainty handling, multi-period models, other mathematical models and heuristics-metaheuristics. These categories are chosen to reflect the general characteristics of the studies in the literature. However, it should be noted that these categories may overlap, and a document given in one category may show characteristics of other categories as well. To avoid repetition, the document mentioned in detail in one category may not be mentioned again in the other.

1.4.1 Linearity and Integerness of Models

Although it is not always easy to reflect real-life problems linearly, it is seen that most researchers propose linear models. Researchers in one of the earliest studies [1] developed a linear model for the optimization of the petroleum supply chain and solved the proposed formulations using GAMS. Sahebjamnia (2020) [41] integrated fuzzy DEMATEL and fuzzy ANP methods with mixed integer linear programming for resilient supplier selection and order allocation purposes. Preston et al. [37] developed a mixed integer linear program to design

Table 1.4: Papers with highest citation counts.

Documents	Year	<2019	2019	2020	2021	2022	2023	Total
[16] Marrying supply chain sustainability and resilience: A match made in heaven	2016	32	28	36	29	45	25	195
[30] Supply chain resilience for single and multiple sourcing in the presence of disruption risks	2018	7	14	26	53	51	26	177
[1] Supply chain optimization of petroleum organization under uncertainty in market demands and prices	2008	73	11	13	9	15	2	123
[6] Vulnerable options in supply chains: Effects of supplier competition	2006	62	3	5	3	2	2	77
[51] Development of a mathematical model for a sustainable closed-loop supply chain with efficiency and resilience systematic framework	2019			2	17	23	10	52
[42] A capacity planning approach for sustainable-resilient supply chain network design under uncertainty: A case study of vaccine supply chain	2021				4	29	17	50
[21] The design of resilient food supply chain networks prone to epidemic disruptions	2021				11	20	12	43
[20] A robust location-inventory model for food supply chains operating under disruptions with ripple effects	2021				3	20	16	39
[46] Sustainable, resilient, and responsive mixed supply chain network design under hybrid uncertainty considering COVID-19 pandemic disruption	2022					12	16	28
[22] Study on supply chain Disruption risk management strategies and mode	2008	17	2	2	1	1	2	25

a resilient hub and spoke network and addressed the problem as a network flow problem. Several other studies [21, 4, 17, 25] also proposed mixed integer linear programming models that solved their problems using GAMS and CPLEX. However, Larimi et al. (2022) [25] and Gholami-Zanjani et al. [21] attracted attention as one of them integrated GIS into their model (2022) [25] and others used Bender's decomposition approach to lessen the complexity of mixed integer model (2021) [21]. It is seen that some of the researchers linearised their proposed model by implementing some techniques. For example, Namdar et al. [30] first, introduced a scenario-based mixed integer programming model with a nonlinear constraint and then linearised that constraint by introducing continuous nonnegative variables. Zare Mehrjerdi and Lotfi (2019) [51] also proposed a mixed integer nonlinear model with objective functions that contain absolute value and maximum functions. They linearised the model by removing absolute value functions and defining covariates. Gholami-Zanhani et al. (2021) [20] also introduced a mixed integer linear programming model by linearising a nonlinear equation by presenting a new variable. Foroozesh et al. (2023) [18] can also be named as other researchers who proposed an integer linear programming model by linearising an equation by defining a new variable. Wang et al. (2016) [48] introduced a nonlinear mathematical model for supplier selection and order allocation considering conditional redirection between suppliers and MATLAB® is utilized for computations. Resilience under disruption was the only objective that the researchers aimed for. Shrivasta et al. (2017) [43] proposed a quadratic mixed integer model to locate and allocate facilities. Another nonlinear model was created for emergency decision-making of a resilient supply chain [27]. Dutta and Shrivasta (2020) [13] formulated a nonlinear mixed integer mathematical model and reduced it to a continuous and piece-wise linear model. Using an approximation approach based on partial relaxation was another method to tackle the NP-hardness of the mixed integer nonlinear model [49]. Aldrighetti et al. (2021) [39] handle the complexity of the mixed integer nonlinear model by using a two-stage solution approach and creating a subproblem. When the studies are examined in general, it is seen that integer programming is generally not preferred and mixed integer programming is preferred. Mixed integer programming seems more realistic because of the wide variety of raw materials and materials that are the subject of supply chains. In addition, it is seen that linear modeling is chosen in general, and in some cases where this is not entirely possible, various techniques are applied for linearization.

1.4.2 Objective Functions of the Models

There is not enough difference between the number of single-objective and multi-objective models to significantly outweigh the other. Single objective models are often focused on either minimizing cost [1, 2, 13, 23, 25, 30, 39, 43, 50, 53], or maximizing profit [6, 20, 21, 35, 45]. As a different approach, Bora et al.

(2014) [8] proposed a mathematical model for constructing a resilient supply chain considering readiness, responsiveness, and recovery. They considered several disruptions that affected the nodes of the supply chain and calculated the monetary effect on nodes. The objective of the model was established as a min-max function and aimed to find a node with the highest monetary effect and minimize it. Zheng et al. (2014) [53] addressed supply disruptions and assumed recovering supplier cooperated with a fellow supplier during the recovery process and introduced a model to minimize the total cost of an interrupted supplier. Fresh food supply chains are vulnerable to contamination and investigation of safety stock and demand under disruption can help decision makers to maximize the total amount of produce transported considering contamination [28]. To ensure a resilient supply chain proactive and reactive strategies must be assessed. Every operation of a supply chain affects the other and procurement as an initial activity has a special importance for a supply chain. Therefore, researchers [23] proposed a multi-period, multistage optimization framework for a resilient humanitarian supply chain. First, the resilience of the suppliers was evaluated using DEMATEL and fuzzy TOPSIS methods and then, a proactive and a reactive mathematical model with the objective of minimizing total procurement cost was introduced. As a proactive strategy, order allocations to the least resilient suppliers were avoided, and the reactive strategy considered disruptions and unmet demand penalties. The efficiency of the proposed model was evaluated using five different case scenarios. The objectives of the documents are given in Table 1.5. When multi-objective models are examined, it is seen that environmental and social impacts are also included in the scope of the studies within the framework of the sustainability approach. Environmental aspects such as emissions, pollution, and resource consumption [4, 16, 17, 18, 31, 32, 34, 38, 40, 42, 46, 51] and social dimension such as employment rates, social awareness, responsibility [4, 16, 18, 32, 34, 38, 40, 42, 46, 51] are addressed as well as the economic dimension such as cost or profit. It can be said that all dimensions of the sustainability approach, social, environmental, and economic, are generally attempted to be evaluated simultaneously in the papers dealing with sustainability in the supply chain within the resilience concept. Some papers [3, 27, 37, 42, 44, 46] examined resilience directly with their objective functions. The objectives of the documents are presented in Table 1.5. There are different methods for dealing with the complexities of multi-objective models such as goal programming or turning models into single-objective ones. Methods used in the studies for handling multi-objective models are given in Table 1.6.

1.4.3 Uncertainty Handling

Uncertainty handling is performed in the majority of the studies by robust programming, stochastic programming, fuzzy programming, or a combination of these approaches (mixed uncertainty). Babich (2006) [6] considered market and

Table 1.5: Objectives of the mathematical models.

Objectives	Documents
Minimizing cost	[39], [2], [30], [1], [25], [43], [13], [53], [50], [23], [38], [16], [51], [49], [41], [37], [42], [44], [4], [40], [32], [24], [17], [46], [27], [18], [31], [19]
Maximizing profit	[35], [6], [21], [20], [45], [22], [3], [34], [10]
Maximizing resilience	[48], [37], [42], [3], [27], [41]
Maximizing total produce	[28]
Minimizing maximum expected cost	[8]
Minimizing loses	[22]
Maximizing environmental scores	[16], [40]
Maximizing social scores	[16], [40]
Minimizing emissions	[38], [51], [42], [4], [17]
Minimizing energy	[51]
Maximizing the employment	[38], [51], [4]
Minimizing risk	[49]
Maximizing crop	[38]
Minimizing loss of resilience	[44]
Minimizing phosphorus loss	[38]
Minimizing time to recover	[24]
Maximizing facility utilization	[17]
Minimizing delivery time	[10]
Minimizing lost business days	[10]
Minimizing environmental impact	[32], [46], [34], [18], [31]
Maximizing social impact	[42], [32], [46], [18]
Minimizing spread of COVID	[34]
Minimizing dissatisfaction	[19]
Minimizing route deviation	[19]

Table 1.6: Methods for handling multi-objectivity.

Method	Document	Method	Document
Fuzzy goal programming	[16]	Augmented lexicographic weighted Tchebycheff (ALWT)	[32]
Multi-choice goal programming	[42]	e -constraint method	[40], [3]
Importance weights	[22], [19]	The augmented e-constraint method	[44]
The LP-metric method	[51], [34], [31]	AUGMECON2	[17], [34], [18]
The weighted average method	[38]	SAUGMECON	[49]
Linear weighting method	[27]	Zimmermann's max-min approach	[4]

supply uncertainty in a stochastic game setting (Nash game and Stackelberg game) between suppliers and manufacturers. This study is the only one that tackles the uncertainty with game theory. Two-stage stochastic programming is a common approach in stochastic programming. An oil supply chain management requires decision-making at various levels such as crude oil production, refining, and petrochemical production. A two-stage stochastic program with fixed recourse (scenario analysis) can be used for crude oil production decisions in the first stage and refinery and petrochemical production decisions in the second stage under demand and price uncertainties [1]. This approach also can be used for supplier selection [2, 30, 45], order allocation [30], location decisions [2, 13, 21, 37, 45, 51], and flow decisions [2, 13, 21, 37, 45, 51]. Some techniques for scenario analysis such as the Monte Carlo procedure to develop scenarios [21], and the scenario reduction method to select scenarios [21, 37] are also used.

Stochastic optimization refers to uncertainty with a known probability distribution and uniform distribution can be used as a way to explore the result of uncertainties such as demand uncertainty [43]. Two-stage stochastic programming implies that uncertainties are faced after the first decision stage, but in real life, it is possible to go through more decision stages and encounter uncertainties. This situation can be modeled with multi-stage stochastic programming [49].

Robust programming approaches are used when probability distributions cannot be forecasted. These methods can be used for various optimization targets for example supplier selection and order allocation under supply uncertainty [44], design [3] and redesign [34] of a supply chain. Resilient supply chain design requires resilient supplier selection, and this can be solved with fuzzy AHP and fuzzy MOORA methods [3].

Fuzzy programming is another approach to address the uncertainty. Supply and process uncertainties [41], cost, emission, and employment uncertainties [4], and uncertain aspects in data [14, 15, 33] can be optimized with a fuzzy programming approach. A combination of uncertainty approaches is also quite common. Stochastic programming for supply uncertainties and fuzzy goal programming for importance weights of objective functions is one of the mentioned combinations [16]. Stochastic modeling and chance-constrained programming can be given as an example of another [35]. Other hybrid approaches can be listed as robust-stochastic programming [20, 25, 38, 46], robust-fuzzy-stochastic programming [32, 40], and robust possibilistic programming [17, 18, 31, 42].

It is seen that stochastic programming is common as well as mixed programming approaches. It is also observed that supply and demand uncertainties are the parameters that are mostly evaluated. The uncertainty parameter and the articles evaluating them can be viewed in Table 1.7.

1.4.4 Multi-Period Models

When all studies are examined, it is seen that most of the studies do not model multi-periods. Multi-period modeling is closer to real life in terms of evaluating changing conditions. Most multi-period models evaluate common changes

Table 1.7: Uncertainty parameters.

Parameter	Document	Parameter	Document
Demand	[22], [1], [43], [51], [13], [49], [37], [42], [21], [3], [20], [38], [25], [17], [34], [10], [35], [18]	Process	[43], [30], [13], [41], [42]
Supply	[6], [16], [30], [13], [41], [44], [40], [38], [35], [2]	Risk	[49]
Price	[6], [1]	Perishability	[49]
Cost	[4], [17], [46], [18]	System	[9]
Emission	[4], [17]	Time	[3]
Employment	[4]	Budget	[3]
Capacity	[20], [45], [35], [18], [2], [31]	Inventory	[20]
Data	[14], [33], [15]	Disruption	[25]

over periods, such as demand, cost, price, and capacity [1, 3, 21, 23, 25, 35, 45, 46, 49, 51, 53]. Disruptions [2, 38, 39, 41] lead time [42], deterioration [20], acceptable distance to response [17], technology, pollution, lost workdays [10], social rating, environmental impact and emission [18] can be shown as the notable variability parameters. Multi-period modelling can be considered as one of the shortcomings of the literature and further variations should be considered in future studies.

1.4.5 Other Mathematical Models

Babich (2006) [6] is the first researcher who evaluates vulnerability in a supply chain from the supplier perspective. Unlike many others, instead of demand, price was assumed to be random, and suppliers' competition was investigated in a stochastic game setting between supplier and manufacturer. The deferment option was defined as vulnerability. Nash game for no deferment scenario and Stackelberg game for deferment scenario were solved for the manufacturer to decide on order quantities under uncertain prices.

Supply chain resilience can be defined as learning abilities and recovery abilities [26]. A supply chain that consists of two members was modeled and researched for its time-varying deformation. Each member was viewed as a cell and the system was analyzed based on the cell resilience model considering resilience abilities. Maximal coverage location problems can enable a resilient demand covering in a supply chain. Researchers gave a theoretical concept for a resilient supply chain design. The maximal coverage location problem with a fuzzy nature was reduced to two Knapsack-like problems using a combination of the Lagrange relaxation technique and a combinatorial algorithm [5]. The Markov chain method stands out because it is different from the methods in other studies. Thanks to this method, the relationship between the supplier and the customer can be studied stochastically, and supply chain resilience can be increased with optimum fund allocation [9]. Multi-criteria mathematical models are another mathematical modeling approach rarely encountered in the resilient supply chain literature. Supply chain vulnerabilities [14] and capabilities [15] can be assessed by using a fuzzy synthetic evaluation method for supply chain resilience. These mathematical models aim to identify vulnerabilities and their criticality level, capabilities, and importance based on expert opinion by applying a fuzzy approach. Supplier selection is one of the primary decisions in supply chain design. A different mathematical modeling approach used for supplier selection can be named as Z-number DEA model [33]. Optimizing the service operations supply chain is not very common among the reviewed studies. A mathematical model based on a multi-channel queuing system can help with decisions and information like optimal team size, and average maintenance time [50].

1.4.6 Heuristics – Metaheuristics

When all the 48 review papers were analyzed, it was seen that the majority of the papers didn't utilize heuristic or metaheuristic methods. It was also observed that the majority of the proposed meta-heuristics belong to the studies published in the last two years and are based on genetic algorithms. In addition, one of the studies [33] draws attention to the use of artificial neural networks to measure the effectiveness of the proposed study. Supply disruption/uncertainty is one of the widely examined aspects of a resilient supply chain as mentioned in the previous subsection. Choosing backup suppliers or fellow suppliers as outlined in [13] may give a chance to recover the interrupted supplier while continuing to meet customer demand. Developed combinatorial optimization problems can ensure minimization the total cost of interrupted suppliers and enhance the chains' ability, for large-scale problems genetic algorithm (GA) with hybrid coding run with MATLAB as proposed in the study [53] can be utilized. Another paper [38] that utilized a genetic algorithm coded in MATLAB also dealt with supply uncertainty along with other uncertainties in a fertilizer supply chain to design a resilient network. Also, Non-dominated Sorting Genetic Algorithm (NSGA-II) [10, 24, 34] and subpopulation genetic algorithm (SPGA-II) [24], Multi-Objective Evolutionary Algorithm (MOEA) [10] are other evolution-inspired and genetic algorithm-based meta-heuristics that researchers employed for resilient design. In addition, a unique solution approach based on enhanced multi-operator differential evolution (EDEcon) is presented [35]. Swarm-inspired algorithms are other meta-heuristic methods applied in research. Particle swarm optimization (PSO) for a responsive and flexible supply chain configuration [31] and a hybrid algorithm of GA and PSO for a competitive network design [45] are other solution directions scientists are pursuing. Also, for the purpose of redesigning a supply chain under disruption caused by COVID sanctions, multi-objective particle swarm optimization (MOPSO) [34] and NSGA-II are implemented. Forecasting emergency probabilities Predicting emergency probabilities provides added value for emergency logistics distribution, while measures of resilience can make the system more effective. In this context, ant colony which is another swarm-inspired algorithm stands out as a method that can be used to reduce customer dissatisfaction, route deviation, and cost [19]. Using the Multi-Echelon Technique for Recoverable Item Control (METRIC) approach and METRIC-based heuristic is another striking method to measure resilience and related financial effect [52]. Other utilized heuristic approaches can be listed as a heuristic based on Benders decomposition and local search [2] and constructive heuristic with the Lagrangian relaxation method [46].

1.5 Conclusion

Even though mathematical models have been applied to supply chain problems widely, utilizing these methods for resiliency of the supply chain is still limited This paper evaluates mathematical models that have been used in vulnerable and resilient supply chain concepts and presents a bibliometric and descriptive review of the related literature. In our view, the limitations of mathematical model-based resilient supply chain studies stand out primarily due to the relatively low number of articles. The increasing publication trend indicates that more studies will be made to close this gap. It is also seen that these studies are more popular in some countries, but since supply chains are now a global phenomenon, it is thought that more countries should give importance to working on this issue. As was said in the previous section, most of the models are shaped as linear and mixed integers. Some of the nonlinear models were linearized with various techniques. However, it is seen that solution methods are not recommended for large-scale problems and remain a research gap throughout the years. However, it should be noted that the majority of the proposed heuristics and metaheuristics have been developed in the last two years. However, more solution methods need to be developed, such as heuristics–metaheuristics, exact solution methods, or hybrid methods. It is seen that the models with multi-objectives do not show superiority in numbers and the studies are generally focused on cost and profit and there are not many studies that target resiliency and recovery as an objective. It is thought that in order to be sure of the resiliency of a supply chain, it should be tended in this direction as an objective function. In multi-objective models, it is seen that environmental and social parameters, which are the other pillars of sustainability along with economics, are also included. While continuing to research these parameters, it is recommended to include global operations and lean and agile logistics approaches. Other suggestions include gathering more data for conducting case studies, further investigating different kinds of disruptions such as ripple effects, contemplating multiple disruptions simultaneously, and including all the possible supply chain members in the model. Of course, this review is not without limitations. Using only the Scopus database is considered a limitation for this study, also keywords for the search are chosen so broadly. More inclusive search words that reflect resiliency might be selected for deep evaluation.

References

[1] Al-Othman, W.B.E., Lababidi, H.M.S., Alatiqi, I.M. and Al-Shayji, K. 2008. Supply chain optimization of petroleum organization under uncertainty in market demands and prices. *European Journal of Operational Research*, 189(3): 822–840.

[2] Aldrighetti, R., Battini, D. and Dmitry Ivanov, D. 2023. Efficient resilience portfolio design in the supply chain with consideration of preparedness and recovery investments. *Omega*, 117: 102841.

[3] Arabsheybani, A. and Khasmeh, A.A. 2021. Robust and resilient supply chain network design considering risks in food industry: Flavour industry in Iran. *International Journal of Management Science and Engineering Management*, 16(3): 197–208.

[4] Ayvaz, B., Kusakci, A.O., Aydin, N. and Ertas, E. 2021. Designing reverse logistics network for end-of-life vehicles: A sustainability perspective in a fragile supply chain. *International Journal of Industrial Engineering*, 28(3).

[5] Azhmyakov, V., Fernández-Gutiérrez, J.P., Gadi, S.K. et al. 2016. A novel numerical approach to the mclp based resilent supply chain optimization. *IFACPapersOnLine*, 49(31): 137–142.

[6] Babich, V. 2006. Vulnerable options in supply chains: Effects of supplier competition. *Naval Research Logistics (NRL)*, 53(7): 656–673.

[7] Barbosa-Póvoa, A.P., da Silva, C. and Carvalho, A. 2018. Opportunities and challenges in the sustainable supply chain: An operations research perspective. *European Journal of Operational Research*, 268(2): 399–431.

[8] Bora, S., Lim, G.J., Biobaku, T.O., Cho, J. and Parsaei, H.R. 2014. Assessing the resiliency and importance of a supply chain network. *CIE44 & IMSS'14 Proceedings*, pp. 1530–1540.

[9] Čech, M. and Lenort, R. 2021. Modelling of financial resource allocation for increasing the supply chain resilience using markov chains. *Acta Logist*, 8: 141–151.

[10] Chen, G., Peivandizadeh, A. et al. 2022. Resilient supply chain planning for the perishable products under different uncertainty. *Mathematical Problems in Engineering*, 2022.

[11] Christopher, M. and Peck, H. 2004. Building the resilient supply chain. *The International Journal of Logistics Management*, 15(2): 1–14.

[12] Denyer, D. and Tranfield, D. 2009. Producing a systematic review.

[13] Dutta, P. and Shrivastava, H. 2020. The design and planning of an integrated supply chain for perishable products under uncertainties: A case study in milk industry. *Journal of Modelling in Management*, 15(4): 1301–1337.

[14] Ekanayake, E.M.A.C., Shen, G., Kumaraswamy, M. and Owusu, E.K. 2022. A fuzzy synthetic evaluation of vulnerabilities affecting supply chain resilience of industrialized construction in Hong Kong. *Engineering, Construction and Architectural Management*, 29(6): 2358–2381.

[15] Ekanayake, E.M.A.C., Shen, G.Q.P. and Kumaraswamy, M.M. 2023. A fuzzy synthetic evaluation of capabilities for improving supply chain resilience of industrialized construction: A Hong Kong case study. *Production Planning & Control*, 34(7): 623–640.

[16] Fahimnia, B. and Jabbarzadeh, A. 2016. Marrying supply chain sustainability and resilience: A match made in heaven. *Transportation Research Part E: Logistics and Transportation Review*, 91: 306–324.

[17] Foroozesh, N., Karimi, B. and Mousavi, S.M. 2022. Green-resilient supply chain network design for perishable products considering route risk and horizontal collaboration under robust interval-valued type-2 fuzzy uncertainty: A case study in food industry. *Journal of Environmental Management*, 307: 114470.

[18] Foroozesh, N., Karimi, B., Mousavi, S.M. and Mojtahedi, M. 2023. A hybrid decision-making method using robust programming and interval-valued fuzzy sets for sustainable-resilient supply chain network design

considering circular economy and technology levels. *Journal of Industrial Information Integration*, 33: 100440.

[19] Geng, L., Xiao, R. and Chen, T. 2017. An integrated resilient management approach to emergency logistics based on soc. *International Journal of Computing Science and Mathematics*, 8(4): 301–312.

[20] Gholami-Zanjani, S.M., Jabalameli, M.S., Klibi, W. and Pishvaee, M.S. 2021. A robust location-inventory model for food supply chains operating under disruptions with ripple effects. *International Journal of Production Research*, 59(1): 301–324.

[21] Gholami-Zanjani, S.M., Klibi, W., Jabalameli, M.S. and Pishvaee, M.S. 2021. The design of resilient food supply chain networks prone to epidemic disruptions. *International Journal of Production Economics*, 233: 108001.

[22] Ji, G. and Zhu, C. 2008. Study on supply chain disruption risk management strategies and model. *In 2008 International Conference on Service Systems and Service Management*, pp. 1–6. IEEE.

[23] Kaur, H. and Singh, S.P. 2022. Disaster resilient proactive and reactive procurement models for humanitarian supply chain. *Production Planning & Control*, 33(6-7): 576–589.

[24] Khezeli, M., Najafi, E., Molana, M.H. and Seidi, M. 2022. A mathematical model for sustainable and resilient supply chain by considering synchronization in the production and distribution network. *International Journal of Industrial Engineering & Production Research*, 33(2): 1–34.

[25] Larimi, N.G., Azhdari, A., Ghousi, R. and Du, B. 2022. Integrating gis in reorganizing blood supply network in a robust stochastic approach by combating disruption damages. *Socio-Economic Planning Sciences*, 82: 101250.

[26] Li, Y. and Zhao, L. 2010. Analyzing deformation of supply chain resilient system based on cell resilience model. *In International Conference on Intelligent Computing for Sustainable Energy and Environment*, pp. 26–35. Springer.

[27] Lv, Z., Qiao, L., Mardani, A. and Lv, H. 2022. Digital twins on the resilience of supply chain under covid-19 pandemic. *IEEE Transactions on Engineering Management*.

[28] MacKenzie, C.A. and Apte, A. 2017. Modeling disruption in a fresh produce supply chain. *The International Journal of Logistics Management*, 28(2): 656–679.

[29] Mallak, L. 1998. Putting organizational resilience to work. Industrial Management-Chicago then Atlanta, pp. 8–13.

[30] Namdar, J., Li, X., Sawhney, R. and Pradhan, N. 2018. Supply chain resilience for single and multiple sourcing in the presence of disruption risks. *International Journal of Production Research*, 56(6): 2339–2360.

[31] Nasrollah, S., Esmaeil Najafi, S., Bagherzadeh, H. and Rostamy-Malkhalifeh, M. 2023. An enhanced pso algorithm to configure a responsive-resilient supply chain network considering environmental issues: A case study of the oxygen concentrator device. *Neural Computing and Applications*, 35(3): 2647–2678.

[32] Nayeri, S., Sazvar, Z. and Heydari, J. 2022. A global-responsive supply chain considering sustainability and resiliency: Application in the medical devices industry. *Socio-Economic Planning Sciences*, 82: 101303.

[33] Nazari-Shirkouhi, S., Tavakoli, M., Govindan, K. and Mousakhani, S. 2023. A hybrid approach using z-number dea model and artificial neural network for resilient supplier selection. *Expert Systems with Applications*, 222: 119746.

[34] Nikian, A., Zare, H.K., Lotfi, M.M. and Nezhad, M.S.F. 2023. Redesign of a sustainable and resilient closed-loop supply chain network under uncertainty and disruption caused by sanctions and covid-19. *Operations Management Research*, 16(2): 1019–1042.

[35] Paul, S.K., Chowdhury, P., Chakrabortty, R.K., Ivanov, D. and Sallam, K. 2022. A mathematical model for managing the multi-dimensional impacts of the covid-19 pandemic in supply chain of a high-demand item. *Annals of Operations Research*, pp. 1–46.

[36] Ponomarov, S.Y. and Holcomb, M.C. 2009. Understanding the concept of supply chain resilience. *The International Journal of Logistics Management*, 20(1): 124–143.

[37] Preston, J.F., Cox, B.A., Rebeiz, P.P. and Breitbach, W. 2021. Developing a resilient, robust and efficient supply network in africa. *Journal of Defense Analytics and Logistics*, 5(2): 224–241.

[38] Rabbani, M., Molana, S.M.H., Sajadi, S.M. and Davoodi, M.H. 2022. Sustainable fertilizer supply chain network design using evolutionary-based resilient robust stochastic programming. *Computers & Industrial Engineering*, 174: 108770.

[39] Riccardo, A., Daria, B. and Dmitry, I. 2021. Increasing supply chain resilience through efficient redundancy allocation: A risk-averse mathematical model. *IFAC-PapersOnLine*, 54(1): 1011–1016.

[40] Sadeghi, Z., Boyer, O., Sharifzadeh, S. and Saeidi, N. 2021. A robust mathematical model for sustainable and resilient supply chain network design: Preparing a supply chain to deal with disruptions. *Complexity*, 2021: 1–17.

[41] Sahebjamnia, N. 2020. Resilient supplier selection and order allocation under uncertainty. *Scientia Iranica*, 27(1): 411–426.

[42] Sazvar, Z., Tafakkori, K., Oladzad, N. and Nayeri, S. 2021. A capacity planning approach for sustainable-resilient supply chain network design under uncertainty: A case study of vaccine supply chain. *Computers & Industrial Engineering*, 159: 107406.

[43] Shrivastava, H., Dutta, P., Krishnamoorthy, M. and Pravin Suryawanshi, P. 2017. Designing a resilient supply chain network for perishable products with random disruptions. *In Proceedings of the International MultiConference of Engineers and Computer Scientists*, 2: 870–875.

[44] Solgi, O., Gheidar-Kheljani, J., Dehghani, E. and Taromi, A. 2021. Resilient supplier selection in complex products and their subsystem supply chains under uncertainty and risk disruption: A case study for satellite components. *Scientia Iranica*, 28(3): 1802–1816.

[45] Vali-Siar, M.M. and Roghanian, E. 2022. Designing a multi-period and multi-product resilient mixed supply chain network under chainto-chain competition. *Kybernetes*, (ahead-of-print).

[46] Vali-Siar, M.M. and Roghanian, E. 2022. Sustainable, resilient and responsive mixed supply chain network design under hybrid uncertainty with considering covid-19 pandemic disruption. *Sustainable Production and Consumption*, 30: 278–300.

[47] Van Eck, N. and Waltman, L. 2010. Software survey: Vosviewer, a computer program for bibliometric mapping. *Scientometrics*, 84(2): 523–538.

[48] Wang, X., Herty, M. and Zhao, L. 2016. Contingent rerouting for enhancing supply chain resilience from supplier behavior perspective. *International Transactions in Operational Research*, 23(4): 775–796.

[49] Zahiri, B., Suresh, N.C. and de Jong, J. 2020. Resilient hazardous materials network design under uncertainty and perishability. *Computers & Industrial Engineering*, 143: 106401.

[50] Zalozhnev, A.Y. and Peremezhko, D.V. 2022. It project management: Supply chain optimization for service operations. *IFAC-PapersOnLine*, 55(10): 2505–2508.

[51] Zare Mehrjerdi, Y. and Lotfi, R. 2019. Development of a mathematical model for sustainable closed-loop supply chain with efficiency and resilience systematic framework. *International Journal of Supply and Operations Management*, 6(4): 360–388.

[52] Zavala, A., Nowicki, D. and Ramirez-Marquez, J.E. 2019. Quantitative metrics to analyze supply chain resilience and associated costs. *Proceedings of the Institution of Mechanical Engineers, Part O: Journal of Risk and Reliability*, 233(2): 186–199.

[53] Zheng, G., Yang, Y. and Zhou, L. 2014. Model and genetic algorithms for resilient supply chain under supply disruption. *In CICTP 2014: Safe, Smart, and Sustainable Multimodal Transportation Systems*, pp. 821–829.

Chapter 2

Operational Optimization of Vectors for Logical Networks

Yavuz Can

2.1 Introduction

Verification is a critical step in the design process of logical networks, as it enables the detection and correction of faults early on. With great effort, the compliance of a Boolean network is checked. An early discovered fault in the design process can be corrected more easily and cost-effectively, which can be adequately expressed by the "Rule of Ten" regarding to the fault costs per product phase. Therefore, verification is performed to detect faults in logic networks. For this purpose, certain test vectors are calculated, which are used to verify a combinatorial network.

As test vectors a series of determined vectors are called which cause changes in the function value at the output of a Boolean network. If there is at least one fault in a logic network this is detected by applying the test vectors, which change the output value. These vectors are formally obtained by the Boolean Differential Calculus. On the one hand, this work demonstrates the reduction of the set of test

Haliç University, Faculty of Engineering, Computer Engineering, Eyüpsultan/Istanbul, Türkiye.
Email: yavuzcan@halic.edu.tr

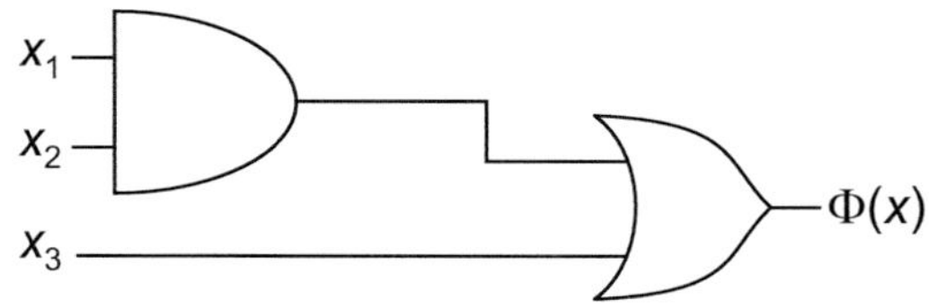

Figure 2.1: Example of the logic network of gates.

vectors and on the other hand the positive influence of an orthogonal form of a Boolean function on the number of test vectors, which are determined by the Boolean Differential Calculus.

2.2 Mathematical Representation

A logic network is a model which represents different kinds of systems, e.g., combinatorial circuits, fault tree diagrams, etc. Consequently, with these models analysis and synthesis of the corresponding system are made.

A logic network, as shown in Figure 2.1, consists of several gates that are interconnected, and its inputs and output(s) can take the values 0 or 1. This kind of logic network is mathematically represented by a Boolean function $\phi(\underline{x})$ containing Boolean expressions [1, 5] and is defined as the mapping $\phi(\underline{x}) : \mathbf{B}^n \rightarrow \mathbf{B}$, whereby the set $\mathbf{B} = \{0, 1\}$ consists of two elements. A Boolean function contains terms - conjunctions and/or disjunctions - which includes one or more variables $x_i = \{x_1, x_2, \ldots, x_n\}$ that are either negated $\bar{x}_n$ or not-negated x_n [1, 5, 9]. A conjunction is also named as product term and a disjunction as clause.

Definition 2.1 Product Term

A product term is a conjunction ($\bigwedge$) of Boolean variables x_i which appear either negated $\bar{x}_n$ or not-negated x_n. The sets A_k and B_k include the number of indices of the consisting variables in a product term $p_k(\underline{x})$.

$$p_k(\underline{x}) = \bigwedge_{i \in A_k} x_i \bigwedge_{j \in B_k} \bar{x}_j \quad \text{whereby} \quad A_k \cap B_k = \emptyset. \tag{2.1}$$

There are several basis forms of Boolean function [2, 7]. One of them is the Disjunctive Form (DF) or by another name Sum of Products (SOP) which is composed of disjunctions ($\bigvee$) of at least two product terms $p_k(\underline{x})$ as defined in Equation 2.2. The synthesis and the analysis of a Boolean network are realized both on a formal level.

Definition 2.2 Disjunctive Form
Disjunctive Form (DF) is the disjunction of at least two product terms $p_{i,j}(\underline{x})$. DF with $m \in \mathbb{N}$ is defined by:

$$\Phi_{DF}(\underline{x}) = \bigvee_{k=1}^{m} p_k(\underline{x}) \tag{2.2}$$

Here, $\bigvee$ denotes the disjunction of at least two product terms. Thus, by substitution, it applies:

$$\Phi_{DF}(\underline{x}) = \bigvee_{k=1}^{m} \Big(\bigwedge_{i \in A_k} x_i \bigwedge_{j \in B_k} \bar{x}_j \Big). \tag{2.3}$$

in which i and j show the indices of non-negated and negated variables, respectively. Hereby, X_i and X_j represent the set of non-negated and negated variables in order. The canonical representation of a product term is called as minterm and a canonical form of a disjunctive form is called as disjunctive normal form (DNF).

For the case, that the product terms of a DF are disjointed to each other in pairs at least in one variable x_i then this DF has the special characteristic of orthogonality. Consequently, this specific DF is named the orthogonal Disjunctive Form (ODF).

Definition 2.3 Orthogonal Disjunctive Form
Thus, a DF as given in Equation 2.2 is orthogonal if it applies $(A_k \cap B_l) \cup (A_l \cap B_k) \neq \emptyset$ for all $k, l \in \{1, 2, \ldots, m\}$ where $k \neq l$. An ODF consists of disjunctions of at least two product terms which are disjointed to each other in pairs by x_i and $\bar{x}_i$. The intersection for every pair of product terms $k, l \in \{1, 2, \ldots, m\}, k \neq l$ results in 0:

$$p_k(\underline{x}) \wedge p_l(\underline{x}) = 0, \tag{2.4}$$

$$\Big(\bigwedge_{i \in A_k} x_i \bigwedge_{j \in B_k} \bar{x}_j \Big) \wedge \Big(\bigwedge_{i \in A_l} x_i \bigwedge_{j \in B_l} \bar{x}_j \Big) = 0. \tag{2.5}$$

In Equation 2.4 $\underline{x}$ shows the tuple of consisting variables in a product term, $\wedge$ represents the conjunction, and finally A_k, B_k, A_l and B_l stand for the sets of the number of indices of the consisting non-negated and negated variables in their corresponding product term. Subsequently, for the cardinalities between these sets following is valid [5]. The number of elements in A_l is greater than or equal to the cardinality of A_k and the same is valid between B_l and B_k:

$$|A_l| \geq |A_k| \quad \text{and} \quad |B_l| \geq |B_k|. \tag{2.6}$$

This special form of orthogonality has several benefits as well as the transforming of an orthogonal DF in an equivalent antivalence form, which means

Exclusive-Or Sum of Products (ESOP). Another benefit is that the number of test vectors, which are calculated by Boolean Differential Calculus (BDC) of an ODF, is fewer. To do this, however, a DF must first be orthogonalized. In the following section, a more recent method is presented for the orthogonalization of every DF.

2.3 Orthogonalization

By Equation 2.7, which is based on the combining technique of the orthogonalizing difference-building $\ominus$ (Equation 2.8) an ODF of every DF consisting of at least two product terms ($m > 1$) can be determined:

$$
\begin{aligned}
{}^{o}\Phi_{DF}(\underline{x}) &:= \bigvee_{k=0}^{m-1} \left(\bigominus_{i=k+1}^{m} p_i(\underline{x}) \right) = \qquad (2.7)\\
&= (p_1(\underline{x}) \ominus \ldots \ominus p_m(\underline{x})) \vee \ldots \vee (p_{m-1}(\underline{x}) \ominus p_m(\underline{x})) \vee p_m(\underline{x}) \\
&= \begin{bmatrix} p_1(\underline{x}) \ominus p_2(\underline{x}) \ominus \ldots \ominus p_{m-1}(\underline{x}) \ominus p_m(\underline{x}) \\ p_2(\underline{x}) \ominus \ldots \ominus p_{m-1}(\underline{x}) \ominus p_m(\underline{x}) \\ \vdots \\ p_{m-1}(\underline{x}) \ominus p_m(\underline{x}) \\ p_m(\underline{x}) \end{bmatrix}.
\end{aligned}
$$

in which $\ominus$ and $\bigvee$ denote the technique of orthogonalizing difference-building and the disjunction of each calculation part, in order. Both the combining $\ominus$-technique and Equation 2.7 for orthogonalization are generally valid, as proved by the mathematical induction in [3, 4]. The orthogonalizing difference-building takes place between two product terms $p_m(\underline{x}) \ominus p_s(\underline{x})$. The intersection between the subtrahend $p_s(\underline{x})$ and the minuend $p_m(\underline{x})$ is removed from the minuend $p_m(\underline{x})$ in a way, that means $p_m(\underline{x}) - (p_m(\underline{x}) \wedge p_s(\underline{x}))$, so that the difference consists of product terms which are disjointed to each other in pairs. For that, it applies the Equation 2.8, with $n, n' \in \mathbb{N}$ as the number of the variables in $p_m(\underline{x})$ and $p_s(\underline{x})$, which merely indicates the order of the variables in the product terms which have to be linked.

Definition 2.4 (Orthogonalizing Difference-Building)
Orthogonalizing difference-building $\ominus$ is the composition of two calculation steps - the usual difference-building and the subsequent orthogonalization - in one step and is defined in Equation 2.8 as long as the subtrahend is not the subset of minuend ($p_s(\underline{x}) \not\subseteq p_m(\underline{x})$). The formula ($\bigvee_{i=1}^{n_j} \bar{x}_i = \bar{x}_{1_j} \vee x_{1_j}\bar{x}_{2_j} \vee \ldots \vee x_{1_j} x_{2_j} \cdot \ldots \cdot \bar{x}_{n_j}$) out of [5]

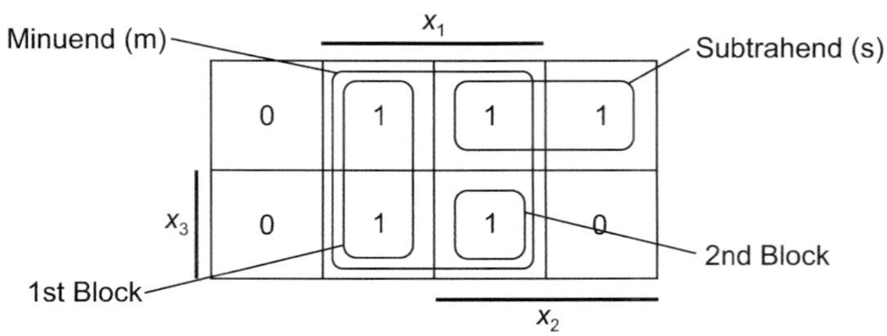

Figure 2.2: Method $\ominus$ presented in a K-map.

is used to describe this alternating technique in a mathematically easier way:

$$
\begin{aligned}
p_m(\underline{x}) \ominus p_s(\underline{x}) &= \bigwedge_{m=1}^{n} x_m \ominus \bigwedge_{s=1}^{n'} x_s \qquad (2.8)\\
&:= \bigwedge_{m=1}^{n} x_m \wedge \bigvee_{s=1}^{n'_j} \bar{x}_{s_j} = \\
&= \left(x_1 \cdot \ldots \cdot x_{n-1} x_n\right)_m \wedge \left(\bar{x}_{1_j} \vee x_{1_j} \bar{x}_{2_j} \vee \ldots \vee x_{1_j} \cdot \ldots \cdot x_{(n'-1)_j} \bar{x}_{n'_j}\right)_s.
\end{aligned}
$$

This technique of $\ominus$ is clarified by the following explanation of the calculation steps and a subsequent example, which is also illustrated in a K-map in Figure 2.2.

It is given two product terms $p_m(\underline{x}) = x_1$ and $p_s(\underline{x}) = x_2\bar{x}_3$. Their orthogonal difference has to be calculated and displayed in a K-map as shown Figure 2.2.

$$
\underbrace{x_1}_{\text{Minuend}} \ominus \underbrace{x_2\bar{x}_3}_{\text{Subtrahend}} = \underbrace{\underbrace{x_1\bar{x}_2}_{1^{st}\text{Block}} \vee \underbrace{x_1 x_2 x_3}_{2^{nd}\text{Block}}}_{\text{Difference}}
$$

The following points explain the calculation steps:

- The first literal of the subtrahend, here x_2, is taken as complement and builds the intersection with the minuend, here x_1. Consequently, the first term of the difference is $x_1\bar{x}_2$.
- Then the second literal, here $\bar{x}_3$, is taken as a complement and builds the intersection with the minuend and the first literal x_2 of the subtrahend. Therefore, the second term is $x_1 x_2 x_3$.
- This process is continued until all literals of the subtrahend are singly complemented and linked by building the intersection with the minuend in a separate product term.

The product terms of the difference are pairwise orthogonal to each other so that the result is already available in an orthogonal form. The following rules must be followed to get correct results when using $\ominus$:

1. If the subtrahend is already disjoint to the minuend ($p_s(\underline{x}) \perp p_m(\underline{x})$) the result is the minuend itself:

$$p_m(\underline{x}) \ominus p_s(\underline{x}) = p_m(\underline{x}). \tag{2.9}$$

2. The orthogonal difference between 0 and the subtrahend is the subtrahend itself:

$$0 \ominus p_s(\underline{x}) = p_s(\underline{x}). \tag{2.10}$$

3. The result between 1 and the subtrahend is the complement of the subtrahend which results in an orthogonal disjunctive form:

$$1 \ominus p_s(\underline{x}) = {}^{o}f(\underline{x}) \mid_{\overline{p_s(\underline{x})}} . \tag{2.11}$$

4. The other way around, if minuend is the subset of the subtrahend ($p_m(\underline{x}) \subseteq p_s(\underline{x})$), the result is 0:

$$p_m(\underline{x}) \ominus p_s(\underline{x}) = 0. \tag{2.12}$$

Out of this $\ominus$-technique Equation 2.7 is established, which allows the orthogonalization of a DF with m as the number of the product terms. Starting with the first product term, the $\ominus$-linkage is calculated up to the last product term. The same procedure is then continued starting with the second product term. This procedure ends with the last product term. The linkages are linked by disjunction ($\vee$). The following example demonstrates this procedure of orthogonalization.

Function $\Phi_1(\underline{x}) = x_1x_2 \vee x_3$ has to be orthogonalized that is illustrated in Figure 2.1. Both forms are then visualized in a K-map (Figure 2.3). The number of product terms is $m = 2$.

$${}^{o}\Phi_1(\underline{x}) = (x_1x_2 \ominus x_3) \vee x_3 = x_1x_2\bar{x}_3 \vee x_3$$

The DF $\Phi_1(\underline{x})$ and the ODF ${}^{o}\Phi_1(\underline{x})$ are represented in K-maps (Figure 2.3). Both functions are equivalent because they cover the same 1s, but differ in their form of coverage.

2.4 Boolean Differential Calculus

The Boolean Differential Calculus (BDC) is a dynamic operation of Boolean functions and is used to describe the behavior of combinatorial networks depending on the change of its input vectors [1, 6, 8]. By the BDC, diagnosis of

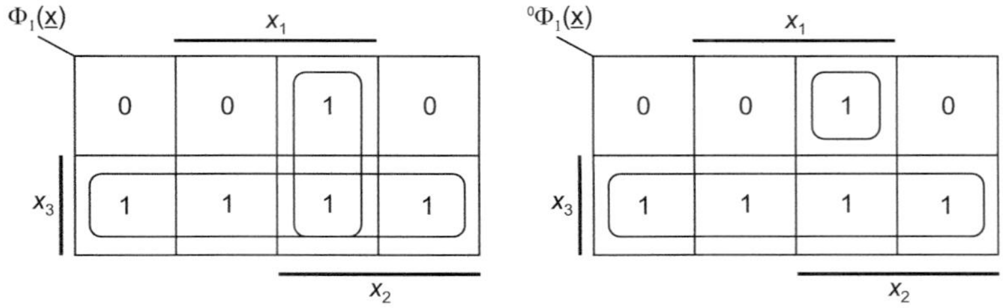

Figure 2.3: $\Phi_1(\underline{x})$ and $^o\Phi_1(\underline{x})$ in K-maps.

faults can be performed. The arranged vectors, which are determined by BDC, cause a change in the output of the faultless network. In the case of a faulty network, no change can be detected. Exactly this adjustment behavior is described by $\frac{d\Phi(\underline{x})}{d\underline{x}}$ and its solution provides all possible test patterns in the form of $\underline{x} \wedge d\underline{x}$ whereby $\underline{x} = \{x_1, x_2, \ldots, x_n\}$ is the tuple of all input variables. Furthermore, $d\underline{x} = \{dx_1, dx_2, \ldots, dx_n\}$ is also binary and is the tuple of the corresponding change parameters. Here, the differential is a parametric derivation from single to multiple differentials and is defined as follows:

$$\frac{d\Phi(\underline{x})}{d\underline{x}} := \Phi(\underline{x}) \oplus \Phi(\underline{x} \oplus d\underline{x}). \tag{2.13}$$

Accordingly, the question of interest is as follows:

> *"At which input vectors and subsequent change of a certain or all variables does the function value of $\Phi(\underline{x})$ change?"*

A variation regarding to a variable x_i is marked, then an additional component dx_i is marked. If a change depending on an input variable x_i is to realize, dx_i is equal to 1, in other circumstances $\overline{dx_i}$ is equal to 0 as shown below:

$$dx_i = \begin{cases} 1, & \text{change of } x_i \\ 0, & \text{no change of } x_i. \end{cases} \tag{2.14}$$

In the case that no particular change variable is contained in a term, the change of the corresponding variable within the product term has no effect on the function value if the remaining variables in the product term change. Overall, the equality below applies to the differential of a function:

$$\frac{d\Phi(\underline{x})}{d\underline{x}} = 1. \tag{2.15}$$

In this following example, the application of Equation 2.15 is demonstrated. Given was a function $\Phi_2(\underline{x})$ which was differentiated and, thus, is its result:

Table 2.1: Example of the application of the differentiation.

	variables			changes	$\Phi_2(\underline{x})$
	x_1	x_2	x_3		
vector	0	1	1		1/0
	$\downarrow$	$\downarrow$	$\downarrow$	$dx_1\overline{dx_3}$	$\downarrow$
vector	1	1/0	1		0/1

$F_2(\underline{x},d\underline{x}) = x_1x_2x_3\ dx_1\overline{dx_3}$. The change variable dx_1 forces a transition of x_1 and at the same time $\overline{dx_3}$ means no transition for x_3. For the variable x_2 both a 0 and a 1 may be used, because of the fact that the change variable dx_2 does not exist, and has no effect regarding the change of the value of the function. When modifying the vectors, e.g., $x_1 = 0$, $x_2 = 1$, $x_3 = 1$ to $x_1 = 1$, $x_2 = 1/0$, $x_3 = 1$, due to the change variables the value of function $\Phi_2(\underline{x})$ alternates from 1 to 0 or vice versa, as listed in Table 2.1.

2.4.1 Grouping of Vectors

By transforming Equation 2.13 according to the following rule

$$a \oplus b = \bar{a}b \vee a\bar{b}, \tag{2.16}$$

the following derivative is obtained. In this way, the $\oplus$-operation is converted into a disjunction of two product terms which specify the separate group of changing types:

$$\frac{d\Phi(\underline{x})}{d\underline{x}} := \underbrace{\overline{\Phi(\underline{x})}\cdot\Phi(\underline{x}\oplus d\underline{x})}_{F_{\alpha^{0\to 1}}(\underline{x},d\underline{x})} \vee \underbrace{\Phi(\underline{x})\cdot\overline{\Phi(\underline{x}\oplus d\underline{x})}}_{F_{\alpha^{1\to 0}}(\underline{x},d\underline{x})} \tag{2.17}$$

In the above expression, $\bar{a}b$ and $a\bar{b}$ represent the conjunctions of both variables a and b in alternating negated form, which are build up due the transformation of the $\oplus$-operation in a disjunction. $\overline{\Phi(\underline{x})}$ and $F_{\alpha^{0\to1}}$ as well as $F_{\alpha^{1\to0}}$ show the complement of a given function $\Phi(\underline{x})$, the part of the differential containing the vectors which lead on to a 0 to 1 change of the function value and the part of the differential containing the vectors which lead on to a 1 to 0 change of the function value.

By $F_{\alpha^{0\to1}}(\underline{x},d\underline{x})$ the set of vectors is marked, which brings to a change of the value of the function $\Phi(\underline{x})$ from 0 to 1. Hereby, $\overline{\Phi(\underline{x})}$ maps all 0-values of $\Phi(\underline{x})$, which pass over to 1 by the differential of $\Phi(\underline{x}\oplus d\underline{x})$. Against this, by $F_{\alpha^{1\to0}}(\underline{x},d\underline{x})$ the set of vectors is signified that cause a transition from 1 to 0. Since $\Phi(\underline{x})$ maps all 1-values of the corresponding network which allow a change to 0 by $\overline{\Phi(\underline{x}\oplus d\underline{x})}$.

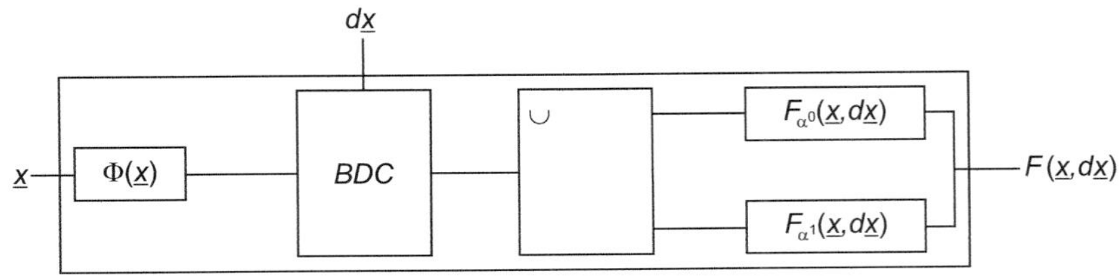

Figure 2.4: Grouping of vectors by BDC.

This decomposition of both vector groups is visualized in Figure 2.4. The result of the BDC leads to the set of all possible test patterns. Here, they are decomposed into two groups regarding their type of change effect those which cause either a 0-1 or a 1-0 modification of the value of the function. On the basis of this decomposition, it can be sufficient to use one of the groups in order to verify a given combinatorial network. Due to the fact it is no longer relevant whether you have a 0-1 or 1-0 change, however the relevance here is a change that can be generated by one of the vector group. In this case, this leads to a reduction of the necessary test pattern for checking the correct functionality of a logic network. Consequently, the set of vectors can be reduced to half approximately.

2.4.2 Reducing Vectors

Here we consider the BDC regarding to DF which can be both orthogonal and non-orthogonal. Corresponding equation is set up and examined. In addition to the benefit of grouping the vectors, further reduction in the set of test vectors can be achieved if an ODF is presented. To justify this statement, first, Equation 2.2 is substituted into Equation 2.17 and then analyzed.

$$\frac{d\Phi(\underline{x})}{d\underline{x}} := \bigvee_{k=1}^{m} \Big(\bigwedge_{i\in A_k} x_i \bigwedge_{j\in B_k} \bar{x}_j \Big) \oplus \bigvee_{k=1}^{m} \Big(\Big(\bigwedge_{i\in A_k} x_i \bigwedge_{j\in B_k} \bar{x}_j \Big) \oplus dx_{i,j} \Big)_k \tag{2.18}$$

$$= \underbrace{\underbrace{\overline{\bigvee_{k=1}^{m} \Big(\bigwedge_{i\in A_k} x_i \bigwedge_{j\in B_k} \bar{x}_j \Big)}} \cdot \bigvee_{k=1}^{m} \Big(\Big(\bigwedge_{i\in A_k} x_i \bigwedge_{j\in B_k} \bar{x}_j \Big) \oplus dx_{i,j} \Big)_k} \vee \bigvee_{k=1}^{m} \Big(\bigwedge_{i\in A_k} x_i \bigwedge_{j\in B_k} \bar{x}_j \Big) \cdot \overline{\bigvee_{k=1}^{m} \Big(\Big(\bigwedge_{i\in A_k} x_i \bigwedge_{j\in B_k} \bar{x}_j \Big) \oplus dx_{i,j} \Big)_k}$$

$$= \underbrace{\underbrace{\bigwedge_{k=1}^{m} \Big(\bigvee_{i\in A_k} \bar{x}_i \bigvee_{j\in B_k} x_j \Big)}_{T^1} \cdot \underbrace{\bigvee_{k=1}^{m} \Big(\Big(\bigwedge_{i\in A_k} x_i \bigwedge_{j\in B_k} \bar{x}_j \Big) \oplus dx_{i,j} \Big)_k}_{Z^1}}_{F_{\alpha^{0\to1}}(\underline{x},d\underline{x})} \vee \underbrace{\underbrace{\bigvee_{k=1}^{m} \Big(\bigwedge_{i\in A_k} x_i \bigwedge_{j\in B_k} \bar{x}_j \Big)}_{T^2} \cdot \underbrace{\bigwedge_{k=1}^{m} \Big(\Big(\bigvee_{i\in A_k} \bar{x}_i \bigvee_{j\in B_k} x_j \Big) \oplus dx_{i,j} \Big)_k \Big)_k}_{Z^2}}_{F_{\alpha^{1\to0}}(\underline{x},d\underline{x})},$$

where $\overline{\bigvee}(.)$ refers to the complement of the corresponding first or second operands. Supplementary, the two vector groups are subdivided separately. $F_{\alpha^{0\to1}}(\underline{x},d\underline{x})$ is graduated in T^1, Z^1 and $F_{\alpha^{1\to0}}(\underline{x},d\underline{x})$ in T^2, Z^2, so that they can be analyzed isolated. Corresponding designations ${}^oT^1, {}^oZ^1, {}^oT^2, {}^oZ^2$ are also assigned for the orthogonal form. The investigation shows that an ODF affects positively the number of test vectors that will be obtained. Therefore, here the substituted form is taken to prove the following purported relation whereby the number of vectors, that are calculated by BDC, is fewer for ODF in comparison to the DF:

$$\underbrace{{}^{o}T^{1}\cdot{}^{o}Z^{1}}_{{}^{o}F_{\alpha 0\to 1}} \vee \underbrace{{}^{o}T^{2}\cdot{}^{o}Z^{2}}_{{}^{o}F_{\alpha 1\to 0}} < \underbrace{T^{1}\cdot Z^{1}}_{F_{\alpha 0\to 1}} \vee \underbrace{T^{2}\cdot Z^{2}}_{F_{\alpha 1\to 0}}. \tag{2.19}$$

Theorem 2.1

Let the number of terms determined in the group $F_{\alpha 0\to 1}$ be identical for both forms - orthogonal and non-orthogonal - and on account of this the following relationship applies.

$${}^{o}T^{1}\cdot{}^{o}Z^{1} = T^{1}\cdot Z^{1}. \tag{2.20}$$

The validity of the relation in Equation 2.20 is established by the following analysis of both ${}^{o}T^{1}$, T^{1} and ${}^{o}Z^{1}$, Z^{1}:

► T^{1}: The complement of an orthogonal form is identical to the complement of its non-orthogonal form. Thus, it applies $T^{1} = {}^{o}T^{1}$. Due to the isomorphism, its proof is taken from the set theory and transferred to the Boolean algebra:

Proof 2.1 There are two overlapping sets A and B, which union is noted once disjoint (orthogonal) with $A \cup \bar{A}B$ (Figure 2.5) and once in non-orthogonal form is illustrated by $A \cup B$ (Figure 2.6).

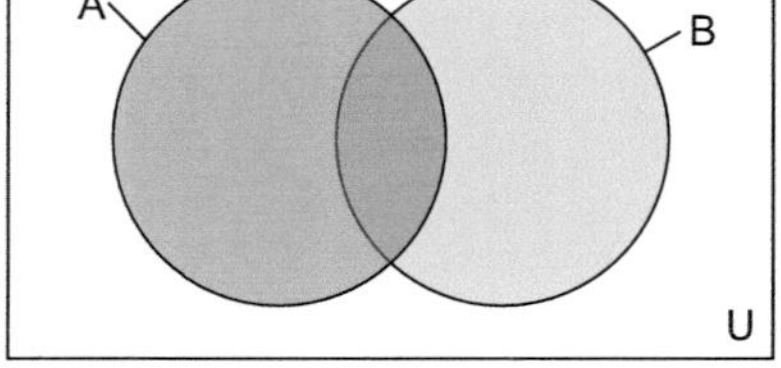

Figure 2.5: Orthogonal - $A \cup \bar{A}B$.

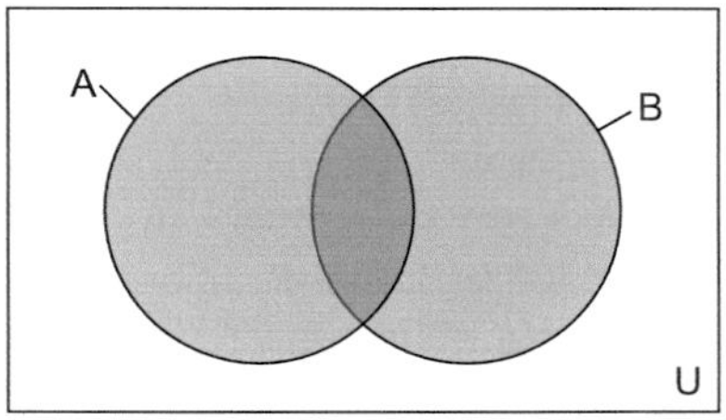

Figure 2.6: Non-orthogonal - $A \cup B$.

After that, building the equality of the complement of both forms is demonstrated. The complements correspond to the difference between the universal set U and the union set:

$$\begin{aligned}
\overline{A \cup \bar{A}B} &\stackrel{?}{=} \overline{A \cup B} \\
1 - (A \cup \bar{A}B) &= 1 - (A \cup B) \\
1 \cap (\overline{A \cup \bar{A}B}) &= 1 \cap (\overline{A \cup B}) \\
\bar{A} \cap (A \cup \bar{B}) &= \bar{A} \cap \bar{B} \\
(\bar{A} \cap A) \cup (\bar{A} \cap \bar{B}) &= \bar{A} \cap \bar{B} \\
\bar{A} \cap \bar{B} &= \bar{A} \cap \bar{B}
\end{aligned}$$

as the result confirms that the left side is equal to the right side. Consequently, the complement of a DF is equivalent to the complement of its ODF via:

$$\overline{\Phi(\underline{x})} = \overline{{}^{o}\Phi(\underline{x})}. \tag{2.21}$$

Here,

$$Z^1$$

is equivalent to ${}^{o}Z^1$. They only differ in their form in such a way that ${}^{o}Z^1$ will be orthogonal.

It is manifested that ${}^{o}T^1 \cdot {}^{o} Z^1 = T^1 \cdot Z^1$ is valid. The number of terms that arises is also equal. However, even after the conjunction ${}^{o}T^1 \cdot {}^{o} Z^1$, the left side is still orthogonal. Because the property of orthogonality continues in this case. An advantage can already be taken here because the number of derived vectors from mutually orthogonal terms is lower than the non-orthogonal terms. With the property of orthogonality, the number of variables in every single term also increases. In addition, this reduces the number of test vectors that are derived from it. The number of test vectors corresponds to 2^k with k as the number of missing variables in a product term according to the given tuple. The more variables there are in the product term, the smaller the number of vectors that are derived from them. Accordingly, it can be noted that

$$^{o}F_{\alpha^{0\to1}}(\underline{x},d\underline{x}) \quad \leq \quad F_{\alpha^{0\to1}}(\underline{x},d\underline{x}). \tag{2.22}$$

In contrast, the number of terms determined in the groups $F_{\alpha^{1\to0}}$ and $^{o}F_{\alpha^{1\to0}}$ are not identical. The number of terms in the orthogonal form is already reduced. Therefore, to prove this claim, the following relation must be demonstrated which is investigated in the following points:

$$^{o}T^{2} \cdot {}^{o}Z^{2} \quad < \quad T^{2} \cdot Z^{2} \tag{2.23}$$

in which T^2 refers to the evidence of the following relation that has already been brought under Z^1. Hence, T^2 is equivalent to $^{o}T^2$. They only differ in their form, whereby, $^{o}T^2$ is orthogonal.

Accordingly, T^2 and $^{o}T^2$ can be neglected on both sides in Equation 2.23. With this, the remaining relation has to be verified $^{o}Z^2 < Z^2$. Thereby, this relation is transformed to $\overline{^{o}Z^2} \vee Z^2 = 1$ because of derivation out of the truth table:

$^{o}Z^2$	Z^2	$^{o}Z^2 < Z^2$
0	0	0
0	1	1
1	0	0
1	1	0

$$\Rightarrow \quad \overline{^{o}Z^2} \vee Z^2 = 1 \tag{2.24}$$

which must be fulfilled so that the advantages of orthogonality can be witnessed in Proof 2.2.

Proof 2.2

$$\overline{{}^{o}Z^{2}} \vee Z^{2} \overset{?}{=} 1$$

$$\overline{\overline{\bigvee_{k=1}^{m}\Big(\big(\bigwedge_{i\in A_k} x_i \bigwedge_{j\in B_k} \bar{x}_j\big)\oplus dx_{i,j}\Big)_k}} \vee \overline{\bigvee_{k=1}^{m}\Big(\big(\bigwedge_{i\in A_k} x_i \bigwedge_{j\in B_k} \bar{x}_j\big)\oplus dx_{i,j}\Big)_k} = 1$$

$$\bigvee_{k=1}^{m}\Big(\big(\bigwedge_{i\in A_k} x_i \bigwedge_{j\in B_k} \bar{x}_j\big)\oplus dx_{i,j}\Big)_k \vee \bigwedge_{k=1}^{m}\overline{\Big(\big(\bigwedge_{i\in A_k} x_i \bigwedge_{j\in B_k} \bar{x}_j\big)\oplus dx_{i,j}\Big)_k} = 1$$

$$\underbrace{\Big[\big((\bigwedge_{i\in A_k} x_i \bigwedge_{j\in B_k} \bar{x}_j)\oplus dx_1\big)\cdot\ldots\cdot\big((\bigwedge_{i\in A_k} x_i \bigwedge_{j\in B_k} \bar{x}_j)\oplus dx_n\big)\Big]_1}_{a} \vee \ldots \vee \underbrace{\Big[\big((\bigwedge_{i\in A_k} x_i \bigwedge_{j\in B_k} \bar{x}_j)\oplus dx_1\big)\cdot\ldots\cdot\big((\bigwedge_{i\in A_k} x_i \bigwedge_{j\in B_k} \bar{x}_j)\oplus dx_n\big)\Big]_m}_{z} \vee$$

$$\underbrace{\Bigg(\underbrace{\Big[\overline{\big((\bigwedge_{i\in A_k} x_i \bigwedge_{j\in B_k} \bar{x}_j)\oplus dx_1\big)} \vee \ldots \vee \overline{\big((\bigwedge_{i\in A_k} x_i \bigwedge_{j\in B_k} \bar{x}_j)\oplus dx_n\big)}\Big]_1}_{\bar{a}} \wedge \ldots \wedge \underbrace{\Big[\overline{\big((\bigwedge_{i\in A_k} x_i \bigwedge_{j\in B_k} \bar{x}_j)\oplus dx_1\big)} \vee \ldots \vee \overline{\big((\bigwedge_{i\in A_k} x_i \bigwedge_{j\in B_k} \bar{x}_j)\oplus dx_n\big)}\Big]_m}_{\bar{z}}\Bigg)}_{\bar{a}\bar{b}\cdot\ldots\cdot\bar{z}} = 1$$

Theorem 2.2
Let

$$a \vee b \vee \ldots \vee z \vee \bar{a}\bar{b} \cdot \ldots \cdot \bar{z} = 1 \tag{2.25}$$

applying to confirm $\overline{{}^{o}Z^{2}} \vee Z^{2} = 1$. *Here,* $\bar{a}\bar{b} \cdot \ldots \cdot \bar{z}$ *shows the conjunction of the complement of each variable* $a, b, \ldots, z$. *In the following Proof 2.3 the validity of Equation 2.25 for 3 variables is illustrated, whereby it can be resolved similarly for z-number of variables.*

Proof 2.3

$$a \vee b \vee \underbrace{c \vee \bar{a}\bar{b}\bar{c}}_{c \vee \bar{a}\bar{b}} = 1 \qquad \textit{(law of absorption)}$$

$$a \vee c \vee \underbrace{b \vee \bar{a}\bar{b}}_{b \vee \bar{a}} = 1 \qquad \textit{(law of absorption)}$$

$$c \vee b \vee \underbrace{a \vee \bar{a}}_{1} = 1 \qquad \textit{(complementation)}$$

The disjunction of a variable with its complement always results in 1.

This analysis confirms that the number of terms using an ODF is fewer. On the one hand, the number of terms determined is less in ${}^{o}T^{2} \cdot {}^{o}Z^{2}$ and the terms are already in orthogonal form so that the test vectors which will be derived from it are fewer. It applies:

$${}^{o}F_{\alpha^{0\to1}}(\underline{x}, d\underline{x}) \quad < \quad F_{\alpha^{0\to1}}(\underline{x}, d\underline{x}). \tag{2.26}$$

2.5 Analytic Result and Discussion

Orthogonalization increases the number of variables in a product term. Variables appear as x_i and $\bar{x}_i$ in pairs. The subsequent conjunction increases the probability that two product terms including their change variables that match each other will disappear due to the calculation rule $x_i \wedge \bar{x}_i = 0$. This reduces the number of calculated vector(s) in the first part $F_{\alpha^{1\to0}}(\underline{x}, d\underline{x})$. Orthogonalization increases the number of variables within the product terms. The intersection of the set of these product terms with the set of product terms including their change variables results in terms from which the test vectors are derived. Since the product terms in the orthogonal form can contain a higher number of variables, which is evidenced by the cardinal relationship in Equation 2.6, the possibility of a common intersection that arises from the coincidence of two product terms will be reduced as a consequence. The number of variables in the product terms increases by orthogonalization. Therefore, the probability increases that the intersections with

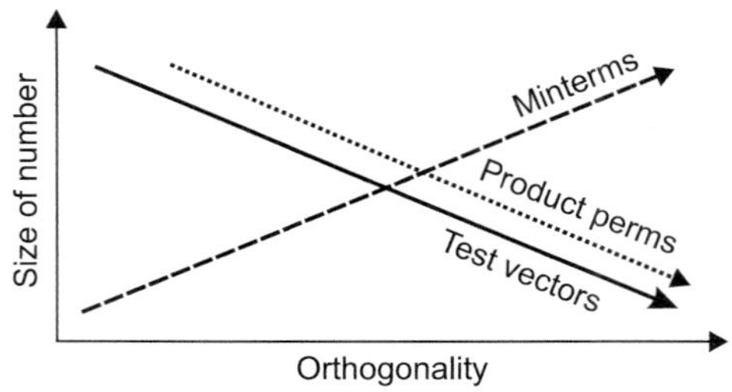

Figure 2.7: Quantitative proportion of orthogonality and number of test vectors.

the subsequent change term will result in minterms. If the number of minterms in the result **increases**, the number of product terms and, proportionally to this, the number of calculated test vectors **decreases** as quantitatively depicted in Figure 2.7. Finally, the vectors of an ODF determined by BDC turn out to be smaller than an equivalent DF.

2.6 Conclusion

This work illustrated that an ODF brings advantages regarding the number of its test vectors, which are determined by BDC, in comparison to its equivalent non-orthogonal DF. Thus, the characteristic of a DF's orthogonality is important in specific Boolean network applications to obtain optimizations. For this purpose, a generally valid method for the orthogonalization of every DF was also presented. Furthermore, the set of test vectors determined by the BDC was analyzed. This led to the realization that the set of test vectors can be separated into two subsets of vector types. One group led to a 0-1 change, and the other led to a 1-0 change. This means, that one subset is sufficient to verify a corresponding function. Consequently, the number of test vectors is optimized by reduction. For future investigations, measurements at the equivalent corresponding algorithms can be performed on the computational level in order to get further detailed outputs regarding their computation time and memory usage.

References

[1] Bochmann, D. 2006. *Binäre Systeme - Ein Boolean Buch*. LiLoLe-Verlag, Hagen, Germany, February 2006.

[2] Can, Y. 2017. Extended forms of switching functions. *In 4th International Conference on Electrical and Electronics Engineering ICEEE*, Ankara, Turkey, 8–10. April 2017. IEEE Support.

[3] Can, Y. 2020. Novel technique for disjointed sum of products. *Turkic World Mathematical Society TWMS - Journal of Applied and Engineering Mathematics*, 10(2): 489–470.

[4] Can, Y., Yaz, O. and Fey, D. 2020. Disjoint Sum of products by orthogonalizing difference-building A. *Open Engineering*, 10(1): 586–594, July 2020.

[5] Crama, Y. and Hammer, P.L. 2011. *Boolean Functions. Theory, Algorithms, and Applications*. Encyclopedia of Mathematics and its Applications 142. Cambridge University Press, New York, USA, May 2011.

[6] Fujiwara, H. 1990. *Logic Testing and Design for Testability*. The MIT Press, Cambridge, MA, USA.

[7] Steinbach, B. and Posthoff, C. 2007. An extended theory of boolean normal forms. *In 6th Annual Hawaii International Conference on Statistics, Mathematics and Related Fields*, pp. 1124–1139, Honolulu, Hawaii.

[8] Steinbach, B. and Posthoff, C. 2017. *Boolean Differential Calculus*. Synthesis Lectures on Digital Circuits and Systems. Morgan & Claypool Publishers, May 2017.

[9] Thayse, A. 1981. *Boolean Calculus of Differences*. Springer, Berlin, Heidelberg, February 1981.

Chapter 3

Application of Various Modelling Techniques into Consumer Confidence Index

Betül Kalaycı,[1,*] *Vilda Purutçuoğlu,*[1] *Gerhard Wilhelm Weber,*[2] *Ömür Uğur*[1] and *Özlem Defterli*[3]

3.1 Introduction to Consumer Confidence Index

The principles of sentiment are presented by economic conditions such that most of the variance in consumer sentiment is caused by economic situations, either directly or indirectly. There are political and economic repercussions to investors' economic optimism or pessimism. Customer expenditures and hence, the direction of the economy's future are predicted by the degree of consumer confidence. Investors, who are in a similar way, are positioned economically, but, have different biased thoughts bringing forward quite different sentiments about the future of economics.

[1]Middle East Technical University.
[2]Poznan University of Technology.
[3]Çankaya University.
* Corresponding author: betulkalayci@gmail.com

Accordingly, when the current situation appears favorable, people are more upbeat about the economy both now and in the future. Basically, economic judgments become more negative, particularly, when inflation or unemployment rates rise. On the contrary, the general public's expectations about the economic future rise when major economic indicators display favorable times that may arrive soon [6].

In the underlying sentimental decision, the Consumer Confidence Index (CCI) plays a significant role. Because it informs the decision-makers and economic forecasters about the current and upcoming state of the economy and serves a special significance in influencing both commercial and governmental policy. Moreover, the CCI measures how optimistic consumers feel about the state of the economy based on their spending and saving habits, which contribute to the national economic expansion. Positive improvements in consumer confidence should be the source of economic growth, while negative changes should prevent it. Hereby, numerous researchers have attempted to determine how macroeconomic factors and CCI are related. Consumers' replies to inquiries regarding the present and future state of the national and personal economy form the foundation of CCI [12]. Following financial and macroeconomic indicators ensures advanced knowledge of the future direction that indicators will monitor. Because of this, CCI which reflects expectations operates as the primary indicators in addition to economic indicators for policy makers, participants in the financial markets, and representatives of the real estate sector. Keynes links *situation of long-term expectations* and *confidence situation* with being vulnerable to changes in the economy's feeling. According to the analysis of Keynes, the susceptibility of consumers and producers to economic improvements plays a crucial part in the statement's economic variations. Hence, consumer confidence and macroeconomic factors have a direct relationship. Considerations, feelings, and the economic decision-making process are not only impressed by numerous macroeconomic variables but, also affected by psychological, political, and sociological choices. As a result, the expectations and actions of economic decision-making units are influenced by the confidence index, which is a key indicator for the economy [3].

There are many studies about the effect of macroeconomic variables and consumer confidence index to each other. Afshar et al. (2007) [1] use quarter data for the USA from 1980 to 2005 to evaluate the links between the consumer, investor, and business confidence indices and economic oscillation. They apply the vector autoregression (VAR) model and the vector error correction model. From their analyses, it is seen that stock returns, purchasing managers' and consumer confidence index all reveal the enormous gap in the Gross Domestic Product (GDP). The results generally approve the considerations of economic swings in such a way that confidence indices play a significant impact [1, 3].

Korkmaz et al. (2009) [14] investigates the causal relationship between the BIST 100 index and Real Sector Confidence Index return by a two-stage dynamic association test. They first use the exponential general autoregressive conditional heteroskedastic (EGARCH) model to estimate the relationship between the variables, and then, they investigate the causal link between the mean and the variance of the model's standardized error terms [3, 14]. In order to determine the impact of the consumer and real sector confidence indices on the Turkish economy, Arısoy (2012) [2] created two different VAR models by using monthly data for the variables Consumption Expenditures, Industrial Production Index, Consumer Confidence Index, Real Sector Confidence Index, Employment Rate, and BIST Index between January 2005 and January 2012. The study's findings reveal that advances in industrial production and stock index are influenced by the Real Sector Confidence Index and the Consumer Confidence Index, respectively [2, 3].

Later, Dees et al. (2013) [7] studied the connection between the variables by using VAR and Threshold models by using the CCI for the USA and the Eurozone, consumption expenditures, real disposable income, financial and real estate wealth, real stock prices, short-term interest rates, unemployment rates, and quarter data from the January 1985–December 1985 and January 2010–December 2010 periods, They indicate that the US consumer confidence index serves as a *security channel* that ensures the shocks' transitivity and that it influences the Eurozone CCI [3, 7].

From all the underlying studies, it is seen that it is beneficial to understand the connections between some macro and financial variables and the CCI, a measure of consumer confidence in economies. A gain in investor and consumer confidence in the nation's economy may result from positive improvements in the factors that may impact such confidence levels. Recently, Başarır et al. (2019) [3] proposed to employ a VAR model to analyze the relation between the chosen macroeconomic and financial variables and CCI in Turkey. As a consequence of the study, it is discovered that there is a causal relationship between the CCI and the industrial production index as well as between the CCI and the BIST100, CPI, and USD Exchange Rates. Furthermore, when the VAR model's consequences are reviewed, it is observed that the USD exchange rate shock has a detrimental impact on the CCI, which in turn negatively affects the dollar exchange rate and BIST100 index [3].

Thereby, in our study, we aim to see the effect of economic variables on consumer confidence levels by using different modeling and operation research techniques. Firstly, we perform machine learning approaches and determine the optimism, and pessimism level of consumers via a hidden Markov model. Then, we apply volatility models to forecast the CCI.

3.2 Hidden Markov Model

A hidden Markov model (HMM) is attributed to a statistical model that is conceived to capture the hidden states of a system and their evolution conducted by a Markov process [22] where the observations are based on several states. Because of its capability of removing patterns and suitable calculation, HMM is one of the most well-known methods in the area of time series modeling [11], speech recognition, bioinformatics, semiconductor malfunction, engineering, image processing and the areas of physical as well as biological sciences [16, 22].

Moreover, this model has an extensive application to forecast stock market data. For instance, it is used for training models in order to forecast the next day's close value of the stock market, for which randomly produced transition probability matrices, emission probability matrices and previous probability matrices are thought. To improve the prediction accuracy and to overcome the overfitting problem, Hassan et al. (2005) [10] employ HMM for a better optimization success. They present a mixed model of HMM with a neural network for stock market forecasting. They also bring together the HMM and fuzzy logic rules to develop the prediction accuracy on non-stationary stock data sets [10, 16]. Then, Lee et al. (2009) [15] model the stock return as a mixture of Gaussian and discrete Markov chains to enhance the predictability of the stock model. They introduce another economic data in order to show Double HMM which works with the Markov Chain of the economic states separately. They affirm that the Double HMM forecasts better than the single HMM [15].

In general, the HMM approach framework is an unsupervised learning technique that allows us to explore new patterns without having to impose a template during the learning process.

The financial time series is produced by some underlying stochastic process that is most likely related to the market environment and investment decisions unknown to the public. That's why, there is a good match between sequential data and HMM in which the forecasting for the next state only depends on the present state not on the whole history of the past process [16]. Accordingly, Li et al. (2016) [16] apply the Hidden Markov Expert Model to forecast the underlying state change of the S&P 500 Index at monthly intervals. They identify the regimes as being two states, bull market and bear market. The underlying states can indirectly guide the direction of market price and thus, are important at the same level. To illustrate, the bear/bull state indicates a clearer temporal pattern than the original market price based on their analysis [16].

Based on the mathematical description, HMM has a sequence of observations and a sequence of states that produce them. Observations' sequence is denoted as $O = (O_1, O_2, \ldots, O_T)$ where each observation is an object from the set $o = \{o_1, o_2, \ldots, o_M\}$. Here, T shows the length of the observation and state sequences, and M denotes the number of possible observations. The hidden states which produce these observations are shown by $S = (S_1, S_2, \ldots, S_T)$ where each state is an object from a set of states $s = \{s_1, s_2, \ldots, s_N\}$. In these expressions, N shows the number of possible states [5]. There are also conditional independence assumptions as shown below.

- $P(O_k|S_1,\ldots,S_T,O_1,\ldots,O_T) = P(O_k|S_k)$ for any $1 \leq k \leq T$.
- $P(O_i,O_j|S_i,S_j) = P(O_i|S_i,S_j)P(O_j|S_i,S_j) = P(O_i|S_i)P(O_j|S_j)$ for $1 \leq i,j \leq T$.
- $P(S_k|S_1,\ldots,S_k-1) = P(S_k|S_k-1)$ for any $2 \leq k \leq T$, i.e., states from a Markov chain [5].

Because of these assumptions, the joint probability of the system can be written as [5]:

$$\begin{aligned} P(O_1,\ldots,O_T,S_1,\ldots,S_T) &= P(O_1,\ldots,O_T|S_1,\ldots,S_T)P(S_1,\ldots,S_T) \\ &= P(O_1|S_1)P(O_2|S_2)\ldots P(O_T|S_T)P(S_1)P(S_2|S_1)P(S_3|S_2)\ldots P(S_T|S_T-1).\ [5]. \\ &= (\textstyle\prod_{i=1}^{T} P(O_i|S_i))P(S_1)(\prod_{i=2}^{T} P(S_i|S_i-1)). \end{aligned} \tag{3.1}$$

For that reason, we require the probabilities below in order to define an HMM [5].

- *Transition probabilities*: $a_{ij} = P(S_t = j|S_t - 1 = i)$ for $1 \leq i, j \leq N$.
- *Emission probabilities*: $a_{ij} = P(O_t = j|S_t - 1 = i)$ for $1 \leq i \leq N, 1 \leq j \leq M$.
- *Initial probabilities*: $\pi = P(S_1|s_i)$ for $1 \leq i \leq N$.

Transition and emission probabilities can also be written in a matrix form. If we say A, B and initial probabilities as a vector π, HMM can be denoted by $\lambda = (A,B,\pi)$. By modeling via HMM, we focus on solving the following three basic problems [5]:

- Finding likelihood of an observation sequence given a HMM with parameters with λ.
- Finding the most probable state sequence given the model parameters and the observation sequence.
- Estimating the model parameters given sequences of states and observations [5].

3.2.1 Calculation of Likelihood

In order to find the likelihood of an observation sequence given the model parameters while the state sequence is hidden, there exists three major approaches.

These are the naive approach, forward algorithm, and backward algorithm. Here, the last two algorithms are developed to gain in computational cost.

Naive Approach

In this method, we find the likelihood given a specific state as shown previously and then, we sum over all possible states as below:

$$P(O|\lambda) = \sum_{S} P(O,S|\lambda), \tag{3.2}$$

where λ is the model parameter. In this description, there are N^T possible states. Therefore, if N and T are large, this approach will become computationally demanding in the order of $\mathcal{O}(N^T)$ to calculate the likelihood [5].

Forward Algorithm

The forward algorithm is a dynamic programming example where we separate the problem into subproblems and use the earlier results in a recursion. In this manner, it is found that the inference problem runs faster than the naive approach. In the below expression, we represent the likelihood of the observation sequence and a specific state at the last position of the state sequence given model parameters which are summed over all possible states [5]:

$$P(O|\lambda) = P_\lambda(O) = \sum_{i=1}^{N} P(S_T = s_i, O|\lambda). \tag{3.3}$$

Therefore, to find the term in the summation that is conditional on the model parameters λ, we define

$$\alpha_k(S_k) = P_\lambda(S_k = s_i, O_1, \ldots, O_k). \tag{3.4}$$

Lastly, the forward algorithm has the complexity $\mathcal{O}(N^2T)$. For large values of N and T, this complexity is lower than the complexity of the naive approach [5].

Backward Algorithm

The Backward algorithm is similar to the forward algorithm, except for the starting point of the calculation. Hereby, we find the likelihood by the expression below:

$$\begin{aligned} P_\lambda(O) &= \textstyle\sum_{i=1}^{N} P_\lambda(S_1 = s_i, O) \\ &= \textstyle\sum_{i=1}^{N} P_\lambda(S_1 = s_i)P_\lambda(O_1|O_2, \ldots, O_T, S_1 = s_i)P_\lambda(O_2, \ldots, O_T|S_1 = s_i) \\ &= \textstyle\sum_{i=1}^{N} P_\lambda(S_1 = s_i)P_\lambda(O_1|S_1 = s_i)P_\lambda(O_2, \ldots, O_T|S_1 = s_i). \\ &= \textstyle\sum_{i=1}^{N} \pi(s_i)b_{s_i}, O_1 P_\lambda(O_2, \ldots, O_T|S_1 = s_i). \end{aligned} \quad [5].$$

To obtain the solution, the last term in the sum is calculated by the following expression [5].

$$\left.\begin{aligned}\beta_k(S_k) &= P_\lambda(O_{(k+1)},\ldots,O_{(N)}|S_k)\\ &= \Sigma_{S_{(k+1)}=s_1}^{S_N} P_\lambda(O_{(k+1)},\ldots,O_T,S_{(k+1)}|S_k)\\ &= \Sigma_{S_{(k+1)}=s_1}^{S_N} P_\lambda(O_{(k+2)},\ldots,0_T|S_{(k+1)},S_k,O_{(k+1)})P_\lambda(O_{(k+1)}|S_{(k+1)},S_k)P_\lambda(S_{(k+1)}|S_k)\\ &= \Sigma_{S_{(k+1)}=s_1}^{S_N} P_\lambda(O_{(k+2)},\ldots,O_T|S_{(k+1)})P_\lambda(O_{(k+1)}|S_{(k+1)})P_\lambda(S_{(k+1)}|S_k)\\ &= \Sigma_{S_{(k+1)}=s_1}^{S_N} \beta_{(k+1)}(S_{(k+1)})b_{(S_{(k+1)})}O_{(k+1)}a_{(S_k,S_{(k+1)})}\end{aligned}\right\} \tag{3.5}$$

while $1 \le k \le N-1$. For $\beta_T(S_T)$, we cannot use the definition which is shown above since it involves $O_{(N+1)}$ which does not exist. Thus, if we use the recursion formula for $k = T-1$, we will get

$$\begin{aligned}\beta_T - 1(S_T - 1) &= \Sigma_{S_T=s_1}^{S_N} P_\lambda(O_T,S_T|S_{(T-1)})\\ &= \Sigma_{S_T=s_1}^{SN} \beta_T(S_T)P_\lambda(O_T|S_T)P_\lambda(S_T|S_{(T-1)}).\end{aligned} \quad [5]. \tag{3.6}$$

However, $P_\lambda(O_T,S_T|S_{(T-1)})$ can be also written as,

$$\begin{aligned}P_\lambda(O_T,S_T|S_{(T-1)}) &= P_\lambda(O_T,S_T|S_{(T-1)})P_\lambda(S_T|S_{(T-1)})\\ &= P_\lambda(O_T|S_T)P_\lambda(S_T|S_{(T-1)}).\end{aligned} \tag{3.7}$$

For that reason, for Equation 3.7 to hold, $\beta_T(S_T) = 1$ [5]. By using Equation 3.1, we obtain

$$P_\lambda(O) = \sum_{i=1}^{N} \pi(s_i)b_{(s_i),o_1}\beta_1(S_1 = s_i). \tag{3.8}$$

3.2.1.1 *Inference of the Most Probable Path and Inference of the Model Parameters*

By using one of the approaches stated below for the calculation of the likelihood, we can calculate the most plausible path for the observations and then, estimate the model parameters of HMM. There are three main methods for this purpose. These are the Viterbi algorithm, expectation-maximization algorithm, and Baum-Welch algorithm. In the following parts, we explain the mathematical details of each approach.

Viterbi Algorithm

This algorithm is a recursive algorithm that is used to find the most probable sequence, which is also named the path given the observation sequence and parameters. In the calculation, after initialization of the state, in every step, we use the previous paths that are found so that the following expression can be computed [5].

$$S^* = argmax_S P(S|O). \tag{3.9}$$

In Equation 3.9, if $f(a) \ge 0$ for all a and $g(a,b) \ge 0$ for all a,b, we will have

$$max_{a,b} f(a)g(a,b) = max_a\{f(a)max_b g(a,b)\}, \tag{3.10}$$

and we get

$$argmax_S P(S|O) = argmax_S P(S|O) \quad (3.11)$$

since $P(O)$ does not include any element from hidden states [5].

Expectation-Maximization Algorithm

The expectation-maximization (EM) algorithm is the method generally used for incomplete or missing data while we need to find the maximum-likelihood estimate of the parameters [4].

The EM algorithm consists of two main steps. The first step, which is called *expectation*, happens when some missing values or latent variables are obtained in the data, because of the problems or limitations that occur in the observation process [4]. To solve this problem, for each of these latent variables, the expected value is computed. The second step happens when the optimization of the likelihood function is analytically challenging. However, by assuming the existence of values for hidden parameters, the likelihood function can be simplified [4]. The complete data is used which is obtained in the previous (expectation) step and the parameters are updated in this step.

In the mathematical representation of this approach, We define the density function $p(x|\theta)$ where θ is a set of parameters. So, our distribution of interest becomes $\mathscr{X} = \{X_1, \ldots, X_N\}$ for the data size N. Here, we assume an incomplete data set $\mathscr{X}$ that is observed and generated by distribution. On the other hand, the complete data set is defined as $\mathscr{Z} = (\mathscr{X}, \mathscr{Y})$ and the joint density function is computed as below [4].

$$p(z|\theta) = p(x,y|\theta) = p(y|x,\theta)p(x|\theta). \quad (3.12)$$

The joint density in Equation 3.12 stems from the marginal density function $p(x|\theta)$ and the assumption of latent variables and parameter value estimates [4]. By using this novel density function, a new likelihood function which is named the *complete-data likelihood*, can be calculated as

$$\mathscr{L}(\theta|\mathscr{Z}) = \mathscr{L}(\theta|\mathscr{X},\mathscr{Y}) = p(\mathscr{X},\mathscr{Y}|\theta). \quad (3.13)$$

In Equation 3.13, the underlying likelihood function is a random variable because missing information $\mathscr{Y}$ is unknown and random. Thus, the likelihood function $\mathscr{L}(\theta|\mathscr{X},\mathscr{Y}) = h_{\mathscr{X},\theta}(\mathscr{Y})$ for some functions where $\mathscr{X}$ and θ are constant and $\mathscr{Y}$ is a random variable. In this expression, the incomplete-data likelihood function is referred as the original likelihood $\mathscr{L}(\theta|\mathscr{X})$ [4].

With more details, the EM algorithm initially computes the expected value of the complete-data log-likelihood $\log p(x,y|\theta)$ with respect to the unknown data $\mathscr{Y}$ given the observed data $\mathscr{X}$ and the current parameter estimates, which is defined as [4]:

$$Q(\theta, \theta^{(i-1)}) = E[\log p(\mathscr{X},\mathscr{Y}|\theta)|\mathscr{X},\theta^{(i-1)}]. \quad (3.14)$$

Here, $\theta^{(i-1)}$ are the present parameter estimates that we use to interpret the expectation and θ are the new parameters that we optimize to raise Q. At this point, to clarify this statement, $\mathscr{X}$ and $\theta^{(i-1)}$ are constants, θ is a normal variable which needs to be adjusted and $\mathscr{Y}$ is a random variable conducted by the distribution $f(y|\mathscr{X},\theta^{(i-1)})$. For that reason, the right side of the Equation 3.14 can be rewritten as [4]

$$E(\log p(\mathscr{X},\mathscr{Y}|\theta)|\mathscr{X},\theta^{(i-1)}) = \int_{y\in\mathscr{Y}} \log p(\mathscr{X},y|\theta)f(y|\mathscr{X},\theta^{(i-1)})dy, \quad (3.15)$$

where $f(y|\mathscr{X},\theta^{(i-1)})$ represents the marginal distribution of the unobserved data and is dependent on both the observed data $\mathscr{X}$ and on the present parameters. $\mathscr{Y}$ refers to the space of the values y.

Suppose that there is a function $h(.,.)$ of two variables and $h(\theta,Y)$ where Y denotes a random variable having a distribution $f_Y(y)$ and θ shows the constant. Then, $q(\theta) = E_Y[h(\theta,Y)] = \int_y h(\theta,Y)f_Y(y)dy$ is a deterministic function that could be maximized if preferred [4].

The interpretation of this expectation is defined as *E-step* of the algorithm. If we realize the meaning of the two arguments in the function $Q(\theta,\theta')$, the first argument θ relates to the parameters that eventually will be optimized to maximize the likelihood and the second argument θ' relates to the parameters which is used to utilize the expectation. On the other side, the second step, *M-step*, of the EM Algorithm is to maximize the expectation that is found in the first step, which is [4]:

$$\theta^{(i)} = \underset{\theta}{\operatorname{argmax}} Q(\theta,\theta^{(i-1)}). \quad (3.16)$$

These steps are iterated as required. Each repetition is assured to raise the log-likelihood and the algorithm is undertaken to converge to a local maximum of the likelihood function [4].

Baum-Welch Algorithm

For an initial HMM and a given sequence of observations, the Baum-Welch algorithm induces optimal parameters to the HMM. Since the Baum-Welch algorithm is a special case of the EM algorithm, the algorithm converges to a local solution which could not be the global optimum [18].

Here, we begin the computation with a starting estimation of the parameters λ. Then, the estimation is iteratively repeated until λ converges. In these calculations, for the predictor of the transition probability between ith and jth variables, the following expression is employed [5]:

$$\hat{a}_{ij} = \frac{Expected\ number\ of\ transitions\ from\ i\ to\ j}{Expected\ number\ of\ transitions\ from\ i}. \quad (3.17)$$

In order to obtain these expected values, the below equation is applied [5].

$$\begin{aligned} \xi(i,j) &= P(S_t = i, S_{t+1} = j|O,\lambda), \\ &= \frac{P(S_t=i,S_{t+1}=j,O|\lambda)}{P(O|\lambda)}, \\ &= \frac{\alpha_t(S_t=s_i)a_{ij}b_{S_{t+1}=s_j;o_{t-1}}\beta_{t+1}(S_{t+1}=s_j)}{\sum_{j=1}^{N}\alpha_t(S_t=s_j)\beta_t(S_t=s_j)}. \end{aligned} \quad (3.18)$$

To predict the emission probability matrix B, similar to Equation 3.18, the following expression is computed.

$$\hat{b}_j(o_k) = \frac{\textit{Expected number of times being in state } s_j \textit{ observing } o_k}{\textit{Expected number of times being in state } s_i} \quad (3.19)$$

In these equations, γ_t is defined as

$$\gamma_t(j) = P(S_t = j|O,\lambda) = \frac{P(S_t = j, O|\lambda)}{P(O|\lambda)} = \frac{\alpha_t(S_t = s_j)\beta_t(S_t = s_j)}{\sum_{j=1}^{N}\alpha_t(S_t = s_j)\beta_t(S_t = s_j)}. \quad (3.20)$$

Here, the estimate for b_j can be written as follows.

$$\hat{b}_j(o_k) = \frac{\sum_{t=1 st O_k=o_k}^{T}\gamma_t(j)}{\sum_{t=1}^{T}\gamma_t(j)}. \quad (3.21)$$

In addition, the estimate of the initial probability π can be expressed as

$$\hat{\pi}_i = \gamma_1(i). \quad (3.22)$$

3.2.2 Application of HMM

The consumer confidence index shows the future developments of households' consumption and saving, based on answers concerning their expected financial situation, their sentiment about the general economic situation, unemployment and capability of savings. In our analyses, we use an indicator above 100 as the signal of an increase in the consumers' confidence towards the future economic situation. As a result, households are less prone to save, and eager to spend money on major purchases in the next 12 months. On the other hand, we accept that Values below 100 show a pessimistic attitude towards future developments in the economy, probably resulting in a leaning to save more and consume less.

In Turkey, the goal of the Consumer Tendency Survey is to measure current situation assessments and future period expectations of consumers on personal financial standing and general economic course and to determine consumers' expenditure and saving tendencies for the near future. The survey covers a randomly chosen sample of all individuals at the age of 15 and above having a job in urban and rural areas. The index is evaluated between 0 and 200. The reading above 100 means consumers are optimistic and below 100 are pessimistic.

Application is done by using the economic variables which are the Consumer Confidence Index (CCI), Unemployment Index, Consumer Price Index, and USD/TRY Index. The CCI, Unemployment Index and CPI are taken from the Turkish Statistical Institute (TÜİK) and USD/TRY exchange rates are taken from the Central Bank of Turkey (TCMB, EVDS Data Central). All these data sets are monthly, starting from 2005 to up to 2023. For that reason, we have 206 variables in total.

On the other side, the Markov model is a probabilistic process that looks at the current state to forecast the probability moving to the next state. HMM attends to play when the complexity is based on not knowing the probability of each regime shift and how to explain these probabilities modifying over time. So, they are able to estimate the transition probabilities for each regime and then, based on current conditions, the output of the most possible regime can be estimated [16]. In this analysis, we apply HMM to forecast the underlying state change of the CCI at a monthly interval. The regimes are defined as being three states, optimistic investor, pessimistic investor, and neutral investor. The aim is to see and analyze the switch between these states and to define a path of these states [16]. We have 206 variables at total. Unfortunately, only one value is higher than 100, which is seen in 2006. That is why, at this point we require a region so that we could make an analysis about the optimism and pessimism level of investors. To construct this region, firstly we take the mean of our CCI data, which is found as 88.9. Afterwards, we calculate the standard deviation of the data set, which is calculated as 6.91. Lastly, by using *Chebyshev's Inequality* given in Equation 3.23, we determine our region.

$$P(|x-\mu| \geq k) \leq \frac{\sigma^2}{k^2}, \tag{3.23}$$

where μ is mean and σ represents the standard deviatio while $\sigma^2 > 0$ [21]. Here, to determine a significant interval that provides us with the most effective region, we set the constant $k = 0.5$. After the calculation, our region within the interval is observed between 85.45 and 92.36. Thus, we evaluate the optimism level of investors if the variables are higher than 92.36, and the pessimism level of investors if the variables are lower than 85.45. Between these intervals are called *neural* or *irresponsive* investors.

After all these determinations, we acquire the probability matrices which are initial, transition, and emission probabilities. According to the initial probability, there is a 0.36 probability for investors to be optimistic, a 0.41 chance to be pessimistic, and 0.23 to be neutral. Furthermore, the transition matrix tells us the probability of moving from one state to each of the states. Here, the transition matrix shows us that there is a 74.3% chance that it stays in the optimistic state, there is a 16.2% chance that it moves to the pessimistic state, and a 9.5% chance to move to the neutral state based on the current data set. On the other hand, there is a 15.3% chance for pessimistic investors to become optimistic and an 84.7%

chance to stay in the pessimistic state. Moreover, for the neutral investors, only 13% chance to move to the optimistic state and an 87% chance to stay in the neutral state.

Accordingly, in order to see the posterior probabilities of being in each state at each time point for a given sequence of observations and a given HMM, we calculate the posterior odds over the entire data set. For the emission probability matrix, we determine an observation. One observation spreads from a state at each time step. Observation symbols are denoted as *H (High), M (Medium)* and *L (Low)*. While constructing this observation, to detect reasonable results, we get the $k = 1$ in Chebyshev's inequality. That is, if the results are higher than 95.81, our observation will be named as *H*. On the other hand, if the value is less than 81.99, our observation will be called *M*, and if the value is between these intervals, then the observation will be denoted as *L*. Hereby, our results show that the probability of optimistic investors staying in the *M* interval is 100.0%. However, the probability for pessimistic investors to be observed at the *H* level is found as 30.0% and to be observed at the *M* level is computed as 70.0%. Also, for the neutral investors, the probability of staying in the *L* interval is calculated as 95.0% and the *M* level is estimated as 5.0%. All these estimations are tabulated in Tables 3.1 and 3.2.

Table 3.1: Transition probabilities in HMM.

	To		
From	Optimistic	Pessimistic	Neutral
Optimistic	0.74	0.16	0.10
Pessimistic	0.15	0.85	0.00
Neutral	0.13	0.00	0.87

Table 3.2: Emission probabilities of HMM.

	To		
From	H	L	M
Optimistic	0.00	0.00	1.00
Pessimistic	0.29	0.00	0.71
Neutral	0.00	0.96	0.04

Additionally, from the analyses, it is observed that the Viterbi training involves much less computational effort than the Baum-Welch algorithm with the same or even a little worse performance. Thus, it is widely preferred among designers of speech recognition systems. On the other hand, the the Baum-Welch algorithm indicates some curious properties: In the case of discrete HMM, it does not require any model initialization, just non-zero random values confirming the stochastic constraints; it thoroughly uses all the available data to generate robust and optimal estimates [19]. From the side of Viterbi training, it is shown that even in the case of discrete HMM, it requires some reasonable initialization by employing the models found for other databases or by training initial models on a hand-labeled subset of the training database [19].

In the following paragraphs, we present the application of our model to observe the comparison and similarities between these algorithms and to analyze the probability of our states.

In order to make an inference based on a trained model and observed data. The most likely states based on our estimates from observed data, we use the *viterbi algorithm* [16]. According to the Viterbi algorithm, we obtain the same initial probabilities for optimistic, pessimistic, and neutral investors. However, for the transition and emission probability matrices, we see different results. For the results of the Viterbi algorithm, the transition matrix shows us that there is a 93.4% chance that it stays in the optimistic state, there is a 2.5% chance that it moves to the pessimistic state, and a 4.1% chance to move to the neutral state based on the current data set. On the other hand, there is a 10.3% chance for the pessimistic investors to become optimistic and an 89.7% chance to stay in the pessimistic state. Moreover, for the neutral investors, only an 8% chance to move to the optimistic state and a 92% chance to stay in the neutral state. According to the emission probability matrix, the probability of the optimistic investors to stay in the M interval is found as 100.0%. However, the probability for the pessimistic investors to be observed at the H level is computed as 64.1% and to be observed at the M level is 35.9%. Also, for the neutral investors, the probability of staying in the L interval is found as 97.8%, and M level is calculated as 2.2%. Tables 3.3 and 3.4 summarize the underlying findings.

Table 3.3: Transition probabilities found in the Viterbi algorithm.

	To		
from	Optimistic	Pessimistic	Neutral
Optimistic	0.93	0.03	0.04
Pessimistic	0.10	0.90	0.00
Neutral	0.00	0.97	0.03

Table 3.4: Emission probabilities found in the Viterbi algorithm.

	To		
From	H	L	M
Optimistic	0.0	0.0	1.0
Pessimistic	0.64	0.0	0.36
Neutral	0.0	0.98	0.02

From the literature, we can deduce that the most credible algorithm to train the HMM is given as the Baum-Welch algorithm. The Baum-Welch algorithm can train the given HMM by an observation sequence and produce a new HMM for detection [23]. Therefore, apart from the Viterbi algorithm, we also see the values of our probability matrix via the *Baum-Welch algorithm*. There are only a few differences between the results that we obtain with the Viterbi algorithm and the Baum-Welch method. Initial probabilities are taken as the same with the previous algorithms. On the other side, for the transition probability matrix, it is estimated that there is a 92.3% chance that it stays in the optimistic state, there is a 3.6% chance that it moves to the pessimistic state, and a 4.1% chance to move to the neutral state based on the current data set. On the other side, there is a 12.8% chance for the pessimistic investors to become optimistic and an 87.2% chance to stay in the pessimistic state. Moreover, for the neutral investors, only an 8.3% chance to move to the optimistic state and a 91.7% chance to stay in the neutral state.

Moreover, for the emission probability matrix, it is computed that the probability of optimistic investors staying in the *M* interval is 100.0%. However, the probability for the pessimistic investors to be observed at the *H* level is found as 60.6% and to be observed at the *M* level is computed as 39.4%. Also, for the neutral investors, the probability of staying in the *L* interval is observed as 95.9% and the *M* level is calculated as 4.1%. Similar to our previous outcomes, we summarize all the estimates of the transition and emission probabilities via the Baum-Welch algorithm in Tables 3.5 and 3.6.

Table 3.5: Transition probabilities found by the Baum-Welch algorithm.

	To		
From	Optimistic	Pessimistic	Neutral
Optimistic	0.92	0.04	0.04
Pessimistic	0.13	0.87	0.0
Neutral	0.08	0.0	0.92

Table 3.6: Emission probabilities found by the Baum-Welch algorithm.

	To		
From	H	L	M
Optimistic	0.0	0.0	1.0
Pessimistic	0.60	0.0	0.40
Neutral	0.0	0.96	0.04

From all these analyses, it is observed that the essential problem related to HMM is to take a sequence of observations, which is defined as a set of hidden states, and fit the most potential HMM. The Baum-Welch algorithm, which is an EM algorithm is used when the HMM parameters are not immediately (empirically) measurable, which is usually the case in real applications similar to our situation. The second step is the *decoding task* which implies searching for the most probable sequence of hidden states given some observations, namely, getting the hidden states that are produced by the monitored output. The sequence of states obtained by the Viterbi algorithm is then compared with the ground truth to measure the accuracy [17].

On the other side, the investors' general success depends on long-term thinking. However, most of them can not help worrying about the short time adjustment to their portfolios. This worry is brought about by the current risings in volatility over the last few years. The past has indicated that the stock market and the economy move in cycles that repeat over and over. For that reason, understanding the different stages of the economy can help guide investment decisions. in such as way that investors can show immense confidence and can tend to increase their stock holding position. On the contrary, safe-haven assets, like gold and bonds, can drop by the wayside in a bear market, which shows a lack of confidence in the economy [16].

3.3 Financial Volatility Models

While trading in financial markets, one of the key roles is to try to seize the movements of the underlying asset. These movements are called *volatility*. The volatility, represented by the symbol σ_t, is the conditional standard deviation of the return on the underlying asset. The volatility possesses some remarkable properties. Its swings with time are one of those crucial characteristics. However, daily data does not immediately reveal this property since each trading day only includes one observation.

Furthermore, the volatility it seems appears in clusters, stops at particular spans, and does not continue to expand indefinitely. Moreover, when the fundamental asset's value declines, the volatility responds differently than when the price of the underlying asset rises. Additionally, it is contingent upon the trading in each day and between the days (the overnight volatility) [9].

Therefore, the aim of the volatility analysis is to describe the factors that produce and affect the volatility. Here, forecasting benefits greatly from the time series structure. The estimation provided by autoregressive conditional heteroskedasticity (ARCH) and generalized ARCH (GARCH) models can be easily utilized if exogenous or predefined variables are present. Thus, both the variance and the mean can be thought of as parts of the estimation problem. To find the optimum formulation, we can also complete the description analyses and the hypothesis testing. Hence, it shouldn't be a surprise that the volatility might be interpreted as a response to news. However, the timing of the news may not have been unexpected, resulting in the volatility's predictable elements, including economic updates. Other news events have an impact on the amount of news events. For instance, the volatility seen in Asian markets earlier in the day as well as the volatility seen in the United States the day before may have an impact on the magnitude of the return moves on the American stock market [8].

Accordingly, market volatility can cause changes in the way financial assets are distributed in terms of risk [20]. These changes naturally affect the sentiment of investors. If we bind this approach with behavioral finance, financial choices are typically predicated on a trade-off between risk and reward. For that reason, the econometric study of risk is a crucial component of risk management, portfolio optimization, option pricing, and asset pricing. So, the ARCH and GARCH models are carried out to a diversified amount of time series analyses and these applications in finance are particularly successful [8].

In order to investigate the connection between investor sentiment and market volatility on the Johannesburg Stock Exchange, Rupande et al. (2019) employ a daily sentiment composite index built from a collection of proxies and GARCH models. The findings indicate a considerable relationship between investor confidence and stock return volatility, demonstrating how behavioral finance may effectively explain how stock returns behave on the Johannesburg Stock Exchange. Additionally, their study stresses how noise trading activity, which is driven by investor sentiment, increases financial market volatility on the South African market [20].

Below, we present the details of the most well-known volatility models, namely, ARCH, GARCH and EGARCH approaches.

3.3.1 Autoregressive Conditional Heteroskedasticity Model

To capture the return on an asset, the autoregressive conditional heteroskedastic (ARCH) model is employed. The ARCH model is predicated on two assumptions regarding the features of the volatility for any time-based financial series. The former presumption holds that there are clusters of high volatility and that the movement of an asset's return depends on its previous value. However, it is unrelated across the whole time series. According to the latter hypothesis, a quadratic function of the earlier delayed values can explain the distribution of the asset returns (a_t), as it is dependent on earlier values [9].

Thereby, with mathematical notations, the input collection present at time $(t-1)$ is used to build the model. The prior m lag innovations determines the conditional variance [9] via

$$\sigma_t^2 = \omega + \alpha_t a_{t-1}^2 + \ldots + \alpha_m a_{t-m}^2. \tag{3.24}$$

In Equation (3.24), σ_t^2 denotes the conditional variance, a_t shows the error term (also called as return residuals), ω and α are the parameters of the model. By squaring them, we can observe that substantial changes in the return innovation exert a more significant impact on the conditional variance. This implies that significant disturbances have a tendency to occur in succession, which is similar to how volatility clusters behave [9].

Thus, the ARCH (m) model can be described as

$$a_t = \sigma_t \varepsilon_t, \quad \sigma_t^2 = \omega + \alpha_t a_{t-1}^2 + \ldots + \alpha_m a_{t-m}^2, \tag{3.25}$$

where ε_t is a white noise and $\varepsilon_t \sim N(0,1)$ under the assumptions of the independent and identically distribution (iid), $\omega > 0$ and $\alpha_i \geqslant 0$ for $i > 0$.

3.3.2 Generalized Autoregressive Conditional Heteroskedasticity Model

There are various criteria that must be followed in the ARCH model such that the model's prediction of the volatility is accurate. For that reason, it is possible to consider and recommend transforming the ARCH model to create a generalized ARCH (GARCH) model. To explain the GARCH model, we begin with the continuously compounded log return series r_t.

Here, let the novelty at time t (a_t) be denoted by $a_t = r_t - \mu_t$ [9]. At this point, we are able to edit e a_t in a GARCH (m,s) model by

$$a_t = \sigma_t \varepsilon_t, \quad \sigma_t^2 = \omega + \sum_{i=1}^{s} \alpha_i a_{t-i}^2 + \sum_{j=1}^{m} \beta_j \sigma_{t-j}^2, \tag{3.26}$$

where ε_t is iid and ε_t $N(0,1)$, a_t is the model's residual at time t, σ_t denotes the conditional standard deviation (volatility) at time t. Moreover, m represents the order of the ARCH component model and s presents the order of the GARCH component model. Furthermore, ω and $\alpha_1, \ldots, \alpha_m$ are the parameters of the ARCH component model and $\beta_1, \ldots, \beta_s$ are the parameters of the GARCH component model. In addition, $\omega > 0$, $\alpha_i \geqslant 0$, $\beta_j \geqslant 0$ and $\sum_{i=1}^{max(m,s)} (\alpha_i + \beta_i) < 1$ [9].

On the other side, the limitation on the ARCH and GARCH parameters (α_i, β_i) is that the conditional standard deviation increases (σ_t) and the volatility (a_t) is finite. The GARCH parameter (β_i) disappears and the remaining portion transforms into an ARCH (m) model if $s = 0$ [9].

Hereby, Equation (3.26) can be also expressed as

$$a_t = \sigma_t \varepsilon_t, \quad \sigma_t^2 = \omega + \alpha a_{t-1}^2 + \beta \sigma_{t-1}^2. \tag{3.27}$$

in which $0 \leqslant \alpha, \beta \leqslant 1, (\alpha + \beta) < 1$

According to this model, large values for a_{t-1}^2 and σ_{t-1}^2 are likely to produce large values for σ_t^2. When clusters of volatility occur [9].

3.3.3 Exponential Generalized Autoregressive Conditional Heteroskedasticity Model

Even though GARCH is an improved version of ARCH, the management of the financial time series may still provide some challenges for the GARCH model.

Therefore, a more comprehensive model is suggested. This extended model is called the exponential GARCH (EGARCH) model. In the EGARCH model, the proposal change is the inclusion of a measured creation to the model, which can account for the inequalities in the asset's return volatility [9].

Thereby, let a_t still be the novelty of the return at time t. The EGARCH (m,s) model can then be expressed as

$$a_t = \sigma_t \varepsilon_t, \quad ln(\sigma_t^2) = \omega + \sum_{i=1}^{s} \alpha_i \frac{|a_{t-i}| + \theta_i a_{t-i}}{\sigma_{t-i}} + \sum_{j=1}^{m} \beta_j \ln(\sigma_{t-1}^2). \tag{3.28}$$

In Equation (3.28), a_t describes the model's residual at time t, σ_t is the conditional standard deviation (volatility) at time t; s represents the order of the ARCH component model and m defines the order of the GARCH component model. Moreover, ω and $\alpha_1, \ldots, \alpha_s$ are the parameters of the ARCH constituent algorithm and $\beta_1, \ldots, \beta_m$ are the GARCH constituent model's parameters. Thus, EGARCH $(1,1)$ is represented as

$$a_t = \sigma_t \varepsilon_t, \quad \ln(\sigma_t^2) = \omega + \alpha(|a_{t-1}| - E(|a_{t-1}|)) + \theta a_{t-1} + \beta ln(\sigma_{t-1}^2), \tag{3.29}$$

where ε_t and $|a_{t-1}| - E(|a_{t-1}|)$ are iid (identically independent distributed) and have mean zero. When the distribution of the error term in the EGARCH is Gaussian, then, $E(|\varepsilon_t|) = \sqrt{2/\pi}$ which gives

$$\ln(\sigma_t^2) = \omega + \alpha(|a_{t-1}| - \sqrt{2/\pi}) + \theta a_{t-1} + \beta ln(\sigma_{t-1}^2). \tag{3.30}$$

Here, there is a single quality that has to be emphasized. The adverse volatility shocks typically have a larger effect and thus, θ is frequently considered to be negative. Indeed as the model employs logarithms, it gives rise to challenges when attempting to estimate an unbiased forecast [9].

Hereby, in our study, as previously applied in the study of Kalaycı (2017) [13], the stochastic model is converted to a MARS-similar expression. By using this finding, we consider checking whether these volatility models can be written similarly to a MARS structure or as a model within the generalized additive model. Alternatively, we also think to detect whether MARS can be written as a similar structure to the ARCH/GARCH model if the errors are at least defined as the normal (Gaussian) distribution. In this way, we can investigate whether the MARS-based model in time series analyses can be informative in behavioral finance data sets.

3.3.4 Application of Volatility Models

We apply these volatility models to CCI. As stated previously, in this data set there are 206 observations. In the analyses, we initially convert the series to a return series by taking its first difference and checking them for the descriptive

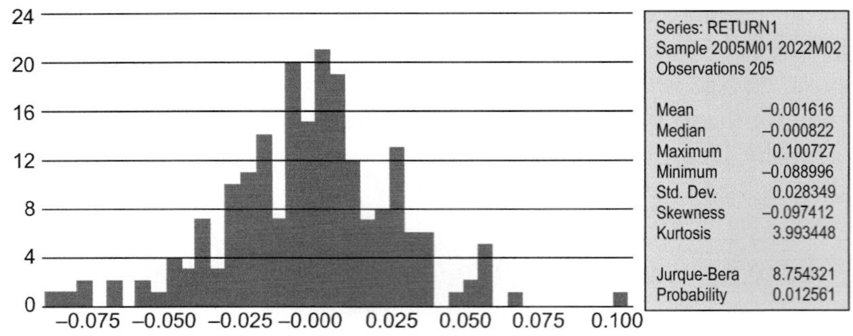

Figure 3.1: The descriptive statistics and the histogram of the return series of CCI.

statistics as shown in Figure 3.1. In this updated data set, we control the outliers by using their scatter plot.

Accordingly, by taking ± 3 standard deviation as the indication of the outliers as used majority of various data sets, we assign the 31st and the 165th observations as outliers. Then, we replace the positive shock by +1 (131st) and the negative shock (165th) by −1 values. Afterward, we check for the stationary case and determine the autoregressive model with the order 2 (AR(2)) and moving average model with the order 1 (MA(1)) as the best-fitted models for the presentation of the variance in the selected volatility model. For the computation, we find ARCH=1 and EGARCH=2 as the optimal model based on Akaike Information Criterion (AIC) and Bayesian Information Criterion (BIC) as seen in Table 3.7.

Table 3.7: Performance of model selection criteria for volatility models with AR(1) and MA(2) model by using CCI.

Fitted Volatility Model	AIC	BIC
ARCH(1)	4.541	4.623
EGARCH(1,2)	4.496	4.626

3.3.4.1 Application for Multivariate Case

For the multivariate volatility case, we apply GARCH model by taking CCI as an output, and other variables (UN, CPI, USDTRY) as an input. We obtain results for ARCH=1. Among our 206 variables, we forecast for the last 20 variables.

In Figure 3.3, the first graph shows us the sample of forecasting for the full data, and at the second graph, we forecast results for the modified sample. This time, the return on CCI is observed as volatile. Accordingly, from the beginning of 2018, when the crisis of currency occurred, the volatility still continued due to the decisions about the macro policy, which made most of the macroeconomic

Table 3.8: Performance of model selection criteria for volatility models with AR(1) and MA(1) model by using CCI for the multivariate case. Here, the errors are assumed as normal (Gaussian) and student-t distributions.

	Under Normal		Under Student-t	
Fitted Volatility Model	AIC	BIC	AIC	BIC
ARCH(1)	-4.6597	-4.5556	-4.6491	-4.5281

variables uncertain and resulted in the volatile level of consumers' confidence level increasing.

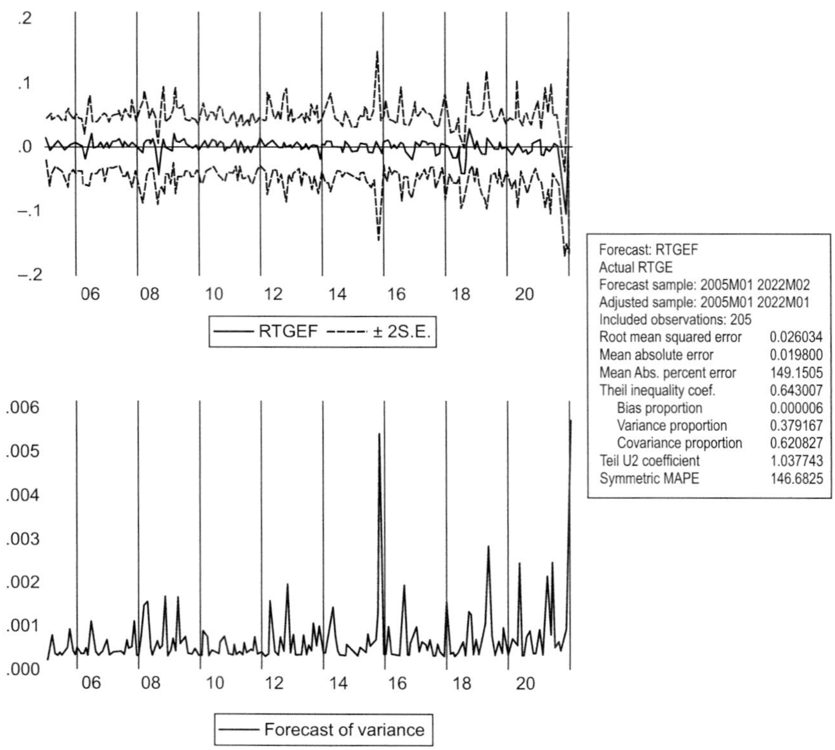

Figure 3.2: Forecasting results of the return series for the entire CCI data.

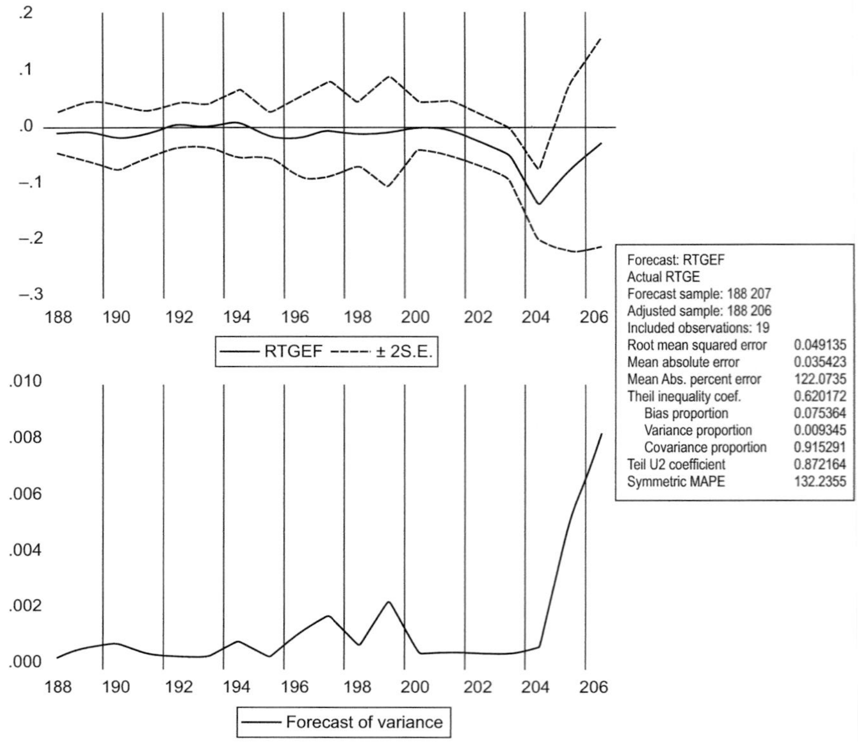

Figure 3.3: Forecasting result of the return series of CCI for the last 20 variables.

3.4 Conclusion

The principles of sentiment are presented by economic conditions such that most of the variation in consumer sentiment stems directly or indirectly from economic conditions. How optimistically or pessimistically investors see the economy has both economic and political consequences. People feel more optimistic about the present and future economy when the current case looks good. Especially, when inflation rates rise or unemployment increases, economic assessments become more pessimistic. When main economic indicators forecast good times ahead, the public's expectations of the economic future improve [6]. Hereby, the consumer confidence index (CCI) identifies the degree of the underlying optimism on the current state of the economy that consumers are expressing through their activities of saving and spending leading to the economic growth of the country.

At this point, we prefer to study the Hidden Markov model (HMM). HMM is attributed to a statistical model conceived to capture the hidden states of a system and their evolution, which are conducted by a Markov process [22]. In this analysis, we apply HMM to forecast the underlying state of change in the CCI at a monthly interval. The regimes are defined as three states, namely, optimistic investor, pessimistic investor, and neutral investor. The aim is to see and analyze the switch between these states and to define a path of these states [16].

Unfortunately, among our variables, only one value is higher than 100, which was seen in 2006. That is why, we need to set a region so that we can identify the level of optimism and pessimism of investors. By using *Chebyshev's Inequality*, we determine our region. Then, in order to see the posterior probabilities of being in each state at each time point for a given sequence of observations and a given HMM, we calculate posterior odds over the entire data set. For the emission probability matrix, we determine an observation. One observation spreads from a state at each time step. Observation symbols are: denoted as *H (High)*, *M (Medium)*, and *L (Low)*. Our results show us that the probability of optimistic investors staying in the *M* interval is 100.0%. However, the probability for pessimistic investors to be observed at the *H* level is 30.0% and to be observed at the *M* level is observed as 70.0%. Also, for the neutral investors, the probability of staying in the *L* interval is calculated as 95.0% and in the *M* level is found as 5.0%. We also see the application of our model to observe the comparison and similarities between Viterbi and Baum-Welch algorithms and to analyze the probability of our states. There are only a few differences between the results that we obtain in both algorithms.

Finally, we obtain forecasting results for the volatility models, starting from the beginning of 2018, when the crisis of currency occurred. The crisis has been still continuing due to the decisions about the macro policy, which makes most of the macroeconomic variables uncertain and results in a volatile level of consumer confidence level to increase.

In conclusion, we observe that the suggested modeling approaches and associated inference techniques of their model parameters on CCI give us a better understanding of the decisions of the consumers and enable us to make an accurate prediction about their future behaviors. Therefore, we believe that the proposal analyses which are the combination of classical time series volatility modeling and HMM can be adapted to other economic indices and more complicated volatility models can be used together with HMM to capture various hidden decisions in economic data sets and to make better planning of economical strategy.

References

[1] Afshar, T., Arabian, G., Zomorrodian, R. et al. 2007. Stock return, consumer confidence, purchasing managers index and economic fluctuations. *Journal of Business & Economics Research (JBER)*, 5(8).

[2] Arısoy, I. 2012. Türkiye ekonomisinde iktisadi güven endeksleri ve seçilmiş makro değişkenler arasındaki ilişkilerin var analizi. *Maliye Dergisi*, (162): 304–315.

[3] Başarır, C., Murat Bicil, I. and Yılmaz, O. 2019. The relationship between selected financial and macroeconomic variables with consumer confidence index. *Journal of Yaşar University*, 14: 173–183.

[4] Bilmes, J.A. et al. 1998. A gentle tutorial of the em algorithm and its application to parameter estimation for gaussian mixture and hidden markov models. *International Computer Science Institute*, 4(510): 126.

[5] Dogan Dar, E., Purutçuoglu, V. and Purutçuoglu, E. 2020. Detection of HIV-1 protease cleavage sites via hidden markov model and. *Numerical Solutions of Realistic Nonlinear Phenomena*, 31: 171.

[6] De Boef, S. and Kellstedt, P.M. 2004. The political (and economic) origins of consumer confidence. *American Journal of Political Science*, 48(4): 633–649.

[7] Dees, S. and Soares Brinca, P. 2013. Consumer confidence as a predictor of consumption spending: Evidence for the united states and the euro area. *International Economics*, 134: 1–14.

[8] Engle, R. 2001. Garch 101: The use of arch/garch models in applied econometrics. *Journal of Economic Perspectives*, 15(4): 157–168.

[9] Grek, A. 2014. Forecasting accuracy for arch models and garch (1, 1) family: Which model does best capture the volatility of the swedish stock market?

[10] Hassan, Md. R. and Nath, B. 2005. Stock market forecasting using hidden markov model: A new approach. *In 5th International Conference on Intelligent Systems Design and Applications (ISDA'05)*, pp. 192–196. IEEE.

[11] Huang, Y. 2020. Estimation and testing of nonparametric hidden markov model with application in stock market. *Communications in Statistics-Theory and Methods*, 49(24): 5917–5929.

[12] Ul Islam, T. and Mumtaz, M.N. 2016. Consumer confidence index and economic growth: An empirical analysis of EU countries. *EuroEconomica*, 35(2).

[13] Kalaycı, B. 2017. *Identification of Coupled Systems of Stochastic Differential Equations in Finance including Investor Sentiment by Multivariate Adaptive Regression Splines*. Ph.D. Thesis, Middle East Technical University.

[14] Turhan Korkmaz, T. and Cevik, E. 2009. Reel kesim güven endeksi ile imkb 100 endeksi arasındaki dinamik nedensellik ilişkisi. *İstanbul Üniversitesi İşletme Fakültesi Dergisi*, 38(1): 24–37.

[15] Lee, J. and Shin, M. 2009. Stock forecasting using hidden markov processes. *ht tp://cs229. stanford. edu/proj2009/ShinLee. pdf*, 2009.

[16] Li, N. 2016. *Hidden Markov Model and Financial Application*. Ph.D. Thesis.

[17] Pérez, O., Piccardi, M., García, J., Patricio, M.A. and Molina, J.M. 2007. Comparison between genetic algorithms and the baum-welch algorithm in learning hmms for human activity classification. *In Workshops on Applications of Evolutionary Computation*, pp. 399–406. Springer, 2007.

[18] Rabiner, L.R. 1989. A tutorial on hidden markov models and selected applications in speech recognition. *Proceedings of the IEEE*, 77(2): 257–286.

[19] Rodríguez, L.J. and Torres, I. 2003. Comparative study of the baumwelch and viterbi training algorithms applied to read and spontaneous speech recognition. *In Iberian Conference on Pattern Recognition and Image Analysis*, pp. 847–857. Springer.

[20] Rupande, L., Muguto, H.L. and Muzindutsi, P.-F. 2019. Investor sentiment and stock return volatility: Evidence from the johannesburg stock exchange. *Cogent Economics & Finance*, 7(1): 1600233.

[21] Saw, J.G., Yang, C.K.M. and Mo, T.C. 1984. Chebyshev inequality with estimated mean and variance. *The American Statistician*, 38(2): 130–132.

[22] Tenyakov, A. 2014. Estimation of hidden markov models and their applications in finance.

[23] Zhang, Y., Zhao, D. and Liu, J. 2014. The application of baumwelch algorithm in multistep attack. *The Scientific World Journal*.

II

OPERATION RESEARCH IN PRODUCTION PLANNING AND INVENTORY MANAGEMENT

Chapter 4

Optimization and Computing Using Intelligent Data-Driven Approaches for Decision-Making

Sourena Rahmani,[1] *Hamed Aghalar*,[1] *Sahel Jebreili*[2] and *Alireza Goli*[3,*]

4.1 Introduction

The onset of the Fourth Industrial Revolution, also known as Industry 4.0 or 4IR, has marked the beginning of an epoch where data has become as valuable as oil. In this era of digitization, a plethora of data is being produced from a variety of sources including the Internet of Things (IoT), business processes, health-

[1]Department of Industrial Engineering, Faculty of Industrial and System Engineering, Isfahan University of Technology, Isfahan, Isfahan, Iran.
[2]Department of Industrial Engineering, Faculty of Engineering, Urmia University, Urmia, West Azerbaijan, Iran.
[3]Department of Industrial Engineering and Futures Studies, Faculty of Engineering, University of Isfahan, Isfahan.
* Corresponding author: Goli.A@eng.ui.ac.ir

care systems, and more. This data, irrespective of whether it is structured, semi-structured, or unstructured, carries enormous potential for influencing strategic decision-making across a wide range of application areas [85]. Data science has risen as a comprehensive field that utilizes statistical methods, data analysis, and related techniques to comprehend and scrutinize phenomena through data. Sophisticated analytics techniques, inclusive of machine learning models, are utilized to derive actionable intelligence or profound understanding from data. This procedure of converting raw data into significant insights is recognized as Data-Driven Decision Making (DDDM) [14, 85].

DDDM, is a methodology that entails gathering data in line with a firm's Key Performance Indicators (KPIs) and converting this data into insights that can be acted upon. During this procedure, Business Intelligence (BI) reporting instruments are frequently employed to streamline the process of data collection and visualization. These tools democratize data analytics, making it accessible to individuals without extensive technical expertise [3]. DDDM is a concept within the realm of decision-making that encompasses a broad spectrum of activities. These activities include the collection of data, the extraction of patterns and facts from this data, and the application of these insights to decision-making processes. DDDM is characterized by an organizational decision-making process that is grounded in empirical data. There are various models of decision-making, and interestingly, even the act of not making a decision is considered a prevalent type or model within these processes [59]. Decision-making is a straightforward process of selecting an option. However, it often poses challenges, especially within an organizational context. Consider the multitude of decisions we face daily. Perhaps you grappled with the decision to hit the snooze button one more time this morning. Or maybe you were overwhelmed by an extensive menu at a restaurant, with the server patiently waiting for your order. You might have even left your wardrobe in disarray after trying on multiple outfits for an important presentation. Even seemingly simple decisions can prove to be difficult. People often go to great lengths and spend considerable amounts of money to avoid making choices. For instance, opting for an expensive tasting menu at a restaurant to avoid choosing from a vast array of dishes, or limiting your wardrobe to black turtlenecks, following the example set by Steve Jobs, to eliminate the need for choosing an outfit [40]. Should you ever find yourself struggling with a decision at your workplace, take comfort in the fact that you are not alone. A study by McKinsey indicates that a significant portion of an executive's time - approximately 40 percent on average - is allocated to decision-making. Alarmingly, they often feel that this time is not used effectively. The burden of decision-making can be so overwhelming that it results in a state of weariness from the constant need to make decisions, a phenomenon referred to as decision fatigue [29]. Yet, decision fatigue is merely one aspect of the repercussions of ineffective decision-making. A study by McKinsey, which included more than 1,200 business leaders worldwide, indicates that inefficient decision-making results in a loss of about

530,000 days of managerial time each year for a typical Fortune 500 company. This translates to approximately $250 million in annual salaries. That's a considerable investment in turtlenecks [40, 60]. Decision-making encompasses a variety of models, including the often-overlooked model of abstaining from making a decision. Individuals employ diverse psychological methods when making decisions, a phenomenon recognized in a notable study as the 'Pentagonal Forms of Decision-Making'. A substantial number of business decisions stem from personal habits and behaviors, which are predominantly intuitive and cognitive. This implies that these decisions are grounded in experience, comprehension, cognition, and education. Conversely, some decisions are not predicated on understanding or experience. Instead, they rely on specific information gleaned from case studies or field research. This approach is commonly known as DDDM. In DDDM, the decision-making process is driven by data rather than direct or implicit knowledge and experience. In this context, information serves as the primary decision-maker [26]. DDDM is not exclusive to the realm of information technology. It has been recognized and utilized extensively across various domains, ranging from business to social and political spheres. In the present era, the ease of data aggregation and accessibility via the internet has amplified the application of this concept, particularly in e-commerce, more effectively than ever before. The successful and optimal utilization of data in business decisions is not necessarily contingent upon the level of technical expertise or the magnitude of an organization's database [59]. Adopting a data-driven approach implies utilizing concrete data to discern trends, draw conclusions, and gain insights that guide the decision-making process. This approach emphasizes making decisions free from personal bias or emotional influence, ensuring that an organization's objectives and strategic plans are grounded in factual evidence and the patterns derived from it. The application of DDDM enables businesses to make enlightened decisions that enhance profitability and sales, foster effective management practices, and streamline operations. This chapter aims to serve as an exhaustive resource for researchers, decision-makers, and application developers interested in leveraging data to address real-world challenges [3, 53]. The degree to which business management depends on decision-making data and guiding statistics plays a pivotal role in shaping the effectiveness and reach of this decision-making approach. While one business may possess an abundance of extensive data repositories and numerous dashboards to transform raw data into actionable information, it may not necessarily exhibit a proactive approach when it comes to analyzing and leveraging this data for decision-making. Conversely, another business, even in the absence of a dedicated data repository on a particular topic, might be dedicated to the collection of information and making decisions guided by relevant indicators [49].

4.2 Optimization and Decision Making: A Brief Overview

4.2.1 Optimization and Decision Making

In the ever-evolving landscape of modern decision-making, the fusion of intelligent data-driven approaches and optimization techniques has emerged as a formidable force, reshaping the way organizations strategize and operate [17, 21]. As businesses and industries grapple with an increasingly complex web of variables and objectives, the synergy between optimization and decision-making has never been more critical [85, 98].

The pivotal role played by these synergistic forces is delved into in this section, with their profound impact on various facets of contemporary problem-solving being unveiled. Through a tapestry of issues that are central to the intersection of optimization and decision-making, navigation is conducted in this exploration. Not only the theoretical underpinnings but also the practical applications that drive innovation across diverse domains are encompassed by these topics. As this section is journeyed through, light is aimed to be shed on the intricate interplay between data-driven insights and optimization strategies, showcasing how decision-makers are empowered to navigate uncertainties, seize opportunities, and engineer efficient solutions through this dynamic duo [45, 55, 75]. The pivotal role of data in shaping decision landscapes is examined us, with the power of intelligent data-driven approaches being emphasized for their ability to uncover hidden patterns, predict outcomes, and inform decisions. It explores how these data-driven insights are fed into optimization models, resulting in their efficacy being amplified in scenarios ranging from resource allocation to supply chain management [54, 85].

The ethical considerations intertwined with data-driven optimization are also explored throughout this section, with the responsible use of algorithms and the importance of transparency and fairness in decision-making processes being discussed. The latest advancements in optimization algorithms and computational techniques that are revolutionizing the way we approach complex decision challenges are additionally spotlighted [84]. As this multifaceted terrain is navigated, the overarching objective is to provide both researchers and practitioners with a comprehensive understanding of the symbiosis between optimization and decision-making. By addressing the key issues, insights, and applications within this realm, a contribution to the ongoing discourse on leveraging intelligent data-driven approaches for enhanced decision-making and problem-solving in the digital age is aimed to be made [41].

Generally, the process of decision-making entails the creation and assessment of various options, ultimately culminating in the choice of the most favorable one that aligns with specified criteria. On the other hand, optimization revolves around the quest for the optimal solution within a range of possible alternatives [55]. At a higher level of abstraction, decision-making can be likened to the concept of optimization. Indeed, in certain scenarios, the decision-making process entails formulating and resolving optimization problems. These scenarios are distinguished by a substantial reservoir of valuable information about the situation at hand and a clear consensus regarding the objective to be achieved. For instance, within the automotive industry, optimization techniques have been harnessed to determine the optimal structural design for automobile frames, with the dual goals of maximizing strength and minimizing weight. Furthermore, multidisciplinary optimization approaches have been deployed during the early stages of vehicle design to identify the most financially lucrative design options [27]. When optimization necessitates the utilization of analysis software, such as finite element analysis, to assess designs, the predominant challenge often resides in the computational workload entailed by solving the optimization problem. Typically, the field of optimization is regarded as a subject of significance within the realms of applied mathematics, operations research, and engineering design.

4.2.2 Deeping in DDDM

The significance of DDDM lies in its ability to facilitate choices grounded on factual information rather than personal biases. For those in leadership roles, the most equitable and balanced approach is to make decisions objectively. The most enlightened decisions are derived from real-time data that aligns with your business objectives. Reporting software can be utilized to compile the necessary data, identify trends, and forecast outcomes. Data can provide valuable insights for various decisions, such as:

1. Strategies for boosting profits and sales.
2. Establishing effective management practices.
3. Streamlining operational processes.
4. Enhancing team performance.

Although not all decisions will be supported by data, they play a crucial role in many of the most significant choices [26, 59].

4.2.2.1 DDDM by 5 Steps

DDDM is a skill that requires practice. To improve your leadership abilities, you must be able to translate raw data into actionable steps that align with your company's goals. The following steps that have been shown in Figure 4.1, can help you make better decisions when analyzing data (DriveResearch, 2023 [39]).

- **Know your vision:** To arrive at well-informed decisions, it is imperative to have a thorough grasp of your company's long-term vision. This understanding serves as the foundation upon which data and strategic planning

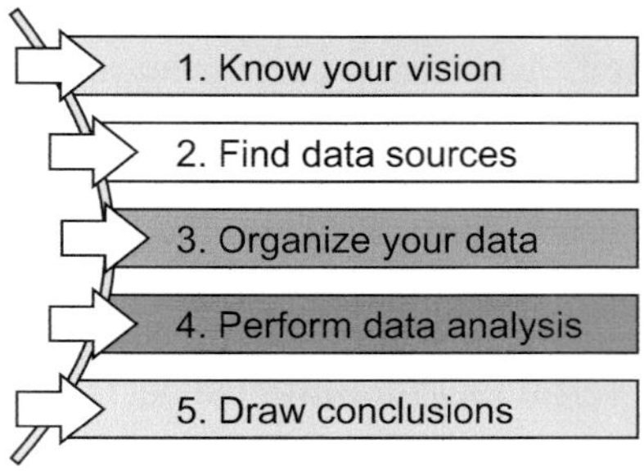

Figure 4.1: How to make a data-driven decision? [39].

can be harnessed to influence your decision-making. Graphs and numerical figures, devoid of the supporting context, lose their significance and effectiveness.

- **Find data sources:** The selection of tools and data resources for your data collection endeavors should be tailored to the specific nature of the data being gathered. When dealing with internal company processes, it is advisable to employ a versatile reporting tool. These tools provide a centralized platform for monitoring the progress of tasks and activities across your organization. An instance of such a tool is Microsoft's Power BI, which facilitates the consolidation of data from diverse external sources. However, when your aim is to analyze marketing trends or competitor metrics, alternative tools designed for these specific purposes are readily available. Evaluating achievement entails the systematic monitoring of an array of performance indicators. Among these key metrics are:

    1. Gross profit margin.
    2. Return on investment (ROI).
    3. Productivity levels.
    4. The overall customer base.
    5. Consistent revenue streams.

- The calculation of gross profit margin involves deducting the cost of goods sold from a company's net sales, allowing for an assessment of profitability. ROI, on the other hand, quantifies the relationship between income and investment, serving as a pivotal gauge to determine the viability of dedicating resources, be it time or capital, to a particular endeavor. Productivity serves as an indicator of how efficiently a company generates goods or services, accomplished by dividing the overall output by the inputs expended. Tracking the total number of customers, a straightforward yet potent metric closely aligns with the business's revenue generation. Additionally, recurring revenue, prevalent among SaaS enterprises, measures the income derived from active subscribers within a defined timeframe, typically on a monthly or annual basis. Beyond these mentioned metrics, numerous other data sets can be explored depending on one's role and organizational objectives. The advent of machine learning has significantly simplified the process of aggregating real-time data, offering a powerful tool for informed decision-making.

- **Organize your data:** Enhancing the effectiveness of data visualization hinges on the meticulous organization of your data, a crucial aspect when it comes to making impactful business decisions. When you are unable to access and comprehend all pertinent data within a unified context, the ability to ensure the highest level of informed decision-making becomes a formidable challenge.
- **Perform data analysis:** After effectively structuring your data, the next step involves delving into its analysis, with the goal of uncovering valuable insights that can inform and enhance the decision-making process. Depending on your specific aims, it may be prudent to juxtapose the data gleaned from your executive dashboard with user-centric research methods such as case studies, surveys, or testimonials. This approach ensures that your conclusions are enriched with insights from the customer's perspective. For instance, if your team is striving to enhance the competitiveness of their SEO tools within the market landscape, you can utilize the following datasets to identify the necessary enhancements:

    1. Data pertaining to the performance of rival companies.
    2. Data concerning the current performance of SEO software.
    3. Present customer satisfaction metrics.
    4. Findings from user research encompassing a range of SEO and marketing tools.

    While a portion of this data originates from your own organization, it may be necessary to acquire certain data from external origins. The value of examining these datasets collectively lies in the fact that you'll derive unique insights compared to the outcomes of analyzing each dataset in isolation.

- **Draw conclusions:** As you engage in the examination of your data, you may find yourself forming interpretations based on your observations. Nevertheless, it is crucial to allocate a distinct section to your interpretations, enabling you to expound upon your insights derived from the data and disseminate your discoveries to others. The fundamental inquiries to pose when crafting these interpretations are:

- What existing knowledge does the data reaffirm?
- What novel insights have I acquired through this data analysis?
- In what ways can I leverage the knowledge acquired to advance my organizational objectives?

Once you have satisfactorily addressed these inquiries, you have effectively completed the process of data analysis and should be well-equipped to make informed decisions guided by data within your business [14].

4.2.2.2 Benefits of DDDM

Employing analytics to inform decision-making is not merely a valuable skill; it is an essential one, especially if you aim to lead by example and foster a culture that values data-driven insights. Through the utilization of data in decision-making, you can establish a business environment that is characterized by fairness, goal-directedness, and an unwavering commitment to growth. In the ensuing sections, Figure 4.2 illustrates a selection of advantages associated with DDDM.

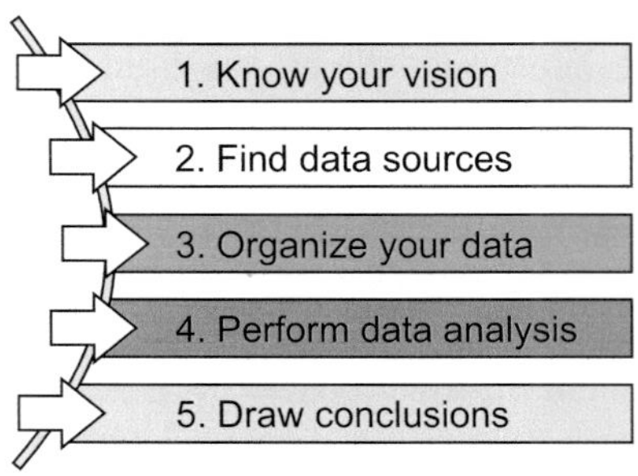

Figure 4.2: Some of DDDM benefits [39].

- **Make confidence decisions:** In the realm of business longevity, enduring enterprises are marked by a resolute belief in their capacity to thrive. When leaders within an organization grapple with uncertainty and hesitation, it can result in errors, increased employee attrition rates, and sub-optimal risk mitigation. Harnessing the power of data to inform critical business choices instills a sense of assurance in these decisions, igniting motivation and determination among both leaders and their teams. This newfound assurance has the potential to cultivate elevated team spirits and enhance overall performance.
- **Guard against biases:** Utilizing data-driven decision-making serves as a safeguard against biases that may infiltrate the judgments of corporate leaders. Biases, often lurking beneath one's awareness, can inadvertently

sway decisions, reflecting personal preferences or values. By grounding decisions in empirical data and numerical evidence, objectivity and impartiality are upheld, thus guaranteeing fairness in the decision-making process. This approach also equips decision-makers with concrete substantiation for their choices, should inquiries arise from team members or stakeholders seeking clarification on the rationale behind specific decisions.

- **Find unresolved questions:** Data plays a pivotal role in elucidating inquiries that might otherwise remain unaddressed. Furthermore, data sets have the capacity to unveil queries that were previously concealed from your awareness. Regardless of the scale or volume of data involved, it confers a valuable advantage to your team, allowing for enhanced insights into facets that remain obscured in the absence of statistical analyses, graphical representations, and charted data. By unearthing these latent questions, you can take solace in the knowledge that your decisions have been meticulously informed, taking into account a comprehensive spectrum of pertinent information.
- **Set measurable goals:** Data plays a pivotal role in addressing previously unanswerable questions, sometimes even revealing inquiries that were not previously within your purview. Irrespective of the volume, data holds the potential to greatly benefit your team, providing a clearer window into dimensions that remain obscured without the aid of statistical analysis, graphs, and charts. Bringing these latent queries to the forefront imbues confidence in your decision-making process, assuring you that every piece of pertinent information has been duly considered. In the realm of modern business strategy, data-driven decision-making stands as an indispensable pillar. It empowers you to establish quantifiable objectives for your team and subsequently guide them towards successful achievement. Through an examination of internal data reflecting past performance, you can identify areas for enhancement and set highly specific targets. For instance, your team can utilize data to delineate objectives such as:

  1. Achieve a 20% year-over-year increase in customer acquisition.
  2. Trim $20,000 from the quarterly budget.
  3. Reduce project budget expenses by $500.
  4. Expand the team by 10 members each quarter.
  5. Lower the cost per hire by $500.

  Without data, gaining insights into expenditure allocation and cost-cutting opportunities would be a formidable challenge for your organization. Establishing measurable goals ultimately guides your decision-making with data, as it prompts you to devise strategies for attaining the set objectives, whether by trimming budgets or expanding customer bases.

- **Improve company processes:** While it is possible to enhance company procedures without relying on data, the utilization of numerical metrics to observe trends in team member performance or the examination of company spending patterns through graphical analysis can significantly augment the process improvement efforts beyond mere observation. Processes that can be refined through the integration of data may encompass:

  1. Enhancing risk management through the analysis of financial data.
  2. Establishing cost estimates based on market pricing data.
  3. Optimizing team member onboarding procedures through insights derived from new hire performance data.
  4. Elevating customer service standards by leveraging customer feedback data. Introducing alterations to company processes can be a challenging endeavor, particularly when the outcome is uncertain. However, the presence of concrete facts and data empowers decision-makers with the confidence needed to navigate these changes effectively.

4.3 Data-Driven Optimization: A Brief Overview

The dynamic realm of Data-Driven Optimization (DDO) is in constant evolution, harnessing the power of data to elevate the efficacy of decision-making across diverse sectors. This section offers a concise introduction to DDO, accentuating its importance, methodological approaches, and wide-ranging practical implementations [107]. Data-driven optimization is predicated on the notion that data, when strategically harnessed, can serve as a powerful catalyst for optimizing complex processes and systems. It leverages vast datasets to inform and enhance decision-making, yielding solutions that are not only efficient but also tailored to dynamic real-world conditions [17, 57]. On the other sight, One of the hallmark features of Data-Driven Optimization is its versatility across a myriad of domains. Whether applied in healthcare, finance, logistics, or engineering, DDO exhibits its prowess in tackling multifaceted challenges. Research findings from various fields have showcased how data-driven optimization can streamline resource allocation, enhance operational efficiency, and yield substantial cost savings [83, 101].

At its core, DDO revolves around the art of making the best decisions when faced with unpredictable circumstances. This entails the utilization of various optimization models and computational techniques capable of handling data that is uncertain, dynamic, structured, or dispersed across vast networks. While traditional mathematical optimization seeks to maximize or minimize a function, denoted as $h(x)$, within a defined domain, DDO introduces a novel dimension by integrating additional input data, denoted as Ψ, thereby transforming the objective into the maximization of an anticipated value [21, 56]. Within the realm of DDO, a primary hurdle lies in effectively managing the unpredictable nature of Ψ. Conventional stochastic optimization necessitates precise information about the specific distribution of Ψ. Nonetheless, relying solely on such assumptions can result in suboptimal outcomes when the real-world distribution differs from these assumptions. In response, innovative approaches known as robust optimization have been introduced, which require only knowledge of the potential range of uncertain parameters. However, it's worth noting that these approaches, while conservative in nature, might not consistently deliver optimal results in practical scenarios [73].

In the pursuit of a harmonious synergy between stochastic and robust optimization techniques, Distributionally Robust Optimization (DRO) has emerged as a compelling solution. DRO operates by optimizing with the utmost consideration for the most adverse potential expected cost value within a predefined set Γ, encompassing various density functions or distributions. By adopting this approach, the optimization process benefits from the incorporation of pertinent problem data while simultaneously fortifying itself against errors stemming from deviations in distribution assumptions [89]. Beyond theoretical constructs, DDO has translated into tangible practical implementations. An illustrative example involves the application of a data-informed robust design optimization technique, aimed at enhancing both the performance and the resilience to aerodynamic challenges of compressor blades. This innovative approach assesses uncertainty by relying on statistical moments derived from limited input data, and it leverages a Gaussian process regression model to enhance the efficiency of robust optimization processes [101]. Furthermore, DRO has been successfully employed in the context of inventory management and financial decision-making, showcasing its adaptability across diverse fields. This approach not only assesses the potential profits for both lenders and sellers but also contributes novel managerial perspectives on the combined facets of inventory and financing choices.

4.3.1 Data-Driven Optimization Approaches

In today's digital age, businesses and organizations are inundated with vast amounts of data. This data, if harnessed effectively, can provide valuable insights and drive informed decision-making. One powerful way to leverage this data is through data-driven optimization approaches. These approaches involve using data analytics and mathematical techniques to optimize processes, operations, and strategies. In this chapter, we will explore the concept of data-driven optimization approaches, their benefits, and some real-world applications [64, 65].

In today's data-rich environment, organizations are leveraging data-driven optimization approaches as sophisticated methodologies that harness the power of data analytics and mathematical modeling to improve decision-making and outcomes across various domains. These methodologies are designed to find the best possible solutions or decisions by considering available data, constraints, and objectives. They play a pivotal role in areas such as supply chain management, marketing, finance, and manufacturing, among others [95, 106]. At the core of data-driven optimization lie powerful mathematical algorithms capable of handling multiple variables, constraints, and objectives. These algorithms come in various forms, including linear programming, nonlinear optimization, and integer programming. Additionally, machine learning techniques, such as regression analysis, neural networks, and genetic algorithms, are commonly integrated into data-driven optimization models to further enhance their adaptability and effectiveness. These methodologies are essential tools that enable organizations to navigate the complexities of the modern business landscape. By using data analytics and mathematical modeling, they empower informed decision-making, leading to enhanced efficiency, cost reduction, increased competitiveness, and ultimately, a competitive edge in today's digital age [6, 33].

- **Linear Programming (LP):** LP is a mathematical technique used to optimize a linear objective function while satisfying linear equality and inequality constraints. It is commonly employed in resource allocation and production planning.

- **Nonlinear Optimization:** In cases where the relationship between variables is not linear, nonlinear optimization techniques come into play. These techniques are used to find optimal solutions in scenarios with nonlinear constraints and objectives.

- **Integer Programming (IP):** IP is an extension of linear programming where decision variables must take integer values. It is used in scenarios where solutions must be whole numbers, such as in discrete optimization problems.

- **Machine Learning:** Machine learning techniques, including regression analysis, neural networks, and genetic algorithms, have found applications in data-driven optimization. These methods can adapt to complex and dynamic data patterns. Let's explore some real-world applications of data-driven optimization approaches in greater detail:

In summary, data-driven optimization approaches are powerful tools that rely on mathematical techniques and data analytics to enhance decision-making across various industries. By taking into account data, constraints, and objectives, organizations can make more informed decisions, increase efficiency, reduce costs, and gain a competitive edge. However, it's essential to navigate the challenges and ethical considerations associated with these approaches to fully realize their benefits in today's data-driven world.

4.4 Optimization Algorithms in Intelligent Decision Making

As the array of optimization methodologies and their practical applications continues to expand, researchers have undertaken systematic comparative analyses to gauge their effectiveness. When conducted rigorously, these comparative assessments serve as invaluable tools, aiding end-users in their selection of the most suitable optimization strategy for specific problem domains. Such investigative endeavors are commonly referred to as "optimization benchmarking" [19]. In a broader perspective, benchmarking entails the systematic evaluation of one or multiple entities in comparison to established industry standards, encompassing a variety of performance criteria. When applied within the context of benchmarking optimization algorithms, the entities scrutinized are specific instantiations of these algorithms. Their performance metrics are derived through rigorous assessments conducted across a diverse range of problem scenarios [19]. This approach brings a certain level of clarity to the process of benchmarking optimization algorithms by establishing a shared understanding of what constitutes 'superior' performance. In an ideal scenario, one approach could be deemed superior to another across all conceivable parameters—faster execution, reduced memory consumption, and the production of superior final results. However, practical reality often diverges from such definitive conclusions. In practice, assessing the outcomes of algorithmic comparisons can pose challenges, as unequivocal judgments are seldom reached.

Nevertheless, when executed correctly, the benchmarking of optimization techniques proves to be an invaluable endeavor. It serves to unveil the strengths and vulnerabilities inherent in an algorithm, thereby guiding more targeted research efforts. Moreover, it aids in validating whether a new iteration of optimization software performs as anticipated. For end-users, it offers guidance in selecting the most appropriate algorithm for specific real-world scenarios.

In contrast, when the process of benchmarking optimization strategies is not carried out correctly, it holds the potential to distort the evaluation of an algorithm's strengths and weaknesses, present fictitious improvements, or recommend an unsuitable algorithmic choice for a particular context. Benchmarking optimization practices involve a multitude of subjective judgments, including factors such as the selection of the test dataset, the determination of the computing environment, the establishment of performance evaluation criteria, and more. Shifting our attention to the sphere of business decisions, the challenge extends far beyond the initial decision-making process; it encompasses the repercussions of those decisions when they fail to yield the anticipated outcomes. Within this context, numerous common challenges surface in the course of making business decisions which include:

- **Insufficient Data:** A fundamental challenge faced by many managers and leaders within organizations is the dearth of information necessary for making well-informed business decisions. When essential information is lacking, those in leadership positions find themselves navigating through a sea of uncertainty, hindering their ability to grasp the required insights and make optimal choices [68].

- **Overwhelmed by Data:** In the past, the prevailing belief was that a surplus of information and data equated to enhanced decision-making capabilities. However, in the contemporary business landscape, an excess of data can actually impede effective decision-making. When confronted with an abundance of data, individuals may find themselves grappling with feelings that overwhelm and over-analysis, which can ultimately hinder or jeopardize the decision-making process. More data doesn't always equate to better outcomes [30].
- **Time Constraints:** Imposed time limitations can exert added pressure, compelling a quicker business decision than initially anticipated. When confronted with the need for swift decision-making, there is a risk that the desired outcomes may be compromised. This haste might inadvertently lead to the omission of crucial details or the oversight of the relative significance of one course of action over another [79].
- **Dealing with Ambiguity:** Another challenge confronting decision-makers within organizations is the inherent ambiguity they encounter. Regardless of their daily exposure to uncertainties, even the most seasoned leaders can find it challenging to surmount the cloud of uncertainty about what lies ahead. This ambiguity can arise from shifting business landscapes, limited access to information, and the potential misinterpretation or misunderstanding of available data [11].

Fortunately, Decision Intelligence, an integral component of the decision-making process at the enterprise level, can offer effective solutions to the frequent challenges and difficulties that businesses encounter when dealing with complex decisions. Decision Intelligence functions as a robust instrument for elevating the effectiveness and accuracy of corporate decision-making processes. This purpose-designed and automated approach enables enterprises to adeptly manage vast datasets, thereby enhancing decision-making capabilities across their entire organizational spectrum through the seamless integration of Artificial Intelligence (AI) and Machine Learning (ML) technologies. Decision Intelligence, a fusion of social science and data science, democratizes analytics by leveraging machine learning on current data, enabling company leaders to make informed decisions. It places no emphasis on who is 'right' or the most knowledgeable in a room but rather takes existing business insights, automates their delivery, and presents them in a manner understandable even to non-technical users. Organizational leaders can harness Decision Intelligence to facilitate well-informed decisions by employing automated knowledge and comprehension [85].

You might wonder why it's beneficial to replace your existing decision models with AI-driven ones. Traditional business models often result in decisions influenced by routine, biases, or a reluctance to acknowledge the shortcomings of these models [32]. In contrast, decision models infused with AI and analytics, commonly known as decision intelligence, empower organizations to align with evolving consumer demands, maintain a competitive edge, and cultivate agility in adapting to change. By integrating machine learning algorithms into the decision-making process, decision intelligence has the potential to enhance decision models. For companies incorporating AI into their daily decision-making processes, the key lies in having confidence in the outputs generated by AI. Decision intelligence platforms are emerging as a solution to expedite decision-making across various industries, fostering trust in AI-driven decision outcomes [24].

Decision intelligence provides several advantages for today's enterprises. Here are a few of the advantages of including decision intelligence in decision-making strategy:

- **Decisions based on data:** To achieve a competitive advantage in today's business environment, your decision-making models must be capable of analyzing existing data, forecasting forecasts or outcomes, and selecting the best alternative for your firm. AI decision-making systems in decision intelligence examine data considerably more deeply, recognizing any unusual or previously unknown patterns. By detecting probable anomalies, AI avoids the possibility of vulnerabilities and interruptions in the data, which might have a significant impact on the company [44].

- **Quicker decisions:** If your business spends too much time attempting to make the correct conclusion, AI decision-making technologies drastically shorten the process by rapidly digesting massive, disparate volumes of data. Reducing the time it takes to analyze vast volumes of data might help your company make better decisions in less time [82].

- **Biases and mistakes are eliminated:** If your leadership team consistently encounters cognitive and emotional biases while making decisions, decision intelligence solutions can assist in eradicating them from having a direct influence on the outcomes of business decisions. Because decision intelligence algorithms look at the given data objectively and are not vulnerable to cognitive and emotional biases, they may entirely avoid these biases [20].

Effective decision-making with intelligence demands essential skills like drawing meaningful conclusions from empirical data, making deductive inferences based on foundational knowledge, and selecting the best option from a set of alternatives. Software designed for Fuzzy Multiple Criteria Decision Making (FMCDM) has been developed and rigorously evaluated in practical, real-world scenarios. Challenges in multicriteria decision-making (MCDM) arise when individuals or a group of decision-makers are tasked with selecting the most suitable option from a multitude of available alternatives [52]. The fundamental stages of MCDM encompass the following key steps:

1. Generating alternative systems to fulfill the defined objectives.
2. Assessing the alternatives based on the chosen criteria.
3. Employing a normative multicriteria analysis technique.
4. Ultimately, designating one alternative as the 'optimal' choice [38].

In essence, decision-makers (DMs) are required to assess both quantitative and qualitative aspects of the criteria when evaluating the available alternatives. Conventional problems in MCDM typically involve precise numerical judgments, necessitating an evaluation of option performance across each criterion. Conversely, information pertaining to alternatives often lacks precision, with decision-makers (DMs) capable of providing only approximations, fragmentary details, or vaguely defined data. Furthermore, the subjective nature of some criteria complicates matters. To tackle these issues, the widespread adoption of fuzzy set theory in MCDM problems has emerged as a promising approach to address common decision-making challenges. Its primary objective is to mitigate the impact of imprecision, encompassing elements such as human judgment and preferences when seeking the optimal solution [7].

Conversely, employing FMCDM techniques can give rise to a range of challenges. Even in relatively straightforward scenarios, the required computations can be time-consuming and prone to errors. Consequently, the development of suitable software for implementing these procedures becomes pivotal. Accordingly, efforts are underway to design software that facilitates the utilization of two distinct FMCDM methodologies. The primary aim of this study is to develop software intended for the application of these two FMCDM approaches, specifically within the context of addressing two practical, real-world challenges. Furthermore, the discussed problems exemplify how this software can assist decision-makers in uncovering intriguing insights.

When it comes to elevating decision-making processes within organizations, the integration of intelligent, data-informed strategies becomes of utmost importance. Decision intelligence stands out as a potent approach that melds data, analytical techniques, and artificial intelligence, with the aim of refining decision-making and revealing fresh pathways for progress and evolution [10, 36].

Decision Intelligence can be defined as the deliberate and strategic application of data, analytics, and artificial intelligence (AI) with the primary goal of enhancing and optimizing the decision-making process. It encompasses a wide array of methodologies, including but not limited to machine learning, predictive analytics, and natural language processing. Instead of depending on instinct or historical precedent, decision intelligence empowers entities, be they businesses or institutions, to make informed choices grounded in valuable insights gleaned from data analysis [10, 58].

Key Benefits of Decision Intelligence: Implementing decision intelligence offers several benefits for businesses in today's complex landscape [10]:

- **Improved Decision-Making:** Executives can utilize analytics to evaluate scenarios and select the most suitable option.
- **Reduced Risks and Uncertainty:** Potential risks can be identified, outcomes investigated, and mitigation strategies developed.
- **Increased Efficiency:** Decision intelligence facilitates faster and more agile decision-making.
- **Enhanced Collaboration:** By aligning stakeholders and providing a single source of truth, communication, and alignment improve [36]. Getting Started with Decision Intelligence: To integrate decision intelligence into an organization, a systematic approach is recommended [10]:
- **Define Decision-Making Criteria:** Identify critical factors for decision-making and establish evaluation methods.
- **Gather and Analyze Data:** Collect relevant quantitative and qualitative data from various sources.
- **Identify Patterns and Insights:** Employ data visualization and analysis tools to identify trends and correlations.
- **Evaluate Options and Decide:** Utilize insights to evaluate options and make informed decisions (Bornet, 2022 [24]). Examples of Decision Intelligence in Action: Decision intelligence finds application across a spectrum of industries and scenarios (Çıtak, 2023 [31]):
- **Resource Allocation and Risk Mitigation:** Airlines optimize flight schedules to minimize delays and maximize profits.
- **Autonomous Decision-Making:** AI systems autonomously handle tasks like music playback through devices such as Amazon's Alexa.
- **Virtual Assistance:** AI assists healthcare professionals in diagnosing diseases and recommending treatments.

4.5 Heuristic Algorithms

The origin of the term 'heuristic' can be traced back to the Greek word 'heuristic,' denoting the act of 'finding' or 'uncovering.' Heuristic algorithms typically consist of practical guidelines or suggestions guiding the exploration of promising segments within the solution space [91]. Heuristic algorithms are problem-solving strategies that discover approximate solutions to complicated problems by employing practical approaches and shortcuts. They are frequently used when exhaustive approaches, such as brute-force, are impracticable due to the large search space or high computing needs. Heuristic algorithms strive to discover a good enough answer in a reasonable period rather than guaranteeing an ideal result [70]. Heuristics are cognitive shortcuts or mental strategies that humans use to make decisions and solve problems quickly and efficiently. These shortcuts rely on past experiences, intuition, and common sense to arrive at solutions, often without engaging in a complex analytical process. They prove highly advantageous in scenarios marked by constraints in terms of both time and resources. Conversely, the term 'Data-driven' pertains to a methodology that leans on real-world evidence and empirical data to guide the process of decision-making and addressing problems [26]. In a context that prioritizes data-driven methodologies, decisions are rooted in concrete facts, statistical analysis, and empirical evidence, rather than relying on mere intuition or gut instincts. When blending heuristics with a data-driven approach, you effectively combine swift, rule-of-thumb strategies with tangible data for decision-making. This fusion can be a potent approach, as it enables the utilization of both your personal insights and historical experiences, while simultaneously benefiting from the impartiality and precision that data brings to the table. For instance, consider a business endeavoring to ascertain the ideal pricing strategy for a new product; they might employ heuristic-based estimates initially, drawing from their industry experience [63]. However, they would then gather and analyze market research data, competitor pricing, and consumer preferences to refine their pricing strategy in a data-driven manner. In general, the concept of "heuristics data-driven" suggests using a combination of quick mental shortcuts and evidence-based data analysis to make informed decisions and solve problems. This approach can help balance the benefits of intuitive decision-making with the reliability of data-backed insights [67]. The features of the heuristic algorithm can be described as follows:

- **Approximation vs. Optimality:** Heuristic algorithms are meant to identify answers that are near the optimum solution but may not always produce the best possible outcome. They trade off optimality for efficiency, making them appropriate for problems with vast or complicated search areas.

- **Search Space and Neighborhood Exploration:** Heuristic algorithms search for nearby solutions from a particular beginning point. The idea of a neighborhood outlines the set of candidate solutions that may be achieved by making minor changes to the existing solution.

- **Heuristic Function:** A heuristic function, also known as a cost function or evaluation function, is a function that quantifies how "good" a solution is. This function is used by the algorithm to direct the search toward better solutions.
- **Greedy vs. Metaheuristic Algorithms:** Within the realm of heuristic algorithms, there exists a category known as 'Greedy algorithms' which operate by making decisions that appear to be the best at each step, aiming to eventually reach a global optimal solution. In contrast, metaheuristic algorithms represent more advanced strategies that oversee and guide the search process, effectively navigating through various sections of the solution space with efficiency.
- **Termination Criteria:** Heuristic algorithms require termination criteria to know when to cease searching. This might be based on a set number of iterations, achieving a certain level of solution quality, or meeting a time constraint.
- **Problem-Specific strategies:** Varied problems often demand distinct heuristic approaches. For instance, certain challenges may find value in the application of genetic algorithms, particle swarm optimization, ant colony optimization, or simulated annealing. Heuristic algorithms are problem-solving strategies that use data or patterns detected in data to make educated judgments and find approximate answers in a "data-driven" setting. These algorithms use data-driven insights to efficiently and effectively direct the search process, especially when dealing with complicated issues with massive datasets or high-dimensional spaces. The notion of heuristic algorithms in data-driven contexts is comparable to the classic definition of heuristics, as previously discussed. Heuristic algorithms, on the other hand, gain from the abundance of knowledge included within the accessible data to improve their decision-making process and solution quality in a data-driven setting. In Table 4.1, the main data-driven aspects of heuristic algorithms are given; In such a way that one can argue a structured view in related applications [62].

4.5.1 Meta-Heuristic Algorithms

Meta-heuristic algorithms are optimization techniques that tackle complex problems with limited time or information, often used in data-driven decision-making contexts. They offer efficient shortcuts to reaching good solutions. These algorithms can be inspired by nature, like the gray wolf optimizer or genetic algorithm, enhancing decision-making processes and employee turnover analysis. Hybrid approaches, combining various strategies, often lead to superior performance. These techniques prove especially valuable when tackling multi-objective optimization challenges, where a solitary solution cannot be deemed

Table 4.1: Data-Driven aspects of heuristic algorithms.

Data-Scarce Optimization [99]	A heuristic decision-making approach driven by data comes into play when optimizing situations grapple with a shortage of available data. For instance, within the realm of materials engineering, one can create Bayesian optimization models to enhance the composition of bio-based composites when data is limited. This approach adjusts meta-heuristic methods based on the data at hand, effectively addressing optimization challenges in scenarios where information is sparse.
Evolutionary Optimization with Data-Driven Approaches [51]	Data-driven evolutionary optimization involves utilizing machine learning techniques and metaheuristics to solve complex optimization problems. This approach combines mathematical programming, evolutionary algorithms, and machine learning to create efficient optimization strategies. It has applications in various fields, including engineering design, process automation, and healthcare.
Heuristic Induction from Human Design Behavior [81]	Data-driven heuristic induction involves deriving heuristic strategies from human design behavior. This approach uses data on how humans make design decisions and translates that into heuristic rules for automated design optimization. It bridges the gap between human creativity and automated optimization by extracting useful heuristics from human expertise.
Instance Feature-Based Algorithm Selection [34]	Instead of manually tuning algorithm parameters, this approach employs historical performance data to automatically configure the parameters of heuristic algorithms. This data-driven optimization enhances the algorithm's effectiveness and efficiency.
Temporal Data-Driven Heuristics [103]	These algorithms leverage time-dependent patterns in historical data to adapt their search strategies. By considering how solutions have evolved, these heuristics can make informed decisions about where and how to explore the search space.

optimal for all objectives, thereby yielding a collection of Pareto-optimal solutions. The incorporation of learning-based methods, such as employing data mining to fine-tune algorithm parameters, plays a pivotal role in augmenting the effectiveness of meta-heuristic methodologies. Overall, these algorithms aid decision-making under uncertainty by efficiently exploring solution spaces and providing actionable insights [92].

The utilization of meta-heuristic optimization algorithms within the realm of intelligent decision-making is experiencing growing prominence across a spectrum of domains, including engineering, finance, and scientific research. These algorithms are tailored to address intricate optimization challenges where obtaining an optimal solution through conventional means proves arduous. Meta-heuristic algorithms form a category of optimization techniques employing heuristic methods to navigate the search space and uncover the most favorable solution. These algorithms draw inspiration from natural phenomena such as evolutionary processes, swarm intelligence, and simulated annealing. The application of these algorithms has yielded encouraging outcomes in the realm of decision-making, encompassing tasks such as scheduling, resource allocation, portfolio optimization, and a multitude of other applications. Given the escalating requirement for streamlined and proficient decision-making procedures, optimization algorithms featuring heuristic methodologies are well-positioned to address the complexities of today's world. Meta-heuristic algorithms constitute a prominent category within the domain of optimization techniques and are extensively harnessed in intelligent decision-making.

The fundamental objective of these algorithms is to ascertain the best possible solution for a given problem by iteratively traversing diverse solution spaces. They bear the moniker 'meta-heuristic' due to their operation at a more abstract level compared to conventional optimization algorithms. Unlike traditional methods, these algorithms do not provide a guarantee of discovering the optimal solution; instead, they strive to identify the most favorable solution within a practical timeframe. Meta-heuristic algorithms have permeated diverse domains, including economics and artificial intelligence, and they are instrumental in tackling intricate challenges that demand resolution. Their remarkable adaptability allows for their application across a broad spectrum of optimization problems, rendering them one of the prevailing and widely adopted algorithmic approaches in contemporary usage [15].

They are designated as 'meta-heuristic' by virtue of their operation at a higher conceptual plane relative to typical optimization algorithms. Unlike conventional techniques, these algorithms do not offer an assurance of uncovering the ultimate solution; their aim is to discern the most advantageous solution within a reasonable timeframe. Meta-heuristic algorithms have made inroads into various domains, spanning economics and artificial intelligence, where they play a pivotal role in addressing complex issues requiring resolution. Their exceptional versatility enables their utilization across a wide array of optimization problems, solidifying their status as one of the foremost and extensively embraced algorithmic methodologies in contemporary practice [69].

Metaheuristic algorithms represent optimization techniques inspired by a diverse range of sources, including natural phenomena, human behavior, and principles from fields like physics. These algorithms have garnered widespread acclaim across multiple domains and have captured the keen interest of both the scientific and industrial communities. However, the proliferation of recent metaheuristic approaches has led to questions regarding the rationale behind introducing new algorithms in an arbitrary manner.

Despite the successful outcomes achieved through metaheuristic algorithms, several challenges continue to persist within this field. Notably, there isn't always a direct and significant correlation between the sources of inspiration for these algorithms and their actual performance. The initialization and control parameters of population-based metaheuristic algorithms hold a pivotal role in enhancing

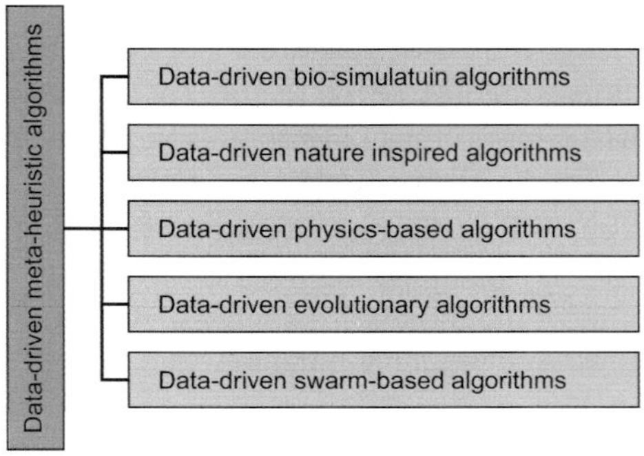

Figure 4.3: Classification of data-driven metaheuristic algorithms.

their effectiveness. Researchers have devoted substantial effort to devising various distribution schemes aimed at augmenting the diversity within initial populations. Achieving the right balance between population size and the number of iterations is crucial to ensure optimal solutions for specific problem sets.

Feature selection represents a critical aspect of machine learning, and metaheuristic algorithms have garnered substantial attention for their prowess in addressing various optimization problems, including feature selection. This chapter offers an extensive literature review spanning a decade (2009–2019) on the utilization of metaheuristic algorithms for solving feature selection challenges. These algorithms are categorized into four groups based on their behavior, and a comprehensive list exceeding a hundred distinct metaheuristic algorithms is presented [2]. In the following, a classification of data-driven metaheuristic headline algorithms is shown in Figure 4.3.

4.5.2 Data-driven Bio-Simulation Algorithms

Bio-simulated metaheuristic algorithms belong to the category of optimization techniques that draw inspiration from various natural processes or phenomena. Their primary objective is to address intricate optimization challenges by emulating the behaviors observed in biological systems or other elements of nature. These algorithms prove especially valuable in scenarios where conventional optimization approaches prove impractical or inefficient. Among the pivotal factors influencing the performance of population-based metaheuristic algorithms, the initial set of solutions assumes a critical role. The placement of these initial solutions can exert a substantial influence on the effectiveness of the search process and the likelihood of discovering the optimal solution [5]. Scholars have invested significant endeavors in devising a range of distribution strategies aimed at enriching the diversity within initial populations while striking an appropriate equilibrium between population size and the number of iterations. In the realm of machine learning, metaheuristic algorithms have garnered substantial interest in addressing the challenge of feature selection. The core objective of feature selection is to trim the dimensionality of the feature set while upholding performance accuracy. Metaheuristic algorithms have found extensive application in this context and a comprehensive examination of the literature has been undertaken to delve into the resolution of feature selection problems through the utilization of these algorithms [4]. Metaheuristic algorithms have the potential to be incorporated seamlessly with artificial neural networks (ANNs) for the purpose of enhancing the optimization of model architecture and feature selection, particularly in the domain of prediction modeling. As an illustration, in a research endeavor focused on forecasting biochar yield through ANNs, six distinct metaheuristic algorithms were subjected to comparative analysis. Their role was to optimize the architecture of the ANN and determine the relevance of individual features. The findings indicated that the ANN model combined with the Rao-2 algorithm exhibited superior performance when contrasted with alternative models [56]. Bio-simulated metaheuristic algorithms present numerous benefits when applied to optimization challenges. They possess the capability to navigate intricate and multidimensional search spaces, demonstrating resilience in the face of noise and uncertainty. Moreover, these algorithms exhibit the capacity to uncover solutions that are close to optimal, even in scenarios where the objective function exhibits non-linearity or discontinuities. Furthermore, their adaptability renders them suitable for diverse problem domains [4].

Over the past decade, significant strides have been made in the development of machine learning algorithms, simplifying the processes of data classification and information extraction. This advancement has led organizations worldwide to recognize the inherent value of their data, resulting in a substantial surge in the demand for data scientists. The establishment of Business Intelligence departments and the adoption of data-driven decision-making have gained prominence. Uncovering valuable insights and concealed patterns from vast datasets can yield substantial benefits for an organization, both in terms of profitability and other aspects. A data-centric approach to machine learning algorithms can facilitate the extraction of valuable knowledge from data and guide well-informed decision-making. Data-driven bio-simulation algorithms encompass the use of data to inform decisions and actions related to optimization problems inspired by natural processes. The integration of artificial neural networks and metaheuristic algorithms, for instance, can enhance the prediction of intricate processes such as biochar yield. Additionally, data-driven techniques can be applied to optimize the initialization approaches for population-based metaheuristic algorithms. Machine learning algorithms play a pivotal role in extracting valuable insights from data, empowering organizations to make informed decisions [5].

4.5.3 Data-driven Nature-inspired Algorithms

Metaheuristic algorithms infused with a data-driven, nature-inspired approach encompass leveraging data to guide decisions and actions pertaining to optimization problems inspired by natural phenomena. This study introduces a fresh wellspring of inspiration for the evolution of metaheuristic algorithms [46]. This newfound wellspring of inspiration, referred to as 'ancient-inspired,' amalgamates the commendable attributes of prevailing sources of inspiration, culminating in the development of highly efficient algorithms. The study reveals that the methodologies, strategies, and technologies employed in ancient times are surprisingly more advanced and optimized than previously envisaged. Consequently, this research contributes a novel classification for metaheuristic algorithms, ushering in a fresh perspective. An approach grounded in data-driven principles applied to ancient-inspired metaheuristic algorithms holds the potential to enhance both their efficiency and effectiveness. Furthermore, the study introduces an innovative algorithm for metaheuristic programming with gene expression, leveraging a diverse array of nature-inspired algorithms as a search engine to tackle the intricate challenge of synthesizing nonlinear models analytically [74].

The algorithm put forth exhibits a marked superiority when compared to the conventional gene expression programming approach. Employing a data-driven strategy in the context of metaheuristic programming with gene expression holds the potential to enhance the efficiency of these algorithms significantly. Notably, previous research has demonstrated the effective utilization of nature-inspired metaheuristics in addressing the complex task of multilevel image thresholding [97].

Multilevel image thresholding plays a vital role in image segmentation, a pivotal step that underpins advanced image analysis. This challenge involves intricate combinatorial optimization with highly complex objective functions, necessitating the application of probabilistic methods for viable solutions. In this concise overview, we explore instances where nature-inspired metaheuristics have demonstrated success in addressing multilevel image thresholding problems. Additionally, we consider how adopting a data-driven approach can enhance the effectiveness of algorithms designed for multilevel image thresholding. Furthermore, we discuss a recent study introducing the 'clique finder algorithm,' which utilizes principles from simulated annealing to address the maximum clique problem (MCP) [8]. The algorithm in question employs a cooling schedule following a logarithmic pattern and dynamically selects two specific moves. To gauge its efficacy, the proposed algorithm underwent evaluation using benchmark graphs sourced from the DIMACS open-source library, and the outcomes indicate a notable success rate. Adopting a data-driven strategy in the context of the clique finder algorithm holds promise for enhancing the performance of such algorithms. Furthermore, a recent study introduces an innovative approach, enhancing the search capability of the traditional cuckoo search algorithm (CSA) for optimization tasks. This method involves a modified local search mechanism accompanied by an adaptive boundary scaling feature that employs a chaotic map [90].

The newly introduced algorithm is subjected to a comparative assessment against both the conventional cuckoo search algorithm (CSA) and a version of CSA without chaos modification, referred to as nonmodified chaos-embedded CSA (CCSA), across a set of 25 benchmark functions. The empirical findings, bolstered by statistical tests, consistently indicate the superior performance of the proposed algorithm over its counterparts. Embracing a data-driven approach in the context of adaptive scaling cuckoo search algorithms holds promise for optimizing their overall effectiveness. Furthermore, the study underscores the pivotal role of initialization control parameters in enhancing the performance of population-based metaheuristic algorithms [5]. Considerable research endeavor has been directed towards devising diverse distribution strategies aimed at enriching the diversity within initial algorithm populations. Simultaneously, there's a concerted focus on achieving the precise equilibrium between population size and the number of iterations, ensuring the attainment of optimal solutions for specific problem sets. Embracing a data-driven methodology for initialization holds significant potential to enhance the overall efficiency and efficacy of population-based metaheuristic algorithms.

The concept of being inspired by ancient wisdom advocates the amalgamation of favorable attributes from existing sources of inspiration to craft metaheuristic algorithms characterized by efficiency [47]. Through a meticulous examination of ancient artifacts and constructions, conceived and executed under constraints and limited resources, scholars can discern methodologies, strategies, and technologies that surpassed initial expectations in terms of advancement and optimization. Integrating these sophisticated techniques into metaheuristic algorithms holds the potential to augment their overall efficiency and effectiveness. The introduction of this ancient-inspired perspective introduces a fresh categorization for metaheuristic algorithms, extending new avenues of exploration for researchers [47]. This categorization serves as a valuable tool for structuring and comprehending the array of algorithms influenced by ancient strategies and technologies.

Through this systematic classification based on ancient-inspired attributes, researchers gain improved clarity in evaluating and contrasting algorithm performance, thereby facilitating the advancement of more effective algorithms. The ancient era furnishes instances of optimization strategies devised to surmount constraints and attain optimal results. Through an examination of these historical strategies, researchers can extract valuable insights on optimizing metaheuristic algorithms. For instance, algorithms inspired by antiquity may integrate methods for efficiently navigating and exploiting the search space, maintaining a delicate equilibrium between convergence and diversity, and adapting to dynamic conditions [47].

4.5.4 Data-driven Physics-based Algorithms

Data-informed algorithms grounded in physics entail the utilization of data to guide decision-making and actions pertinent to optimization challenges within the domain of physics. This resource serves as a valuable reference for engineers, geophysicists, and geoscientists, offering a comprehensive introduction to data science and analytics terminology relevant to subsurface characterization. Furthermore, it illustrates the practical application of data-driven techniques for a range of tasks, including outlier detection, geomechanically and electromagnetic characterization, image analysis, fluid saturation estimation, and pore-scale characterization within the subsurface. The reference showcases real-world case studies that draw upon field observations, laboratory experiments, and simulation data to underscore the effective utilization of data-driven methods for characterizing subsurface phenomena [88]. A doctoral thesis delves into an innovative application platform referred to as a 'transit hub.' This platform serves as a facilitator for the seamless integration of sensor data streams dispersed both spatially and temporally. It also enables the fusion of simulation-based decision support systems and the execution of experiments aimed at elucidating how advanced decision support tools enhance the optimization of transportation infrastructure usage.

Within this framework, data mining and machine learning methodologies were custom-tailored to cater to the context-specific prediction of delays in public transit networks across various time horizons – ranging from long-term to short-term and real-time scenarios. Addressing the challenge of data sparsity, novel approaches were devised, including the creation of shared route segment networks and the development of multitask neural networks. These innovations fostered effective solutions for handling sparse data in public transit contexts. In the broader landscape, the past decade has witnessed the evolution of diverse machine-learning algorithms, designed to streamline the processes of data classification and information extraction [35]. Organizations all over the world have realized the true intrinsic value of their data, and the demand for data scientists has risen tremendously. Setting up Business Intelligence departments and making data-driven decisions has gained popularity. In an extensive research investigation, a comprehensive catalog is compiled, featuring over 500 metaheuristic algorithms. This study proceeds to undertake a rigorous comparative analysis, evaluating the performance of both recently enhanced competitive variants and newly introduced metaheuristics. The benchmark for this evaluation is the CEC2017 benchmark suite [71]. The creators of these algorithms credit their uniqueness to the sources of inspiration they draw from, including biology, human behavior, physics, and various observable phenomena. The overarching concept underpinning 'nature-inspired' algorithms is centered on the idea of developing algorithmic solutions by mimicking natural processes or biological behaviors, all with the aim of efficiently tackling optimization problems [66]. The study provides insights into the performance of nature-inspired metaheuristic algorithms and the need for a rational introduction of new metaheuristics.

Data-driven methods can be used for outlier detection, geomechanically electromagnetic characterization, image analysis, fluid saturation estimation, and pore-scale characterization in the subsurface [48]. Real-world scenarios involving the application of data-driven approaches to subsurface characterization, drawing from field observations, laboratory experiments, and simulation data, can provide concrete illustrations of these methodologies. Such case studies hold relevance for diverse domains like oil and gas exploration, geothermal energy utilization, and groundwater management. In addition, data mining and machine learning methodologies are instrumental in enabling context-aware predictions of delays in public transit networks, spanning various temporal scales, including long-term, short-term, and real-time scenarios [93]. Strategies like the creation of collaborative route segment networks and the implementation of multitask neural networks can be devised to effectively address challenges associated with sparse data. These techniques hold the potential for enhancing the efficiency of transportation infrastructure utilization and mitigating congestion in urban settings. Furthermore, a diverse array of machine-learning algorithms can be employed to streamline the tasks of data categorization and information extraction, simplifying these processes [35].

Foundational machine learning techniques that are essential for anyone enthusiastic about data science encompass decision trees, random forests, support vector machines, k-nearest neighbors, and neural networks. These algorithms serve versatile purposes, including predictive modeling, classification, clustering, and anomaly detection. In a separate realm, metaheuristic algorithms come into play to address optimization challenges, emulating natural processes and biological behaviors. These algorithms find applications in tasks such as automated synthesis of nonlinear models, evaluating performance, and comprehensive cataloging of metaheuristic algorithms (Sun, 2018).

4.5.5 Data-driven Evolutionary Algorithms

Utilizing data as a foundational element, data-driven evolutionary algorithms guide the decision-making and operational aspects of evolutionary optimization challenges. Noteworthy advancements have been witnessed in the field of evolutionary computation theory within the realm of discrete search spaces, particularly in the last decade, signaling a substantial leap forward [37]. Refined and intricate runtime analysis models of evolutionary algorithms find application in the enhancement of submodular function optimization. Recent strides have been taken in addressing the challenges posed by stochastic and dynamic problem domains. At the forefront of this field lies drift analysis, a remarkably potent analytical technique, offering the potential to enhance the effectiveness of evolutionary algorithms. Additionally, non-deterministic parameter control mechanisms can be harnessed to adapt algorithm parameters, leveraging insights gleaned from the evolutionary process [28].

Self-adaptation is widely embraced within the realm of evolutionary strategies, entailing the encoding of parameter configurations within individual chromosomes, which then evolve through processes like mutation and crossover. Theoretical evaluations of runtime performance reveal that in discrete problems with undefined structures, a non-elitist evolutionary algorithm employing self-adaptive mutation rates can surpass the performance of evolutionary algorithms employing fixed mutation rates. In addition, the construction of nonlinear models in an analytical format can be facilitated through the application of metaheuristic programming, utilizing gene expression techniques inspired by various facets of nature-inspired algorithms [74].

The algorithm under consideration demonstrates superior performance in automating the synthesis of nonlinear models when compared to the conventional gene expression programming approach. The optimization of population-based metaheuristic algorithms heavily relies on the effective management of initialization control parameters, serving as a pivotal factor in elevating algorithmic efficiency [5]. Significant research endeavors have been dedicated to the exploration of diverse distribution strategies aimed at enriching the initial populations of algorithms. The goal is to strike the right equilibrium between population size and the number of iterations, ensuring the attainment of optimal solutions for specific problem sets. The realm of multi-objective evolutionary algorithms (MOEAs) showcases the potential to derive highly effective algorithms within a customizable framework, facilitating automated generation [22]. Employing a multi-objective framework for the automated design of Multi-Objective Evolutionary Algorithms (MOEAs) can result in the development of an MOEA that exhibits top-tier performance across a range of predefined metrics.

4.5.6 Data-driven Swarm-based Algorithms

Data-informed swarm-based algorithms encompass algorithmic approaches where data serves as a guiding influence for directing a collective of agents or robotic entities toward predefined objectives. These algorithms have garnered widespread recognition across diverse domains, spanning applications such as exploring and covering uncharted environments, diagnosing faults in physical systems, predicting wind speeds, and optimizing performance within industrial systems. Their design prioritizes the seamless facilitation of data classification and information extraction processes, leveraging localized communication and sensing capabilities among agents for orchestrating the swarm's navigation and search activities. The algorithms' effectiveness is substantiated through empirical assessments in tasks related to coverage and search, with comparative evaluations against alternative swarm strategies [96]. Over the last decade, machine learning algorithms have been meticulously crafted with the aim of simplifying data classification and information extraction procedures.

Machine learning algorithms have been specifically crafted to streamline the process of data classification and the extraction of information [94]. These algorithms leverage data to influence the actions of a collective of agents or robots, thereby enhancing their efficiency in pursuing predefined objectives. In the realm of data-driven swarm-based algorithms, the coordination of navigation and search tasks within the swarm is facilitated through the employment of localized communication and sensing mechanisms among the agents [72]. This enables a more streamlined and proficient synchronization among agents. Data-informed swarm-based algorithms have garnered recognition across a range of domains, encompassing applications such as the exploration and coverage of uncharted terrains, diagnosing faults within physical systems, predicting wind speeds, and optimizing the performance of industrial systems [72]. This adaptability renders them applicable across a broad spectrum of use cases. To gauge their effectiveness, these algorithms are empirically validated through experiments involving tasks related to coverage and search, and their performance is benchmarked against alternative strategies employed by swarms [72]. These experimental findings illustrate that, in specific situations, data-informed swarm-based algorithms can surpass the performance of alternative swarm strategies.

4.6 Integration of Optimization and Computing Approaches

One of the primary drivers behind the surge in big data, cloud technologies, and the Internet of Things (IoT) can be attributed to the widespread access to an array of digital devices and tools. This access has enabled diverse methods of data acquisition, aggregation, and the generation of extensive data volumes. In many scenarios, algorithms endowed with substantial computational capabilities and the capacity to store vast datasets are employed, particularly within the realms of Data Science and Machine Learning. Data-driven computing represents a burgeoning field of computer science where prediction outcomes are directly derived from input data [76, 100].

The term 'data-driven' signifies that data serves as the determining factor shaping the behavior of individuals engaged in an event or activity. This paradigm has played a pivotal role in fueling early developments and fostering a multitude of prospective enhancements and applications within the domain. Computational models offer the means to translate observations into predictions concerning future events, act as experimental platforms for novel concepts, extract value from data, and formulate inquiries pertaining to behavior [105]. The outcomes are subsequently harnessed to gain insights, enhance, oversee, and predict the functioning of intricate systems and procedures, spanning an array of domains from public policy to self-governing systems. The utility of models has transcended traditional domains like engineering and science, extending their application into diverse fields such as finance, economics, corporate administration, public policy, and urban planning. The growing processing capabilities and data accessibility have facilitated the emergence of novel computational models capable of offering more intricate descriptions of the systems under consideration [23].

These computational models effectively enable us to simulate hypothetical scenarios, providing the ability to adjust the factors influencing specific aspects before applying them in real-world situations. Computational modeling, much like any other technological instrument, lacks inherent virtue or harm. Models have the potential to offer insights or lead astray, and their utilization can be either accurate or misguided. Therefore, it is crucial to possess a thorough comprehension of the computational modeling process and a heightened awareness of when and how to use these models safely [104]. This responsibility should not rest solely with the modelers; those who commission and utilize these models must also possess a degree of expertise. Making informed decisions during the model commissioning process, as well as discerning when and how to employ a model, carries equal significance to the more technical aspects of model development. Just as a meticulously engineered hammer may align precisely with its specifications but is ill-suited for driving screws, the appropriateness of model usage hinges on these considerations [87].

Integrating optimization methodologies with computational strategies is a fundamental aspect of data-driven algorithms. Optimization methods are harnessed to discover the best possible solutions for various problems, while computational strategies are employed to manage and analyze vast volumes of data. The synergy of these two approaches enables the development of data-driven algorithms that excel in tackling complex challenges and handling extensive datasets. A prime example of this convergence of optimization and computational techniques is evident in the evolution of machine learning algorithms. In the realm of machine learning, optimization methods are utilized to identify the most suitable model parameters that align harmoniously with the available data, while computational techniques play an indispensable role in processing and scrutinizing large and intricate datasets [33].

This fusion empowers machine learning algorithms to acquire knowledge from extensive datasets and deliver precise predictions. Similarly, this synergy is exemplified in the creation of optimization algorithms tailored for addressing large-scale challenges. While optimization algorithms excel at finding optimal solutions, their computational demands can be prohibitive when dealing with large-scale problems. Employing computational strategies such as parallel computing or distributed computing enables the scalability of optimization algorithms, enabling them to effectively manage expansive datasets and intricate problem domains [101].

The fusion of optimization and computational techniques involves the blending of various tools and approaches to methodically evaluate a wide range of algorithms across complex benchmark scenarios. This method offers a way to address the task of choosing and setting up an algorithm that harmonizes efficiently with the particular problem's structure, optimization goals, and the computational resources at hand [42]. An illustrative instance of this unified approach involves combining the algorithm framework 'Paradise' with the automated algorithm configuration tool 'race' and the experimental platform 'IO profiler.' This integrated process offers several benefits, such as swift assessment times, the creation of comprehensive datasets to support algorithm analysis, and the establishment of a standardized interface suitable for benchmarking various sampling-based optimization techniques. In addition to facilitating methodical investigations into algorithm configuration, this approach also enables the assessment of how new concepts interact with existing methods. This avenue of research holds significant potential in a field that frequently emphasizes comparing fully developed algorithm instances [16]. Optimization and computing approaches can improve the performance of data-driven algorithms in several ways:

- **Improved accuracy:** Optimization methods are instrumental in locating the most optimal solution for a given challenge, whereas computational methodologies are adept at handling and scrutinizing extensive data sets. When these two strategies are harmoniously integrated, it becomes possible to create data-driven algorithms that excel in addressing intricate problems and managing vast data volumes, thereby enhancing precision and efficiency [13].

- **Scalability:** Optimization algorithms often demand significant computational resources when applied to extensive problems. However, employing computational techniques like parallel computing or distributed computing allows for the expansion of optimization algorithms, enabling them to tackle both complex challenges and substantial datasets effectively [25].

- **Efficiency:** Integrating constraints within the framework of a data-driven algorithm can notably enhance the efficiency of making inferences during testing. For example, a novel approach to constrained generative modeling circumvents the need to address an optimization challenge during every training iteration or to manually balance data fidelity and constraint adherence by introducing extra hyperparameters [86].

- **Fairness:** Utilizing computational strategies allows for obtaining additional data pertaining to specific features while keeping within budget constraints, enabling the pursuit of fairness in decision-making processes [18].

- **Optimization of Quality of Experience (QoE):** Enhancing Internet Quality of Experience (QoE) through data-driven methods involves harnessing a centralized, up-to-the-minute perspective on performance across a vast array of endpoints, which typically represent millions of clients [50].

Exact techniques strive to deliver results that are rigorously precise and free from any form of approximation or rounding discrepancies. They rely on meticulous computations and representations of numerical values. Exact methods are typically favored when confronted with scenarios necessitating the utmost accuracy, such as storing precise data in a database or conducting highly accurate calculations. Conversely, approximate methods entail employing estimations or close approximations to obtain outcomes that closely approximate the true values. These approaches may incorporate simplifications or assumptions to simplify calculations or enhance computational efficiency. Approximate methods come into play when exact calculations are impractical or when an approximate solution suffices. They find utility in scenarios where obtaining the exact solution poses difficulties, particularly when managing extensive datasets. Furthermore, approximate methods serve as valuable tools for validating or comparing results against exact methods, thereby assessing their accuracy and computational efficiency [1].

Data-driven mathematical models leverage data to formulate mathematical expressions capable of forecasting or elucidating phenomena. These models find application in diverse fields like epidemiology, the theory of big data, petrochemical planning, expectation forecasting, and control systems for mining equipment. Various techniques, including differential equations, fuzzy logic, and rough set theory, can be employed to construct these models. The overarching objective of data-driven mathematical models is to furnish precise and dependable predictions or explanations grounded in the available data. Data-driven mathematics involves the process of directly deriving mathematical models from data. This approach entails a comprehensive examination of complete datasets through statistical methodologies and the amalgamation of analytical outcomes from each of these complete datasets [12]. The resultant models find application in the interpretation of observed occurrences and the categorization of data. Data-driven mathematical modeling is a versatile tool deployed across various domains, such as mathematical biology and the development of intelligent urban areas. Within the context of smart and sustainable cities, researchers are actively crafting novel mathematical models by harnessing the potential of big data. In the realm of mathematical biology, data-driven modeling plays a pivotal role in constructing phenomenological models from empirical data and formulating explanatory mathematical models rooted in dynamic systems. This method entails the analysis of data, the classification of information, and the elucidation of observed phenomena through the framework provided by these models. In a recent investigation, a data-driven mathematical model was employed to unify disparate datasets, creating a coherent depiction of the interconnected signaling and apoptosis pathways that regulate the survival of plasma cells [61, 80]).

Single-objective optimization pertains to the process of enhancing a solitary objective function within predefined constraints. The primary aim is to identify the most favorable solution that either maximizes or minimizes the objective function, all while ensuring compliance with the given constraints. This approach is extensively applied across a spectrum of disciplines, including engineering, economics, and logistics. On the other hand, multi-objective optimization tackles the optimization of multiple objective functions under constraints. The fundamental objective here is to discover a set of Pareto-optimal solutions, representing the most advantageous compromises among the competing objectives. Multi-objective optimization finds utility in numerous domains where optimal decisions necessitate balancing trade-offs among two or more conflicting objectives, such as in engineering design, financial decision-making, and environmental management [9].

A data-driven method for approximating solutions is a technique that relies on data to derive an estimation for addressing a problem. This approach proves particularly valuable when a precise analytical solution is either unavailable or excessively intricate to compute. In the data-driven methodology, a simplified representation of the solution space is established, achieving this through either a model-based or data-driven approach. This streamlined representation is then applied to expedite the solution process. The data-driven solution is reached by employing an adaptive solver with a dataset, and the accuracy of the solution is assessed relative to the size of the data [78]. As mentioned in the previous section, heuristic and metaheuristic data-driven approaches can be used as approximate approaches. In Figure 4.4, metaheuristic approaches are presented in an expanded category that has a more practical and detailed look at solving real-world problems than Section 4.7, which was classified into five general categories. Evidence demonstrates that the fusion of optimization and computational strategies has found application across a wide spectrum of fields. These include computer vision measurement, enhancements in heating, ventilation, and air conditioning system performance, optimization within tunneling operations, planning assembly sequences, enabling intelligent edge services, optimizing the architecture of multilayer perceptron, fine-tuning parameters in friction stir welding processes, and enhancing economic load dispatch mechanisms. The optimization techniques harnessed in these studies encompass a range of soft computing approaches, such as artificial neural networks, adaptive neuro-fuzzy inference systems, genetic algorithms, and metaheuristic algorithms. The outcomes of these investigations underscore the ability of optimization techniques to deliver more precise results and to curtail error rates effectively. Furthermore, it's notewor-

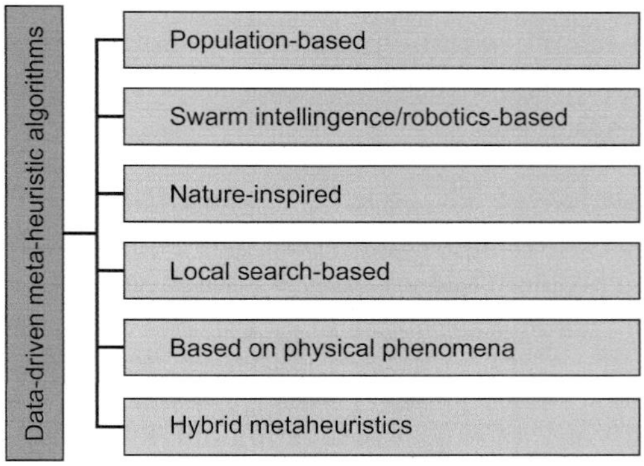

Figure 4.4: A developed classification of data-driven solution approaches.

thy that the optimization techniques employed in these studies exhibit versatility, as they can be effectively employed to tackle non-linear optimization problems encompassing both equality and inequality constraints. Combining optimization and computing approaches can provide several benefits, including [43, 77]:

- **Enhanced Precision:** Optimization techniques have the capability to generate results with a higher degree of accuracy, effectively diminishing error rates.

- **Accelerated Problem Solving:** Soft computing methods, including metaheuristics, can efficiently compute solutions of high quality for challenging problems within reasonable timeframes.

- **Handling Complexity:** Optimization techniques are well-suited to address intricate issues, effectively tackling non-linear optimization problems characterized by both equality and inequality constraints.

- **Adaptability to Uncertainty:** Soft computing techniques, such as metaheuristics, possess the ability to work effectively with imprecise, uncertain, or approximated data. They can manage partial information and provide accurate solutions for optimization problems even when exact data is unavailable.

- **Meeting Realistic Timelines:** Metaheuristics are adept at ensuring that solutions are achieved within practical timeframes, a feature valuable across diverse domains, ranging from informatics to industrial and commercial applications.

- **Innovative Hybrid Approaches:** The fusion of optimization and computing methodologies can foster the creation of innovative hybrid intelligent methods, expanding the problem-solving toolkit.

- **Enhanced System Performance:** Combining optimization and computing approaches can result in substantial improvements in system performance, applicable across fields such as heating, ventilation, and air conditioning systems.
- **Time Efficiency:** The synergy between optimization and computing techniques can lead to a significant reduction in the time elapsed between defining a problem and arriving at a solution.

In conclusion, the amalgamation of optimization and computing strategies yields a multitude of advantages, encompassing enhanced precision, expedited problem-solving, adeptness in tackling intricate challenges, adaptability to vagueness and unpredictability, adherence to practical timeframes, innovation through hybrid intelligent techniques, performance enhancement, and a reduction in time expended. Particularly, soft computing methods like metaheuristics prove adept at effectively addressing complex problems, autonomously adapting to the terrain of the search space, striking a balance between exploration and exploitation, accommodating imprecision and ambiguity, producing high-quality solutions, supporting hybridization with other optimization techniques, and ensuring the fulfillment of realistic resolution timelines in numerous application domains. These studies underscore the significance of hybridizing metaheuristics with other optimization methods and highlight the rapidly evolving nature of the field of metaheuristics.

4.7 Integrated Optimization and Computing Approaches Using Intelligent Data-Driven

This section delves into the synergy between optimization techniques, computing methodologies, and intelligent data-driven approaches to facilitate more informed, efficient, and effective decision-making. By integrating these diverse fields, decision-makers can harness the power of data analytics, machine learning, and optimization to address complex problems across domains such as supply chain management, finance, healthcare, and engineering.

Optimization and computing have long been fundamental components of problem-solving in fields such as engineering, logistics, finance, and more. These techniques involve finding the best possible solution from a set of alternatives, given specific constraints and objectives. Classical optimization algorithms, such as linear programming and dynamic programming, have played pivotal roles in optimizing processes and decision-making. However, as problems grew in complexity and data became more abundant, traditional optimization and computing methods started to encounter limitations. They struggled to handle large-scale, nonlinear, and uncertain scenarios efficiently. This limitation prompted researchers and practitioners to look for alternative approaches.

Integrated optimization and computing approaches involve the seamless integration of mathematical optimization techniques with high-performance computing systems, including cloud computing, supercomputing, and parallel computing. Mathematical optimization is the process of finding the best solution to a problem from a set of possible solutions while adhering to certain constraints. This can include maximizing profits, minimizing costs, optimizing resource allocation, and much more. Computing approaches, on the other hand, encompass a wide range of technologies and methodologies, such as machine learning, artificial intelligence (AI), data analytics, and quantum computing. These computational tools can be used to enhance the effectiveness of optimization processes, leading to more accurate and timely decisions.

4.7.1 Applications Across Industries

Manufacturing and Supply Chain Optimization: Integrated optimization and computing can optimize production schedules, minimize inventory costs, and improve supply chain logistics. By combining optimization algorithms with real-time data analytics and AI-driven demand forecasting, manufacturers can respond swiftly to changes in market demand, reduce wastage, and enhance overall production efficiency.

- **Healthcare and Treatment Planning:** In the healthcare sector, these approaches are used to optimize patient treatment plans, hospital resource allocation, and clinical trial designs. By considering various factors like patient history, treatment options, and resource constraints, healthcare providers can make data-driven decisions to deliver more effective and efficient care.
- **Financial Portfolio Management:** In the world of finance, integrated optimization and computing enable portfolio managers to optimize investment strategies, mitigate risk, and maximize returns. These approaches take into account market data, economic indicators, and individual investor preferences to create optimized investment portfolios.
- **Energy and Sustainability:** The energy industry utilizes optimization and computing techniques to optimize power generation, distribution, and consumption. By leveraging data from smart grids, weather forecasts, and renewable energy sources, energy companies can minimize costs and reduce environmental impact.

4.7.2 Benefits of Integration

Integrated optimization and computing approaches represent a powerful synergy between mathematical optimization and advanced computing technologies.

As industries continue to embrace digital transformation, these approaches will play a crucial role in driving efficiency, reducing costs, and enabling data-driven decision-making. From manufacturing and healthcare to finance and energy, the integration of optimization and computing is poised to reshape industries and create a more efficient and sustainable future. The integration of optimization and computing approaches offers several compelling benefits:

- **Enhanced Decision-Making:** By leveraging computational power and optimization algorithms, organizations can make data-driven decisions quickly and accurately, leading to improved overall performance.
- **Cost Reduction:** Optimizing processes and resource allocation can lead to significant cost savings. Whether it's in supply chain management, energy consumption, or financial investments, efficiency gains translate into financial benefits.
- **Improved Resource Utilization:** Companies can maximize the utilization of resources such as time, labor, and equipment, resulting in higher productivity and reduced waste.
- **Adaptability:** Integrated approaches can adapt to changing conditions in real-time, helping organizations respond swiftly to unexpected events or market fluctuations.
- **Sustainability:** These approaches contribute to sustainability goals by optimizing resource use and minimizing environmental impact.

4.8 Conclusion and Future Directions

In this comprehensive exploration of Optimization and Computing using intelligent data-driven approaches for decision-making, we have delved deep into the transformative potential of these methodologies. Throughout this chapter, we have witnessed the power of leveraging data-driven insights and computational capabilities in various domains, including but not limited to [mention specific domains or applications discussed in your chapter]. Our journey has underscored the versatility and effectiveness of these intelligent data-driven approaches in enhancing decision-making processes. From the deployment of advanced machine learning algorithms for predictive analytics to the fine-tuning of optimization techniques for resource allocation, it is evident that the synergy between data and computation is reshaping the landscape of problem-solving. As we reflect on the rich tapestry of insights and methodologies presented herein, we recognize that this chapter merely scratches the surface of what is possible. The evolution of Optimization and Computing using intelligent data-driven approaches for decision-making is ongoing, and the horizons are filled with potential avenues for research, application, and innovation.

4.8.1 Future Directions

- **Interpretability and Explain ability:** As machine learning models become increasingly complex, there is a growing need to improve their interpretability and explainability. Future research should focus on developing methodologies and tools that enable us to understand and trust the decisions made by AI systems, especially in critical applications.

- **Ethical AI and Fairness:** Ethical considerations surrounding AI and data-driven decision-making are paramount. Researchers and practitioners must continue to address issues related to bias, fairness, and accountability. The development of ethical frameworks and guidelines will be crucial in ensuring responsible AI deployment.

- **Adaptive Learning Systems:** The creation of adaptive learning systems that can autonomously evolve and adapt to changing data distributions and environments is a compelling area of research. Such systems could provide decision support that is not only accurate but also agile in responding to emerging trends and challenges.

- **Human-AI Collaboration:** Exploring ways to foster productive collaboration between humans and AI systems is essential. Developing AI models that can effectively complement human decision-makers and vice versa will be key in harnessing the full potential of intelligent data-driven approaches.

- **Cross-disciplinary Research:** The convergence of knowledge from various fields, including computer science, domain-specific expertise, and behavioral sciences, can yield transformative solutions to complex problems. Encouraging cross-disciplinary collaboration and research will be instrumental in advancing the field.

- **Scalable and Efficient Algorithms:** With the ever-increasing volume of data, there is a growing demand for algorithms that are not only accurate but also scalable and computationally efficient. Future work should focus on developing techniques that can handle large-scale data without compromising performance.

In conclusion, the field of Optimization and Computing using intelligent data-driven approaches for decision-making is on an exciting trajectory. By addressing these future directions and challenges, we can harness the full potential of data-driven decision-making to drive positive societal and economic impact. As researchers and practitioners, we stand at the forefront of a transformative era, and our collective efforts will shape the future of intelligent decision support systems.

References

[1] Aandahl, R.Z., Stadler, T., Sisson, S.A. and Tanaka, M.M. 2014. Exact vs. approximate computation: Reconciling different estimates of Mycobacterium tuberculosis epidemiological parameters. *Genetics*, 196(4): 1227–1230.

[2] Abdollahzadeh, B., Soleimanian Gharehchopogh, F. and Mirjalili, S. 2021. Artificial gorilla troops optimizer: A new nature-inspired metaheuristic algorithm for global optimization problems. *International Journal of Intelligent Systems*, 36(10): 5887–5958.

[3] Abraham, D. 2020. Data-driven decision making and community indicators: towards an integration of DDDM in community development. pp. 199–210. *In*: Ridzi, F., Stevens, C. and Davern, M. (eds.). *Community Quality-of-Life Indicators: Best Cases VIII*. Springer International Publishing.

[4] Agrawal, P., Abutarboush, H.F., Ganesh, T. and Mohamed, A.W. 2021. Metaheuristic algorithms on feature selection: A survey of one decade of research (2009–2019). *IEEE Access*, 9: 26766–26791.

[5] Agushaka, J.O. and Ezugwu, A.E. 2022. Initialization approaches for population-based metaheuristic algorithms: A comprehensive review. Applied Sciences, 12(2), 896.

[6] Aigner, K.-M., Goerigk, M., Hartisch, M., Liers, F. and Miehlich, A. 2023. A framework for data-driven explainability in mathematical optimization.

[7] Al Mohamed, A.A., Al Mohamed, S. and Zino, M. 2023. Application of fuzzy multicriteria decision-making model in selecting pandemic hospital site. *Future Business Journal*, 9(1): 14.

[8] Almuhaideb, S., Altwaijry, N., AlMansour, S., AlMklafi, A., AlMojel, A.K., AlQahtani, B. and AlHarran, M. 2022. Clique finder: A self-adaptive simulated annealing algorithm for the maximum clique problem. *International Journal of Applied Metaheuristic Computing (IJAMC)*, 13(2): 1–22.

[9] Alsheddy, A. 2018. A penalty-based multi-objectivization approach for single objective optimization. *Information Sciences*, 442: 1–17.

[10] Andrew Yeung, S.D.P.M. 2023. Decision Intelligence: Using Data to Drive Better Outcomes. https://www.thoughtspot.com/datatrends/ai/decision-intelligence.

[11] Arend, R.J. 2020. Strategic decision-making under ambiguity: A new problem space and a proposed optimization approach. *Business Research*, 13(3): 1231–1251.

[12] Arridge, S., Maass, P., Öktem, O. and Schönlieb, C.-B. 2019. Solving inverse problems using data-driven models. *Acta Numerica*, 28: 1–174.

[13] Asaad, S.M., Potrus, M.Y., Ghafoor, K.Z., Maghdid, H.S. and Mulahuwaish, A. 2022. Improving positioning accuracy using optimization approaches: A survey, research challenges and future perspectives. *Wireless Personal Communications*, 122(4): 3393–3409.

[14] Asana, T. 2022. Data-driven decision making: A step-by-step guide. Asana Team. https://asana.com/resources/data-driven-decision-making.

[15] Ayyarao, T.S., Ramakrishna, N., Elavarasan, R.M., Polumahanthi, N., Rambabu, M., Saini, G., Khan, B. and Alatas, B. 2022. War strategy optimization algorithm: A new effective metaheuristic algorithm for global optimization. *IEEE Access*, 10: 25073–25105.

[16] Aziz-Alaoui, A., Doerr, C. and Dreo, J. 2021. Towards large-scale automated algorithm design by integrating modular benchmarking frameworks. *Proceedings of the Genetic and Evolutionary Computation Conference Companion.*

[17] Baardman, L., Cristian, R., Perakis, G., Singhvi, D., Skali Lami, O. and Thayaparan, L. 2023. The role of optimization in some recent advances in data-driven decision-making. *Mathematical Programming*, 200(1): 1–35.

[18] Bakker, M., Valdés, H.R., D Tu, P., Gummadi, K.P., Varshney, K.R., Weller, A. and Pentland, A.S. 2020. Fair enough: Improving fairness in budget-constrained decision-making using confidence thresholds.

[19] Beiranvand, V., Hare, W. and Lucet, Y. 2017. Best practices for comparing optimization algorithms. *Optimization and Engineering*, 18(4): 815–848.

[20] Belenguer, L. 2022. AI bias: Exploring discriminatory algorithmic decision-making models and the application of possible machine-centric solutions adapted from the pharmaceutical industry. *AI and Ethics*, 2(4): 771–787.

[21] Bertsimas, D. and Thiele, A. 2006. Robust and data-driven optimization: Modern decision making under uncertainty. *In Models, Methods, and Applications for Innovative Decision Making* (pp. 95–122). Informs.

[22] Bezerra, L.C. and Manuel, L. 2020. Automatically designing state-of-the-art multi-and many-objective evolutionary algorithms. *Evolutionary Computation*, 28(2): 195–226.

[23] Bibri, S.E. 2021. Data-driven smart sustainable cities of the future: Urban computing and intelligence for strategic, short-term, and joined-up planning. *Computational Urban Science*, 1(1): 8.

[24] Bornet, P. 2022. Is Decision Intelligence The New AI? https://www.forbes.com/sites/forbestechcouncil/2022/05/25/is-decisionintelligence-the-new-ai/?sh=3dcac9594e42.

[25] Brownlee, J. 2021. How to Choose an Optimization Algorithm. https://machinelearningmastery.com/tour-of-optimization-algorithms.

[26] Buijsse, R., Willemsen, M. and Snijders, C. 2023. Data-driven decision-making. pp. 239–277. *In*: Liebregts, W., van den Heuvel, W.-J. and van den Born, A. (eds.). Data Science for Entrepreneurship: Principles and Methods for Data Engineering, Analytics, Entrepreneurship, and the Society. Springer International Publishing.

[27] Canonico, P., De Nito, E., Esposito, V., Fattoruso, G., Pezzillo Iacono, M. and Mangia, G. 2022. Visualizing knowledge for decision-making in Lean Production Development settings. Insights from the automotive industry. *Management Decision*, 60(4): 1076–1094.

[28] Case, B. and Lehre, P.K. 2020. Self-adaptation in nonelitist evolutionary algorithms on discrete problems with unknown structure. *IEEE Transactions on Evolutionary Computation*, 24(4): 650–663.

[29] Ceschi, A., Demerouti, E., Sartori, R. and Weller, J. 2017. Decision-making processes in the workplace: How exhaustion, lack of resources

and job demands impair them and affect performance [Original Research]. *Frontiers in Psychology*, 8.

[30] Jingshang, C.H.E., Chenjie Xiao, S.H. and Aimei, L.I. 2019. Why information overload damages decisions? An explanation based on limited cognitive resources. *Advances in Psychological Science*, 27(10): 1758–1768.

[31] Çıtak, E. 2023. Elevating business decisions from gut feelings to data-driven excellence.

[32] Colson, E. 2019. What AI-Driven Decision Making Looks Like. https://hbr.org/2019/07/what-ai-driven-decision-making-looks-like.

[33] De Loera, J.A., Haddock, J., Ma, A. and Needell, D. 2021. Data-driven algorithm selection and tuning in optimization and signal processing. *Annals of Mathematics and Artificial Intelligence*, 89(7): 711–735.

[34] Deniz, N. and Ozceylan, E. 2023. A bibliometric and social network analysis of data-driven heuristic methods for logistics problems. *Journal of Industrial and Management Optimization*, 19(8): 5671–5689.

[35] Deshmukh, R.K. and Kubal, M. 2020. An overview of machine learning algorithms for data science.

[36] Dilmegani, C. 2023. Decision Intelligence: In-Depth Guide for Businesses in 2023.

[37] Doerr, B. and Neumann, F. 2021. A survey on recent progress in the theory of evolutionary algorithms for discrete optimization. *ACM Transactions on Evolutionary Learning and Optimization*, 1(4): 1–43.

[38] Dos Santos Gonçalves, P.V. and Campos, L.M.S. 2022. A systemic review for measuring circular economy with multi-criteria methods. *Environmental Science and Pollution Research*, 29(21): 31597–31611.

[39] DriveResearch. 2023. What is Data-Driven Decision Making? (And Why It's So Important).

[40] Elbanna, S. and Fadol, Y. 2016. The role of context in intuitive decision-making. *Journal of Management Organization*, 22(5): 642–661.

[41] Felländer, A., Rebane, J., Larsson, S., Wiggberg, M. and Heintz, F. 2022. Achieving a data-driven risk assessment methodology for ethical AI. *Digital Society*, 1(2): 13.

[42] Guo, X., Stein, B.V. and Bäck, T. (2019, 6–9 Dec. 2019). A new approach towards the combined algorithm selection and hyper-parameter optimization problem. *2019 IEEE Symposium Series on Computational Intelligence (SSCI)*.

[43] Guo, Z., Moayedi, H., Foong, L.K. and Bahiraei, M. 2020. Optimal modification of heating, ventilation, and air conditioning system performances in residential buildings using the integration of metaheuristic optimization and neural computing. *Energy and Buildings*, 214: 109866.

[44] Gupta, S., Modgil, S., Bhattacharyya, S. and Bose, I. 2022. Artificial intelligence for decision support systems in the field of operations research: Review and future scope of research. *Annals of Operations Research*, 308(1): 215–274.

[45] Hakanen, J. and Allmendinger, R. 2021. Multiobjective optimization and decision making in engineering sciences. *Optimization and Engineering*, 22(2): 1031–1037.

[46] Harifi, S. Ancient-inspired: A novel source of inspiration for the development of metaheuristic algorithms.

[47] Harifi, S., Mohammadzadeh, J., Khalilian, M. and Ebrahimnejad, S. 2021. Giza Pyramids Construction: An ancient-inspired metaheuristic algorithm for optimization. *Evolutionary Intelligence*, 14(4): 1743–1761.

[48] Heße, F., Müller, S. and Attinger, S. 2023. Data-driven estimates for the geostatistical characterization of subsurface hydraulic properties. *Hydrol. Earth Syst. Sci. Discuss.*, 2023: 1–31.

[49] Höchtl, J., Parycek, P. and Schöllhammer, R. 2016. Big data in the policy cycle: Policy decision making in the digital era. *Journal of Organizational Computing and Electronic Commerce*, 26(1-2): 147–169.

[50] Jiang, J. 2017. Enabling Data-Driven Optimization of Quality of Experience for Internet Applications.

[51] Jin, Y., Wang, H. and Sun, C. 2021. Data-driven evolutionary optimization. Springer.

[52] Kahraman, C. 2008. Multi-criteria decision-making methods and fuzzy sets. pp. 1–18. *In*: Kahraman, C. (ed.). Fuzzy Multi-Criteria Decision Making: Theory and Applications with Recent Developments. Springer US.

[53] Kaspi, S. and Venkatraman, S. 2023. Data-Driven Decision-Making (DDDM) for higher education assessments: A case study. *Systems*, 11(6).

[54] Katsaliaki, K., Galetsi, P. and Kumar, S. 2022. Supply chain disruptions and resilience: A major review and future research agenda. *Annals of Operations Research*, 319(1): 965–1002.

[55] Keith, A.J. and Ahner, D.K. 2021. A survey of decision making and optimization under uncertainty. *Annals of Operations Research*, 300(2): 319–353.

[56] Khan, M., Ullah, Z., Mašek, O., Naqvi, S.R. and Khan, M.N.A. 2022. Artificial neural networks for the prediction of biochar yield: A comparative study of metaheuristic algorithms. *Bioresource Technology*, 355: 127215.

[57] Khan, U.A., Bajwa, W.U., Nedić, A., Rabbat, M.G. and Sayed, A.H. 2020. Optimization for data-driven learning and control. *Proceedings of the IEEE*, 108(11): 1863–1868.

[58] Kozyrkov, C. 2019. Introduction to Decision Intelligence.

[59] Krishen, A.S. and Petrescu, M. 2019. Data-driven decision making: Implementing analytics to transform academic culture. *Journal of Marketing Analytics*, 7(2): 51–53.

[60] Levinthal, D.A. and Newark, D.A. 2023. Putting the individual in the organizational context: A carnegie perspective on decision making. *Frontiers in Psychology*, 14: 1165713.

[61] Li, J., Xiao, X., Boukouvala, F., Floudas, C.A., Zhao, B., Du, G., Su, X. and Liu, H. 2016. Data-driven mathematical modeling and global optimization framework for entire petrochemical planning operations. *AIChE Journal*, 62(9): 3020–3040.

[62] Liao, X., Shi, J., Li, Z., Zhang, L. and Xia, B. 2020. A model-driven deep reinforcement learning heuristic algorithm for resource allocation in ultra-dense cellular networks. *IEEE Transactions on Vehicular Technology*, 69(1): 983–997.

[63] Lim, S., Henriksson, A. and Zdravkovic, J. 2021. Data-driven requirements elicitation: A systematic literature review. *SN Computer Science*, 2(1): 16.

[64] Luo, J. 2023. Data-Driven Innovation: What Is It? *IEEE Transactions on Engineering Management*, 70(2): 784–790.

[65] Lyu, S. 2022. Data-driven optimization approaches to the power system planning under uncertainty.

[66] Ma, Z., Wu, G., Suganthan, P.N., Song, A. and Luo, Q. 2023. Performance assessment and exhaustive listing of 500+ nature-inspired metaheuristic algorithms. *Swarm and Evolutionary Computation*, 77: 101248.

[67] Mahadeen, T., Galanakis, K., Samara, E. and Kilintzis, P. 2021. Heuristics and Evidence Decision (HeED) Making: A case study in a systemic model for transforming decision making from heuristics-based to evidenced-based. *Journal of the Knowledge Economy*, 12(4): 1668–1693.

[68] Malak, H.A. 2023. 10 Information Management Challenges and Solutions for 2023.

[69] Mangla, P. and Goyal, S.K. 2017. Heuristic vs meta-heuristic approaches for load balancing in cloud environment. *International Journal of Computer Mathematical Sciences*, 6(4): 297–301.

[70] Martí, R. and Reinelt, G. 2022. Heuristic methods. *In Exact and Heuristic Methods in Combinatorial Optimization: A Study on the Linear Ordering and the Maximum Diversity Problem* (pp. 27–57). Springer Berlin Heidelberg.

[71] Martins, A.F., Machado, M., Bernardino, H.S. and de Souza, J.F. 2021. A comparative analysis of metaheuristics applied to adaptive curriculum sequencing. *Soft Computing*, 25(16): 11019–11034.

[72] Michaloglou, A. and Tsitsas, N.L. 2021. Particle swarm optimization algorithms with applications to wave scattering problems. *Swarm Intelligence.*

[73] Mohajerin Esfahani, P. and Kuhn, D. 2018. Data-driven distributionally robust optimization using the Wasserstein metric: Performance guarantees and tractable reformulations. *Mathematical Programming*, 171(1): 115–166.

[74] Monakhov, O. and Monakhova, E. 2021. Comparative evaluation of algorithms for automatic construction of nonlinear models based on metaheuristic programming with gene expression. *CEURWorkshop Proceedings.*

[75] Morelli, M., Casagrande, M. and Forte, G. 2022. Decision making: A theoretical review. *Integrative Psychological and Behavioral Science*, 56(3): 609–629.

[76] Moumtzidis, I., Kamariotou, M. and Kitsios, F. 2022. Digital transformation strategies enabled by internet of things and big data analytics: The use-case of telecommunication companies in Greece. *Information*, 13(4): 196.

[77] Nesmachnow, S. 2015. Using metaheuristics as soft computing techniques for efficient optimization.

[78] Nguyen, L.T.K., Aydin, R.C. and Cyron, C.J. 2022. Accelerating the distance-minimizing method for data-driven elasticity with adaptive hyperparameters. *Computational Mechanics*, 70: 621–638.

[79] Ordonez, L. and Benson III, L. 1997. Decisions under time pressure: How time constraint affects risky decision making. *Organizational Behavior and Human Decision Processes*, 71(2): 121–140.

[80] Phillips, T., Lenhart, S. and Strickland, W.C. (2021). A data-driven mathematical model of the heroin and fentanyl epidemic in Tennessee. *Bulletin of Mathematical Biology*, 83(10): 97.

[81] Puentes, L., Cagan, J. and McComb, C. 2021. Data-driven heuristic induction from human design behavior. *Journal of Computing and Information Science in Engineering*, 21(2): 024501.

[82] Ross, M. and Taylor, J. 2021. Managing AI Decision-Making Tools.

[83] Saeid, S., Maghsoud, A. and Farzaneh Mansoori, M. 2021. Artificial Intelligence and its application in optimization under uncertainty. *In*: Ciza, T. (ed.). *Data Mining* (pp. Ch. 8). IntechOpen.

[84] Saltz, J.S. and Dewar, N. 2019. Data science ethical considerations: A systematic literature review and proposed project framework. *Ethics and Information Technology*, 21(3): 197–208.

[85] Sarker, I.H. 2021. Data science and analytics: an overview from data-driven smart computing, decision-making and applications perspective. *SN Computer Science*, 2(5): 377.

[86] Schwenzer, M., Ay, M., Bergs, T. and Abel, D. 2021. Review on model predictive control: An engineering perspective. *The International Journal of Advanced Manufacturing Technology*, 117(5): 1327–1349.

[87] Sezer, H.B. and Namukasa, I.K. 2021. Real-world problems through computational thinking tools and concepts: The case of coronavirus disease (COVID-19). *Journal of Research in Innovative Teaching Learning*, 14(1): 46–64.

[88] Shukla, K., Xu, M., Trask, N. and Karniadakis, G.E. 2022. Scalable algorithms for physics-informed neural and graph networks. *Data-Centric Engineering*, 3: e24, Article e24.

[89] Soma, T., Gatmiry, K. and Jegelka, S. 2022. Optimal algorithms for group distributionally robust optimization and beyond.

[90] Song, Z., Song, S., Zhang, B., Qian, J., Tang, C. and Ji, J. 2021. An adaptive scaling cuckoo search algorithm. *2021 4th International Conference on Advanced Electronic Materials, Computers and Software Engineering (AEMCSE)*, 562–566.

[91] Spirov, A.V. and Myasnikova, E.M. 2022. Heuristic algorithms in evolutionary computation and modular organization of biological macromolecules: Applications to *in vitro* evolution. *Plos One*, 17(1): e0260497.

[92] SS, V.C. and HS, A. 2022. Nature inspired meta heuristic algorithms for optimization problems. *Computing*, 104(2): 251–269.

[93] Sun, F. 2018. Algorithms for Context-Sensitive Prediction, Optimization and Anomaly Detection in Urban Mobility.

[94] Tang, J., Liu, G. and Pan, Q. 2021. A review on representative swarm intelligence algorithms for solving optimization problems: Applications and trends. *IEEE/CAA Journal of Automatica Sinica*, 8: 1627–1643.

[95] Tirkolaee, E.B., Sadeghi, S., Mooseloo, F.M., Vandchali, H.R. and Aeini, S. 2021. Application of machine learning in supply chain management: A comprehensive overview of the main areas. *Mathematical Problems in Engineering*, 2021: 1476043.

[96] Tran, V.P., Garratt, M.A., Kasmarik, K., Anavatti, S.G. and Abpeikar, S. 2022. Frontier-led swarming: Robust multi-robot coverage of unknown environments. *Swarm and Evolutionary Computation*, 75: 101171.

[97] Tuba, M. 2014. Multilevel image thresholding by nature-inspired algorithms: A short review. *Computer Science Journal of Moldova*, 66(3): 318–338.

[98] Ullah, I. and Youn, H.Y. 2020. Intelligent data fusion for smart iot environment: A survey. *Wireless Personal Communications*, 114(1): 409–430.

[99] Van Der Schelling, M. 2021. A data-driven heuristic decision strategy for data-scarce optimization: With an application towards bio-based composites.

[100] Wagner, R. and Cozmiuc, D. 2022. Extended reality in marketing—Amultiple case study on internet of things platforms. *Information*, 13(6).

[101] Wang, H., Gao, L., Yang, G. and Wu, B. 2023. A data-driven robust design optimization method and its application in compressor blade. *Physics of Fluids*, 35(6).

[102] Wang, H., Liang, M., Sun, C., Zhang, G. and Xie, L. 2021. Multiple-strategy learning particle swarm optimization for large-scale optimization problems. *Complex Intelligent Systems*, 7(1): 1–16.

[103] Wang, N., Wang, H., Pei, S. and Zhang, B. 2023. A data-driven heuristic method for irregular flight recovery. *Mathematics*, 11(11): 2577.

[104] Winsberg, E. 2003. Simulated experiments: Methodology for a virtual World. *Philosophy of Science*, 70(1): 105-125.

[105] Wirtz, B.W. 2022. Artificial Intelligence, Big Data, Cloud Computing, and Internet of Things. pp. 175–245. *In*: Wirtz, B.W. (ed.). *Digital Government: Strategy, Government Models and Technology*. Springer International Publishing.

[106] Yang, B., Xu, X., Gong, Y. and Rekik, Y. 2023. Data-driven optimization models for inventory and financing decisions in online retailing platforms. *Annals of Operations Research*.

[107] Ye, Y. 2014. Data-Driven Optimization. https://web.stanford.edu/ỹyye/Data-DrivenOPT.pdf.

Chapter 5

Aggregate Production Planning: A Novel Game Model

Ali Hamidoğlu,[1,*] *Selma Gütmen*[2] and *Gerhard-Wilhelm Weber*[2]

5.1 Introduction

Aggregate production planning (APP) is the process of determining a company's production, inventory, and employment levels simultaneously over a set period of time. Assuming that both sales and production capacity are constant, the goal is to minimize the total relevant costs while simultaneously satisfying non-constant and time-varying demand. The process of organizing a scheme for the level of production, storage, and labor to fulfill the demand over a period of time up to one year is referred to as APP. To be more specific, the APP develops a decision-making process that is both medium-term and long-term, and it covers a planning horizon of three to twelve months [1]. As a result, many different activities, such as the capacity of the machinery, the storage level, the environment of the labor, etc., are managed and employed throughout this [2]. The primary objective is to achieve the lowest possible total production cost while simultaneously making the most efficient use of the resources that are at one's disposal.

[1]Department of Mathematical and Statistical Sciences, University of Alberta, Edmonton, Alberta, Canada T6G 2G1.

[2]Poznan University of Technology, Faculty of Engineering Management, Poznan, Poland. Institute of Applied Mathematics, Middle East Technical University, Ankara, Turkey.

* Corresponding author: hamidogl@ualberta.ca

The process of applying for an APP requires numerous phases, including the following:

- *Demand forecasting*: The process of anticipating the future demand for a product or service by looking at historical data, current market trends, and other relevant factors [3].
- *Planning for capacity*: The process of evaluating an organization's resources and determining whether or not it is able to generate enough goods or services to fulfill the anticipated demand [4].
- *Creating a production plan*: Follow a strategy for the manufacture of products or provision of services that satisfies the anticipated level of demand while simultaneously maximizing the use of available resources and reducing expenditures [5].
- *Planning for inventory*: The process of calculating the appropriate level of stock to maintain in order to fulfill both the production schedule and the requirements of the market [6].
- *Budgeting and financial planning*: The process of preparing a budget and financial plan for the production activities to ensure that they are financially viable. This can be done to ensure that the production activities are financially feasible [7].
- *Keeping an eye on things and making any changes*: The process of keeping a close eye on the production plan at all times and making any necessary changes to ensure that it stays on the track [8].

In general, APP is a crucial procedure for businesses that wish to efficiently manage their production activities and optimize their resources to satisfy the demand of their customers while keeping their costs to a minimum and increasing their profits as much as possible.

The APP model was first studied in [9, 10], and since then, it has received a significant amount of attention from researchers [11, 12, 13, 14]. Regarding solution techniques for problems of APP, optimization and meta-heuristic models are frequently utilized in order to build production plans that are efficient and effective while also meeting the business goals and objectives of the firm. These approaches are used to achieve the goals of maximizing earnings, minimizing costs, and optimizing resource utilization.

Mathematical techniques are utilized in optimization models in order to determine the optimal production plan for a given set of constraints, which may include the availability of resources, the capacity of the production facility, and the demand for the product. Linear programming [15], mixed-integer programming [16], and non-linear programming [17] are the three types of optimization models that are utilized in APP the most frequently. In linear programming, creating a linear objective function that expresses the purpose of the company, such as maximizing profits or lowering expenses, while adhering to a set of linear constraints is a necessary step in the process. In mixed-integer programming, discrete choice variables are considered in APP. In other words, this technique is utilized when the planning of production involves decisions that can only be made in whole numbers, such as the number of production units or the number of personnel required. Besides, non-linear programming is implemented whenever there are non-linear objective functions or constraints involved in production planning problems.

Meta-heuristic models, on the other hand, are optimization strategies that make use of iterative algorithms to search for near-optimal solutions to difficult problems. The ant colony algorithm [18, 19], particle swarm optimization [20, 21], fuzzy linear programming [22], goal optimization [23, 24], simulated annealing algorithm [25], generic algorithm [26], all refer to different types of meta-heuristic models. These models are especially helpful in situations in which the problem being solved is either too difficult or has an excessive number of variables for it to be solved using more conventional optimization techniques. These models are powerful tools that are used in APP to produce production plans that are efficient and effective while also meeting the business goals and objectives of the organization. Meta-heuristic models have the potential to offer superior answers to difficult production planning issues, despite the fact that optimization models have been around longer and are utilized more frequently.

The interaction between a number of different agents that are involved in APP can be modeled with game theory. In this context, game theory offers a framework for understanding the behavior of the many agents, including their decision-making processes and the results of their interactions with one another. The idea behind the game theory is that each participant in the process of production planning has their own set of information, objectives, and limitations to work with. The production crew, the sales team, and the finance team may all be considered to be agents in this regard. Every group is working toward its own unique objectives and may have access to unique data regarding the market, production costs, and capacity limits. The goal of modeling these various aims and information in game theory is to determine how the agents will behave and interact with one another so that predictions can be made. In this context, cooperative [27] and competitive [28] game models are the ones that come up most frequently in APP applications of game theory. In games with cooperative play, players work together towards the accomplishment of a shared objective, whereas in competitive games, players pursue their individual objectives without the assistance of other players. In general, game theory offers a robust framework for evaluating the behavior of several actors in an APP. By simulating the interactions between various agents and their decision-making processes, game theory can assist in identifying potential inefficiencies in the production plan and developing methods to maximize the organization's overall profit.

In this chapter, we investigate a two-player game platform where each player's strategy is related to the number of people working regular-time and overtime shifts to satisfy the demand in each period of time. In this game, the first player is given the role of an evader and the responsibility of starting the game. The evader's objective is to meet the specified demand while employing the fewest possible full-time workers and, if possible, accumulating surplus goods for use in the following round. The second player, on the other hand, replicates the actions of the first player, but instead of stockpiling additional items, he employs either the same number of workers as the evader or fewer workers, and he manages the number of overtime hours worked by each worker in order to bring the storage level down to zero or as close to zero as possible during each phase of the game. Once the evader has developed his strategy to employ a sufficient number of workers without requiring them to perform overtime duty in order to meet the demand, the pursuer is able to become aware of the level of workforce and inventory of APP in advance in each period, which enables him to construct a better schedule for the production planning. This is beneficial to both parties. The results of this comparison show that there is a growing disparity in the overall production costs of each participant throughout the course of each period of time. The question then arises as to whether there are any other planning tactics that are more effective than the schedule that was prepared by the pursuer. If the cost of storing items drops to such a low level and the cost of paying workers for overtime rises to such a high level, then this planning strategy might not work. When seen from this angle, one may execute superior production planning, which would result in lower overall production costs than those of the pursuer. Because of this, the study has some inherent limitations. So, we take into consideration a scenario with a high cost of storage in which the planner wishes to meet the demand in each period of time while simultaneously minimizing the number of products that are held for the following period. When the disparity between the total production cost of the evader and that of the pursuer reaches its highest point, the game is over. In this scenario, the production plan of the pursuer is constructed in accordance with the method utilized by the evader, which ultimately determines who wins the game.

There have been several recent works [29, 30, 31, 32] that focus on the construction of game models that simulate real-world challenges. To describe the given problem as a two-person game in order to build chase tactics for one player that either becomes the solution to the given problem or approximates the answer within a defined range is the primary perspective that is shared among those studies. Moreover, we refer to some recent works [33, 34, 35, 36, 37, 38] for various directions and perspectives developed in the context of logistics and supply chain management.

The chapter is organized as follows. In Section 5.2, some preliminary information along with several notifications are provided regarding the APP issue. In Section 5.3, we constructed the strategies for both players in accordance with the proposed algorithm. Here, we see that one of the pursuer's strategies involves making use of the evader's production plan in order to achieve the goal of reducing the total amount of money spent on production. In Section 5.4, an APP problem is studied for a short-term period where the strategy of each player is constructed in accordance with the main result. The paper finishes with several remarks, observations and a discussion of potential future work in Section 5.5.

5.2 Preliminaries and Statement of the Problem

In this part, we provide several notions, a lemma, and a general problem of APP with some remarks.

Definition 5.1 Floor function takes as input a real number x, and assigns output as the greatest integer less than or equal to x, denoted by $\lfloor x \rfloor$. On the other hand, the ceiling function maps x to the least integer greater than or equal to x, denoted by $\lceil x \rceil$.

For example, $\lfloor 1.23 \rfloor = 1$, $\lfloor -1.2 \rfloor = -2$, $\lceil 2.32 \rceil = 3$. Moreover, the fractional part of a real number x is denoted by $\{x\}$ which is defined by the formula: $\{x\} = x - \lfloor x \rfloor$. Obviously, we have $0 \le \{x\} < 1$, for all $x \in \mathbb{R}$. Now, we state the following novel theorem which would play a crucial role in building various strategies for the pursuer to overcome his opponent in the game.

Theorem 5.1
For given any rational number $\mathscr{A}$, there exists at least one couple (α, β) with $\alpha, \beta \in (0,1)$ such that

$$\left\lfloor \frac{\mathscr{A}}{(1+\alpha)} \right\rfloor (1+\alpha) = \left\lfloor \frac{\mathscr{A}}{(1+\beta)} \right\rfloor (1+\beta). \tag{5.1}$$

Here, we omit the proof of this lemma which could be proved by using the theory of numbers [39]. Although we make use of this lemma for rational number $\mathscr{A}$, the hypothesis could be extended to any real number $\mathscr{A}$ in a general sense.

As an example, consider $\mathscr{A} = 10$, then $\alpha = \frac{1}{2}$, $\beta = \frac{4}{5}$ and $\alpha = \frac{2}{7}$, $\beta = \frac{1}{8}$ would work for Theorem 5.1.

In this section, we make note of the fact that the contribution made by workers who put in overtime during the ith period could be considered to be proportional to one worker who puts in normal hours. To be more specific, if an employee works for eight hours on a regular-time basis and one hour for overtime, then his total contribution to the company could be regarded as 0.125 times that of an average worker who works regular hours. Roughly speaking, he does 1.125 times the amount of workouts that a regular-time employee would do. In the following section, we will utilize this perspective in order to combine the workforce that works regular hours with the workforce that works overtime:

$$Cost = \sum_{i=1}^{n} \left((D_i - S_i - S_{i-1})m_i + S_i\beta_i + c_i\omega_i + \alpha_i o_i \ell_i \right) + h_1\omega_1 + \sum_{i=1}^{n-1} V_i|\Delta\omega_i|,$$

where

$$\Delta\omega_i = \omega_{i+1} - \omega_i, \tag{5.2}$$

$$V_i = \begin{cases} h_i, & \text{if } \Delta\omega_i > 0, \\ f_i, & \text{if } \Delta\omega_i < 0, \end{cases} \tag{5.3}$$

and $S_0 = 0$.

Table 5.1: Parameters and their descriptions regarding to APP.

Parameters	Descriptions
ω_i	Number of regular-time workers in i^{th} period of time
c_i	The cost of each worker in i^{th} period
S_i	The number of items kept in the storage in i^{th} period
β_i	Holding cost of each unit in i^{th} period
o_i	Contribution of overtime worker in i^{th} period of time
α_i	Cost of each overtime worker in i^{th} period
h_i	Hiring cost of each worker in i^{th} period
f_i	Firing cost of each worker in i^{th} period
m_i	The cost of each item in i^{th} period
D_i	Demand in i^{th} period
η_i	Production of each employee during i^{th} period
ℓ_i	The number of overtime workers in i^{th} period

In this study, we make the assumption that the planner is obligated to meet all of the demand at each period of time; hence, the APP problem in period n would be a problem of minimizing the costs of the following in terms of manpower and storage, subject to some limitation.

5.3 A Novel Game Construction in APP Problems

In this section, we develop a game theoretical point of view with two players in APP problem. The evader's purpose is to meet all the demand by possibly storing items or products for the upcoming period, but, he does this without posing overtime duty on his employees. On the other hand, the pursuer's purpose is to make use of the evader's plan in order to create a better schedule by using less number of workers with overtime duty and trying to meet the exact demand for APP problem The primary issue at hand is determining the most effective tactic for the pursuer to use in order to cut down on the overall amount of money spent on production while simultaneously widening the gap between himself and his rival. These issues will be covered in the next section.

Denote $e_{w,i}$ as the number of workers the evader uses in i^{th} period. Moreover, let $S_{e,i}$ be the amount of product stored by the evader for the next upcoming period $i+1$. The evader starts his move by choosing his number of employees as

$$e_{w,1} = \lceil e_1 \rceil, \text{ with } e_1 = \frac{D_1}{\eta_1}, \text{ which makes } S_{e,1} = \eta_1 e_{w,1} - D_1. \tag{5.4}$$

Denote $p_{w,i}$ as the number of workers that the pursuer plans to use in i^{th} period and similarly let $S_{p,i}$ be the amount of product stored by the pursuer for $(i+1)^{th}$ period. Then, the pursuer builds his first move as finding a rational number $\delta \in (0,1)$ such that

$$p_{w,1} = \left\lfloor \frac{e_{w,1}}{1+\delta_1} \right\rfloor \text{ and } D_1 = \eta_1(1+\delta_1)p_{w,1}. \tag{5.5}$$

Here, it is shown that there is at least one $\delta \in (0,1)$ satisfying this situation. In this context, the pursuer wishes to use $\delta_1 p_{w,1}$ amount as the contribution of $p_{w,1}$ number of workers in overtime duty which results in $S_{p,1} = 0$. Hence, he eliminates the storage cost of the product, but instead, uses overtime workers to cover this gap.

For $k \geq 1$, evader constructs his $(k+1)^{th}$ step as the following recursive manner:

$$e_{k+1} = \frac{\eta_k}{\eta_{k+1}} e_k + \left\lceil \frac{D_{k+1} - \eta_k e_k}{\eta_{k+1}} \right\rceil, \text{ with } e_{w,k+1} = \lceil e_{k+1} \rceil. \tag{5.6}$$

Here, we find out that

$$S_{e,k+1} = \eta_{k+1} e_{w,k+1} + S_{e,k} - D_{k+1}. \tag{5.7}$$

On the other hand, the pursuer's $(k+1)^{th}$ step then becomes

$$p_{w,k+1} = \left\lfloor \frac{e_{w,k+1}}{1+\delta_{k+1}} \right\rfloor, \text{ and } D_{k+1} = \eta_{k+1}(1+\delta_{k+1})p_{w,k+1}, \tag{5.8}$$

for some rational number $\delta_{k+1} \in (0,1)$.

Here, we have the following main result.

Theorem 5.2

If the evader's workforce meets the recursive relation (5.6) *with initial start* (5.4) *for each period of time, then the pursuer's workforce should be constructed as* (5.8) *with initial start* (5.5), *which becomes a superior plan for approaching the optimal cost of the APP problem.*

Here, we omit the proof of the theorem. From Theorem 5.1, we can say that the following strategy set of the pursuer satisfying (5.5) and (5.8):

$$P = \{p_{w,k} : k \leq n\}, \tag{5.9}$$

contains at least two strategies depending on the choice of $\{\delta_i\}_{i=1}^{i=n}$ from the unit open interval $(0,1)$. Due to this fact, performing a meta-heuristic strategy would be a better option when attempting to choose the most suitable candidate in order to reduce costs associated with a variety of APP problems. Moreover, we open a new discussion for considering overtime workers as a contribution by means of partial regular-time workers such as $\delta_k p_{w,k}$ with $\delta_k \in (0,1)$.

5.4 Applications

In this section, we look at a production planning problem for the textile sector that involves the sophisticated manufacturing of school uniforms over a period of four months (May, June, July, and August). Here, we consider the following Table 5.2.

Table 5.2: The values of the parameters for the four-month demand.

Parameters	May	June	July	August
D_i	2483	2685	7482	1629
η_i	340	420	440	420
c_i	350	350	350	350
β_i	1	2	2	3
m_i	15	15	12	14
h_i	160	140	120	150
f_i	30	30	30	30
α_i	400	400	400	400

In this problem, we assume without loss of generality that all the workers of the pursuer are willing to participate partially in overtime duty, i.e., $\ell_i = p_{w,i}$ for $i = 1,2,3,4$. By using the construction argument presented in the previous section and from Theorem 5.2, we obtain the following APP for both players: In

Table 5.3: Workforce and storage data of both players according to four months demand.

Parameters	May	June	July	August
$e_{w,i}$	8	7	18	5
$S_{e,i}$	237	492	930	1401
$p_{w,i}$	7	5	13	3
$S_{p,i}$	0	0	0	0
δ_i	0.0434	0.279	0.3081	0.2933
o_i	0.0434	0.279	0.3081	0.2933

general, the contribution of each overtime worker is considered as

$$o_i = \frac{\delta_i}{\ell_i} p_{w,i} \tag{5.10}$$

for each period i. Finally, the total cost of APP problem for the evader and the pursuer would be

$$E_{cost} = E_{May} + E_{June} + E_{July} + E_{August} = 45117 + 47564 + 104520 + 35743 = 232944$$

and

$$\begin{aligned} P_{cost} = P_{May} + P_{June} + P_{July} + P_{August} &= 40936,52 + 42643 + 96896,12 + 25382,6 \\ &= 205858,24 \approx 205858, \end{aligned}$$

respectively. The gap would be

$$E_{cost} - P_{cost} = 232944 - 205858 = 27086, \tag{5.11}$$

which shows that the strategy followed by the pursuer gets a better production planning than the one used by his opponent. Here, we have two more strategies followed by pursuer with the same construction, but different $\delta_i \in (0,1)$. In this regard, we consider Table 5.4.

Table 5.4: Workforce and storage data of the pursuer with different $\delta_i \in (0,1)$.

Parameters	May	June	July	August
$p_{w,i}$	4	6	14	2
$S_{p,i}$	0	0	0	0
δ_i	0.8253	0.0657	0.2146	0.9394
o_i	0.8253	0.0657	0.2146	0.9394
Parameters	May	June	July	August
$p_{w,i}$	5	4	12	2
$S_{p,i}$	0	0	0	0
δ_i	0.4601	0.5985	0.4169	0.94
o_i	0.4601	0.5985	0.4169	0.94

Using those data sets, we find two different total costs of the pursuer as follows:

$$\begin{aligned} P_{1,cost} &= P_{1,May} + P_{1,June} + P_{1,July} + P_{1,August} \\ &= 40605,48 + 42812,68 + 96845,76 + 24617,52 \approx 204881. \end{aligned}$$

and

$$\begin{aligned} P_{2,cost} &= P_{2,May} + P_{2,June} + P_{2,July} + P_{2,August} \\ &= 40715,2 + 42902 + 96945,12 + 24558 \approx 205120. \end{aligned}$$

Observe that

$$P_{1,cost} < P_{2,cost} < P_{cost}.$$

Hence, the best strategy that the pursuer should follow is P_1, which makes the gap $E - P_1 = 28063$, which is the best gap so far.

5.5 Conclusion, Remarks and Future Works

In this work, we propose a method for constructing several schedules for the simple APP problem by making use of a game theory aspect of the problem. In this study, we investigate a scenario in which the planner wants to meet the demand for each period of time while also minimizing the number of products that are stored for the next period. In this regard, one pursuer and one evader are considered in the APP problem, where the evader's purpose is to meet the entire demand with the minimum possible storage without having an overtime plan for his employees, and the pursuer attempts to make use of the evader's plan in order to gather the knowledge of his workforce to create a better schedule by means of a smaller number of workers with overtime duty, in addition to meeting the exact demand for the APP problem. If the cost of storage is high enough, the newly constructed APP will finally be useful. If the cost of storing items drops to an extremely low level and the cost of paying workers for overtime rises to an extremely high level, then this planning scheme is likely doomed to fail. In this regard, one could perform superior production planning, which would result in lower total production costs than those of the pursuer. To get around this limitation, we simply switch the roles of the two players: the evader employs regular-time workers in addition to overtime labor without having storage, whereas the pursuer, once again, follows the knowledge of his opponent to plan his storage level as well as his workforce. This allows the evader to avoid the limitation without any additional work on his part. Obviously, the construction could be entirely different, which is something that needs to be investigated further in the work that comes after this. In addition, a new contribution concept has been introduced to combine regular-time workers with overtime workers in order to make the process of firing employees and hiring new employees more flexible. This has been accomplished by combining the two types of workers.

In the numerical example, we have investigated the following maximization problem:

$$\max_{P_w \in P} E - P_w, \tag{5.12}$$

where the set P represents all strategies followed by the pursuer and E is the strategy of the evader. In this regard, we construct three strategies for the pursuer and choose the one which becomes near-optimal. Here, we see that the strategies of the pursuer totally depend on the proper selection of parameters $\delta_k \in (0,1)$. Motivated by this and Theorem 5.1, it may be possible to design a novel metaheuristic perspective to select the best candidate δ_k to solve the maximization problem (5.12). This question also remains open in this study.

References

[1] Mirzapour Al-e-hashem, S.M.J., Aryanezhad, M.B. and Sadjadi, S.J. 2012. An efficient algorithm to solve a multi-objective robust aggregate production planning in an uncertain environment. *The International Journal of Advanced Manufacturing Technology*, 58(5):765–782.

[2] Jamalnia, A., Yang, J.-B., Feili, A., Xu, D.-L. and Jamali, G. 2019. Aggregate production planning under uncertainty: A comprehensive literature survey and future research directions. *The International Journal of Advanced Manufacturing Technology*, 102(1): 159–181.

[3] Ha, C., Seok, H. and Ok, C. 2018. Evaluation of forecasting methods in aggregate production planning: A Cumulative Absolute Forecast Error (CAFE). *Computers & Industrial Engineering*, 118: 329–339.

[4] Zhang, R., Zhang, L., Xiao, Y. and Kaku, I. 2012. The activity-based aggregate production planning with capacity expansion in manufacturing systems. *Computers and Industrial Engineering*, 62(2): 491–503.

[5] Váncza, J., Kis, T. and Kovács, A. 2004. Aggregation-the key to integrating production planning and scheduling. *CIRP Annals*, 53(1): 377–380.

[6] Liu, Z., Chua, D.K.H. and Yeoh, K.W. 2011. Aggregate production planning for ship building with variation-inventory trade-offs. *International Journal of Production Research*, 49(20): 6249–6272.

[7] Yaghin, R.G. 2020. Enhancing supply chain production-marketing planning with geometric multivariate demand function (a case study of textile industry). *Computers and Industrial Engineering*, 140(106220): 1–15.

[8] Dhaya, R. and Kanthavel, R. 2022. Energy efficient resource allocation algorithm for agriculture IoT. *Wireless Personal Communications*, 125(2): 1361–1383.

[9] Holt, C.C., Modigliani, F. and Simon, H.A. 1955. A linear decision rule for production and employment scheduling. *Manag. Sci.*, 2(1): 1–30.

[10] Holt, C.C., Modigliani, F., Muth, J.F. and Simon, H.A. 1961. Planning production inventories and workforce. *American Economic Review*, 51(4): 697–699.

[11] Ning, Y., Liu, J. and Yan, L. 2013. Uncertain aggregate production planning. *Soft Comput.*, 17(4): 617– 624.

[12] Goli, A., Tirkolaee, E.B., Malmir, B., Bian, G.-B. and Sangaiah, A.K. 2019. A multi-objective invasive weed optimization algorithm for robust aggregate production planning under uncertain seasonal demand. *Computing*, 101(6): 499–529.

[13] Jamalnia, A., Yang, J.-B., Xu, D.-L. and Feili, A. 2017. Novel decision model based on mixed chase and level strategy for aggregate production planning under uncertainty: Case study in beverage industry. *Computers and Industrial Engineering*, 114: 54–68.

[14] Jang, J. and Chung, B.D. 2020. Aggregate production planning considering implementation error: A robust optimization approach using bi-level particle swarm optimization. *Computers and Industrial Engineering*, 142, 106367: 1–12.

[15] Campo, E.A., Cano, J.A. and Gómez-Montoya, R.A. 2018. Linear programming for aggregate production planning in a textile company, Fibres and Textiles in Eastern Europe.

[16] Tirkolaee, E.B., Goli, A. and Weber, G.W. 2019. Multi-objective aggregate production planning model considering overtime and outsourcing options under fuzzy seasonal demand. *In Advances in Manufacturing II: Volume 2-Production Engineering and Management* (pp. 81–96). Springer International Publishing.

[17] Leung, S.C. and Wu, Y. 2004. A robust optimization model for stochastic aggregate production planning. *Production Planning and Control*, 15(5): 502–514.

[18] Qamhan, A.A., Ahmed, A., Al-Harkan, I.M., Badwelan, A., Al-Samhan, A.M. and Hidri, L. 2020. An exact method and ant colony optimization for single machine scheduling problem with time window periodic maintenance. *IEEE Access*, 8: 44836–44845.

[19] Zhang, S., Li, X., Zhang, B. and Wang, S. 2020. Multiobjective optimization in flexible assembly job shop scheduling using a distributed ant colony system. *European Journal of Operational Research*, 283(2): 441–460.

[20] Wibawa, A.P., Mahmudy, W.F., Rizki, A.M., Yuliastuti, G.E. and Tama, I.P. 2022. Multi-site aggregate production planning using particle swarm optimization. *Journal of Engineering, Project, and Production Management*, 12(1): 62.

[21] Chakrabortty, R.K., Akhtar Hasin, M.A., Sarker, R.A. and Essam, D.L. 2015. A possibilistic environment based particle swarm optimization for aggregate production planning. *Computers and Industrial Engineering*, 88: 366–377.

[22] Vasant, P., Nagarajan, R. and Yaacob, S. 2004. Decision making in industrial production planning using fuzzy linear programming. *IMA Journal of Management Mathematics*, 15(1): 53–65.

[23] Goli, A., Tirkolaee, E.B., Malmir, B., Bian, G.B. and Sangaiah, A.K. 2019. A multi-objective invasive weed optimization algorithm for robust aggregate production planning under uncertain seasonal demand. *Computing*, 101(6): 499–529.

[24] Khemiri, R., Elbedoui-Maktouf, K., Grabot, B. and Zouari, B. 2017. Integrating fuzzy TOPSIS and goal programming for multiple objective integrated procurement-production planning. *The 2017 22nd IEEE International Conference on Emerging Technologies and Factory Automation (ETFA)*.

[25] Baxendale, M., McGree, J.M., Bellette, A. and Corry, P. 2021. Machine-based production scheduling for rotomoulded plastics manufacturing. *International Journal of Production Research*, 59(5): 1301–1318.

[26] Yuliastuti, G.E., Rizki, A.M., Mahmudy, W.F. and Tama, I.P. 2019. Optimization of multi-product aggregate production planning using hybrid simulated annealing and adaptive genetic algorithm. *International Journal of Advanced Computer Science and Applications*, 10(11): 484–489.

[27] Hafezalkotob, A., Chaharbaghi, S. and Lakeh, T.M. 2019. Cooperative aggregate production planning: A game theory approach. *Journal of Industrial Engineering International*, 15: 19–37.

[28] Karmarkar, U.S. and Rajaram, K. 2012. Aggregate production planning for process industries under oligopolistic competition. *European Journal of Operational Research*, 223(3): 680–689.

[29] Hamidoğlu, A., Taghiyev, M. and Weber, G. 2022. On construction of pursuit-evasion games in discrete control models. *Applied and Computational Mathematics*, 21(1): 52–60.

[30] Hamidoğlu, A. 2021. A novel one target game model in the life insurance market. *International Journal of Management Science and Engineering Management*, 16(3): 221–228.

[31] Hamidoğlu, A., Taghiyev, M.H. and Weber, G.W. 2021. On building two-player games with treatment schedules for the SIR model. *Azerbaijan Journal of Mathematics*, 11(2): 183–195.

[32] Hamidoğlu, A. 2022. A game theoretical approach for finding near optimal solutions of an optimization problem. *Optimization*, 1–23.

[33] Goli, A., Tirkolaee, E.B. and Weber, G.W. 2022. Preface to the special issue on computational performance analysis based on novel Intelligent methods: Exploration and future directions in production and logistics. *Foundations of Computing and Decision Sciences*, 47(2): 107–110.

[34] Tirkolaee, E.B., Goli, A., Gütmen, S., Weber, G.W. and Szwedzka, K. 2022. A novel model for sustainable waste collection arc routing problem: Pareto-based algorithms. *Annals of Operations Research*, 1–26.

[35] Das, S.K., Roy, S.K. and Weber, G.W. 2022. The impact of carbon tax policy in a multi-objective green solid logistics modelling under sustainable development. *In Computational Modelling in Industry 4.0: A Sustainable Resource Management Perspective* (pp. 49–66). Singapore: Springer Nature, Singapore.

[36] Gürbüz, B., Mawengkang, H., Husein, I. and Weber, G.W. 2022. Rumour propagation: An operational research approach by computational and information theory. *Central European Journal of Operations Research*, 30: 345–365.

[37] Savku, E. and Weber, G.W. 2022. Stochastic differential games for optimal investment problems in a Markov regime-switching jump-diffusion market. *Annals of Operations Research*, 312(2): 1171–1196.

[38] Gütmen, S., Weber, G.W., Goli, A. and Tirkolaee, E.B. 2022. Human paradigm and reliability for aggregate production planning under uncertainty. *In Advances of Artificial Intelligence in a Green Energy Environment* (pp. 193–201). Academic Press.

[39] Hamidoğlu, A. and Taghiyev, M.H. 2021. On construction of almost periodic sequences and applications to some discrete population models. *Journal of Difference Equations and Applications*, 27(1): 118–131.

Chapter 6

Balancing Cooperation and Competition in Game Theory for Business

Jinal Parikh,[1] *Jacek Dominik Śledziński,*[2,*] *Sırma Zeynep Alparslan Gök,*[3] *İsmail Özcan*[4] and *Gerhard-Wilhelm Weber*[2,5]

6.1 The Importance of Research on the Phenomena of Cooperation and Competition

There is no doubt that cooperation is a prerequisite for achieving the most ambitious goals of humanity (e.g., space exploration, global activities undertaken to protect the earth's ecosystem, creating buildings and structures of international importance). Following the work of probes on Mars, the operation of the Large Hadron Collider at CERN or the construction of the Eurotunnel connecting England with France, we are aware of the incredible complexity of these projects, the need to take advantage of the knowledge and experience of thousands of people, the need to coordinate the work of these people so as to achieve one common objective. We realize that no single person would ever be able to accomplish on their own even a fraction of the challenges that can be achieved by collaborating with others.

[1] Ahmedabad University, Ahmedabad, India.
[2] Poznan University of Technology, Poznan, Poland.
[3] Suleyman Demirel University, Isparta, Turkey.
[4] Universidad Miguel Hernández de Elche, Alicante, Spain.
[5] Institute of Applied Mathematics METU, Ankara, Turkey.
* Corresponding author: jacek.dominic@gmail.com

At the same time, the need for competition (of which conflict is an integral part) is deeply rooted in human nature. After all, the evolutionary mechanism, one of the foundations of which is constant struggle, in particular the constant effort to gain a better position by harming opponents, is much older than humanity itself. In nature, the simplest organisms compete with each other, it is a mechanism that has accompanied life on Earth basically from the beginning of its existence. Understandably, in the popular understanding, competition means the opposite of cooperation (the online dictionary of antonyms indicates the terms "cooperation" and "competition" as a pair of words with opposite meanings).

In this context, it is fascinating that the phenomenon of competition (and conflict) is also a powerful (perhaps even greater) force driving the progress of civilization. So are we dealing with a paradox? Competition drives progress, but to create something truly great, you need cooperation! Nowadays, after a deeper analysis, it seems that only the tendency of the human mind to oversimplify and the poverty of natural human language meant that the concepts of competition and cooperation are colloquially perceived as extreme positions of the same, proportional linear scale - in the common understanding (which is confirmed by the above-mentioned dictionary of antonyms) the more competition, the less cooperation (by the same amount), and vice versa. However, this is a false picture. In fact, competition and cooperation are much more complex phenomena, with a rich spectrum of properties, interpenetrating and interacting with each other, creating a kind of amalgam. Often we still do not understand the nature of these phenomena well, and therefore we still cannot manage them properly.

To better understand the phenomena of competition and cooperation, a "nonlinear" holistic approach to them is needed. The authors of this essay wonder how to study these phenomena in business and propose game theory as a tool for describing and designing better strategies (the situations considered here, i.e., interactions of entities with different goals whose situation depends on the actions of other entities, will be called strategic situations).

The following paragraphs present some important areas of business activity and discuss the possibility of using game theory to better understand the situation between business entities and to design better strategic solutions for these entities.

6.1.1 Cooperation

In the realm of business strategy, the interplay between cooperation and competition is fundamental. Game theory provides a robust framework for understanding how these dynamics shape decision-making among rational actors. This section explores the significance of both cooperation and competition in game theory for business and their applications.

Cooperation in business often leads to win-win outcomes, where firms collaborate to achieve mutual benefits. This can be seen in various contexts, such as strategic alliances, joint ventures, and partnerships. Cooperation fosters innovation, risk sharing, and the ability to access new markets or technologies collectively.

A real-life example of cooperative game theory applied in business is the formation and operation of airline alliances. Airline alliances are strategic collaborations between multiple airlines to achieve mutual benefits, enhance network connectivity, and compete more effectively in the global aviation industry [1].

One prominent example is the "Star Alliance," which is the world's largest airline alliance [2]. It includes major airlines such as United Airlines, Lufthansa, Air Canada, and many others. Cooperative game theory principles are evident in several aspects of airline alliances:

1. **Code-Sharing**: Airlines within an alliance often engage in extensive code-sharing agreements. This means that one airline can market and sell seats on another alliance member's flights. Passengers benefit from a wider range of destinations and more convenient connecting flights. Cooperative game theory helps determine how the revenue from code-sharing is distributed among the participating airlines, ensuring they share the benefits fairly [3].

2. **Network Expansion**: By forming alliances, airlines can expand their global networks without having to invest heavily in new routes and infrastructure. Cooperative game theory assists in deciding which routes and hubs each member airline will focus on, taking into account their strengths and competitive advantages.

3. **Cost Sharing**: Cooperative game theory is used to allocate the costs of joint marketing campaigns, shared lounges, and other initiatives among the alliance members. This ensures that the burden of these expenses is distributed fairly based on each airline's contribution and benefit.

4. **Antitrust Immunity**: In some cases, airlines seek antitrust immunity to collaborate more closely within an alliance. This legal arrangement allows them to engage in pricing and scheduling coordination, among other activities, without violating antitrust laws. Game theory helps determine the terms and conditions of such immunity, ensuring it doesn't harm competition.

5. **Frequent Flyer Programs**: Many airline alliances allow passengers to earn and redeem frequent flyer miles across multiple member airlines. Cooperative game theory is applied to determine the value of these programs and how the rewards are shared among the alliance partners.

6. **Competitive Positioning**: Airlines in an alliance cooperate to compete more effectively with other global carriers, such as those in rival alliances.

Cooperative game theory guides the strategies and tactics used to position the alliance as a strong alternative in the global aviation market.

Overall, airline alliances exemplify the application of cooperative game theory in business by enabling multiple firms to collaborate strategically, share resources, and compete more effectively on a global scale. These alliances not only benefit the participating airlines but also provide passengers with expanded travel options and greater convenience.

6.1.2 Competition

Competition drives firms to continually improve products, services, and operational efficiency. It incentivizes innovation and keeps prices in check. In a competitive market, consumers tend to benefit from a wider array of choices and lower prices.

A real-life example of competitive game theory applied in business is the smartphone industry, particularly the competition between two major players: Apple Inc. and Samsung Electronics [4]. The rivalry between these companies illustrates various aspects of competitive game theory:

1. *Product Differentiation*: Both Apple and Samsung engage in intensive product differentiation to gain a competitive edge. They continually invest in research and development to introduce new features, designs, and technologies in their smartphones. Competitive game theory comes into play as each company tries to anticipate the other's moves and strategically position their products to capture market share.

2. *Pricing Strategies*: Pricing decisions in the smartphone industry are influenced by competitive game theory. Apple and Samsung engage in price wars, periodically adjusting the prices of their devices to outmaneuver each other. This dynamic competition can lead to Nash equilibriums, where both companies strive to find the optimal pricing strategy that maximizes their own profits while considering their competitor's response.

3. *Advertising and Branding*: Competitive game theory is evident in the aggressive advertising and branding efforts of both companies. They use marketing tactics to position themselves as market leaders and influence consumers' perceptions. Ad campaigns often directly or indirectly compare their products to those of their competitors, creating a competitive strategic environment.

4. *Patent Wars*: The smartphone industry has seen numerous patent disputes and lawsuits between Apple and Samsung. These legal battles can be viewed through the lens of competitive game theory, as both companies

strategically use patents and legal action to protect their intellectual property, gain market advantages, and potentially force the other party into costly litigation.

5. *Product Launch Timing*: Apple and Samsung carefully time their product launches to gain a competitive advantage. They monitor each other's product release schedules and aim to introduce new devices just before or after their competitor to capture consumer attention and market share.
6. *Ecosystems and Lock-In Effects*: Both companies have developed ecosystems around their products, including app stores, cloud services, and proprietary accessories. Competitive game theory factors into their efforts to lock customers into their ecosystems, making it more costly for consumers to switch to the competitor's products.
7. *Strategic Partnerships*: Apple and Samsung enter into strategic partnerships with suppliers, manufacturers, and service providers to secure critical components and distribution channels. These partnerships are influenced by competitive considerations, as both companies seek to gain an advantage in the supply chain and distribution networks.
8. *Global Expansion and Market Share*: The competition extends to global markets where Apple and Samsung vie for dominance. Competitive game theory guides their decisions on market entry, expansion, and market share acquisition, considering factors such as local consumer preferences, regulatory environments, and the actions of their rival.

In sum, the smartphone industry exemplifies competitive game theory in business, where two major players continually strategize and respond to each other's moves to gain a competitive advantage in a dynamic and evolving market. The interactions between Apple and Samsung showcase the complex interplay of pricing, product differentiation, branding, legal tactics, and other competitive strategies that are central to the field of competitive game theory.

6.1.3 Coopetition - Balancing Cooperation and Competition

Coopetition: The concept of coopetition, introduced by Adam M. Brandenburger and Barry J. Nalebuff [5] highlights the simultaneous presence of cooperation and competition among firms. Coopetition is crucial in industries where companies need to cooperate on certain aspects (e.g., industry standards) while fiercely competing in others (e.g., market share).

A real-life example of co-opetitive (cooperative and competitive) game theory applied in business is the partnership between Microsoft and Linux [6], specifically through Microsoft's Azure cloud platform and its collaboration with various Linux distributions and open-source technologies. Here's how competitive game theory is demonstrated in this context:

Cooperation:

1. *Azure and Linux Integration*: Microsoft's Azure cloud platform has actively embraced open-source technologies, including Linux distributions like Ubuntu, Red Hat, and SUSE. Microsoft has worked closely with these Linux providers to ensure their systems run smoothly on Azure. This cooperation benefits customers who can deploy a wide range of applications on Azure, regardless of their operating system preferences.
2. *Open-Source Contributions*: Microsoft has contributed significantly to the open-source community, supporting projects like the Linux kernel and various open-source software projects. This collaborative approach fosters innovation and helps improve the quality of open-source software.
3. *Interoperability*: Microsoft has developed tools and technologies that facilitate interoperability between its products and open-source solutions. For example, the company has created Visual Studio Code, a popular open-source code editor, and PowerShell, which can manage Linux systems, promoting cross-platform compatibility and cooperation.

Competition:

1. **Cloud Services Market**: Microsoft's Azure competes directly with other cloud providers like Amazon Web Services (AWS) and Google Cloud Platform. In this competitive landscape, Microsoft strives to gain a larger market share by offering a wide range of services and features, which includes hosting Linux-based workloads.
2. **Hybrid Cloud**: Microsoft positions itself as a provider of hybrid cloud solutions, allowing organizations to integrate on-premises infrastructure with Azure. This competitive strategy targets businesses that rely on Linux in their data centers but are open to hybrid cloud solutions offered by Microsoft.
3. **Azure vs. Linux Cloud Providers**: While cooperating with Linux distributions, Microsoft also competes with Linux-focused cloud providers like IBM Cloud and Oracle Cloud. This competitive rivalry drives innovation and pricing strategies as they vie for customers who favour Linux-based solutions.

Benefits and Outcomes: The competitive relationship between Microsoft and Linux demonstrates the effectiveness of combining cooperation and competition:

- **Diverse Customer Base**: Microsoft's willingness to cooperate with Linux and open-source technologies has attracted a more diverse customer base, including those who prefer Linux-based solutions. This has increased Azure's market share and revenue.
- **Innovation**: The collaboration between Microsoft and the Linux community has led to the development of innovative solutions and improvements in interoperability between Windows and Linux environments.
- **Market Expansion**: By embracing Linux, Microsoft has expanded its presence in markets that traditionally favoured open-source solutions, such as web hosting and container orchestration.
- **Choice for Customers**: The coopetitive approach provides customers with choices, allowing them to select the best solutions for their needs while avoiding vendor lock-in.

In summary, the partnership between Microsoft and Linux, driven by competitive game theory principles, illustrates how businesses can simultaneously cooperate with potential competitors in certain areas while competing vigorously in others. This strategy has allowed Microsoft to strengthen its position in the cloud services market and better cater to the diverse needs of its customers.

6.2 Game Theory as a Tool

Game theory provides us tools to analyze and optimize strategies in situations involving both cooperation and competition. Mixed-strategy games and sequential games with cooperation phases allow firms to navigate these complex dynamics effectively.

6.2.1 Uber and Lyft – A Real-life Example of Applying Game Theory in Business

A real-life example of mixed-strategy games and sequential games with coopetition in business can be found in the context of ride-sharing services, particularly the coopetition between Uber and Lyft. Both companies compete fiercely for market share while also occasionally cooperating in certain areas [7].

Mixed-Strategy Games in Pricing:

1. *Dynamic Pricing*: Uber and Lyft engage in dynamic pricing strategies, often referred to as surge pricing or Prime Time, where prices for rides increase during high-demand periods. This is a mixed-strategy game where both companies adjust their pricing algorithms to optimize revenue while

responding to competitor actions. For example, if Lyft observes Uber implementing surge pricing in a specific area, it may decide to do the same to attract more drivers and riders.

2. *Pricing Discounts*: Both companies offer promotions and discounts to riders and incentives to drivers. These promotional strategies involve mixed strategies, as they periodically adjust discounts and incentives based on the competitive landscape. For instance, if Uber offers a 20% discount in a particular city, Lyft might respond by offering similar discounts or additional perks.

Sequential Games in Expansion and Market Entry:

1. *Market Entry*: Uber and Lyft have engaged in sequential games when entering new markets. One company typically enters a new city first, and the other follows suit. The second entrant observes the actions and strategies of the first entrant and adjusts its market entry strategy accordingly. This sequential decision-making process reflects a classic example of sequential game theory.
2. *Geographic Expansion*: As both companies expand their services to new regions and countries, they often follow a sequential game strategy. They observe each other's moves in terms of international expansion, carefully selecting the markets where they believe they can gain a competitive advantage.

Coopetition in Driver Partnerships:

1. *Driver Partnerships*: Uber and Lyft often cooperate in recruiting and retaining drivers. While they are competitors, many drivers work for both platforms simultaneously. This coopetition allows drivers to maximize their earnings by switching between Uber and Lyft based on factors like demand and incentives.
2. *Interoperability*: In some regions, Uber and Lyft have integrated their services to allow riders to book rides on either platform through a single app. This interoperability is a form of coopetition that benefits both companies by expanding the customer base and making it more convenient for riders.

Strategic Investments:

1. *Investments in Competing Startups*: Uber and Lyft have made strategic investments in competing ride-sharing startups worldwide. While they compete for market share in their core markets, they also invest in startups in other regions, creating a portfolio of investments to hedge their bets and benefit from the success of emerging competitors.

In this highly dynamic and competitive industry, Uber and Lyft engage in mixed-strategy games, sequential games, and coopetition to gain and maintain market share. Their strategic decisions involve a careful balance of competition and co-operation, demonstrating the complex interplay of game theory in the real world of business.

6.3 The Phenomenon of Uncertainty in Business and Game Theory

6.3.1 Cooperative Games

In crisp cooperative game theory payoffs to coalitions of players are known with certainty. A cooperative game [8], is a pair $< N, v >$ where $N = \{1, 2, ..., n\}$ is the set of players and $v : 2^N \to \mathbb{R}$ is a map, assigning to each coalition $S \in 2^N$ a real number, such that $v(\emptyset) = 0$. We denote by G^N the family of all crisp cooperative games with player set N, and $\mathbb{R}^N$ the set of all payoff vectors.

Let $\Pi(N)$ the set of all permutations $\sigma : N \to N$ of N. The set $P^\sigma(i) := \{r \in N \mid \sigma^{-1}(r) < \sigma^{-1}(i)\}$ consists of all predecessors of i with respect to the permutation σ.

Let $c \in G^N$ and $\sigma \in \Pi(N)$. The marginal contribution vector $m^\sigma(v) \in \mathbb{R}^N$ with respect to σ and v has the i-th coordinate the value $m_i^\sigma(v) := v(P^\sigma(i) \cup \{i\}) - v(P^\sigma(i))$ for each $i \in N$. The Shapley value $\phi(v)$ of a game $v \in G^N$ is the average of the marginal vectors of the game as follows [8]:

$$\phi(v) := \frac{1}{N!} \sum_{\sigma \in \Pi(N)} m^\sigma(v).$$

Next, we give some solution concepts from cooperative game theory such as the centre of gravity of the imputation set value, shortly denoted by CIS-value, the egalitarian non-separable contribution value shortly denoted by $ENSC$-value [9, 10].

The CIS-value assigns to every player its individual worth, and distributes the remainder of $c(N)$ equally among all players and is defined by:

$$CIS_i(v) = v(\{i\}) + \tfrac{1}{|N|}(v(N) - \sum_{j \in N} v(\{j\})) \text{ for all } i \in N.$$

The $ENSC$-value assigns to every game v the CIS-value of its dual game and is defined by:

$$CIS_i(v^*) = ENSC_i(v) = -v(N \backslash \{i\}) + \tfrac{1}{|N|}(v(N) + \sum_{j \in N} v(N \backslash \{j\})), \text{ for all } i \in N.$$

Here v^* is the dual game of the game $v \in G^N$ and defined by:

$$v^*(S) = v(N) - v(N \backslash S).$$

6.3.2 Fuzzy Intervals

A fuzzy set $\mathscr{F}$ in $\mathbb{R}$ is a function $u_{\mathscr{F}} : \mathbb{R} \to [0,1]$, where $u_{\mathscr{F}}$ assigns to each point in $\mathbb{R}$ a degree of membership. For any $\alpha \in [0,1]$, α-level set (α-cut) of $\mathscr{F}$ defined by as follows [11]:

$$[u_{\mathscr{F}}]^{\alpha} = \{x \in \mathbb{R} : u_{\mathscr{F}}(x) \geq \alpha\} = \left[u_{\mathscr{F}}^{-}, u_{\mathscr{F}}^{+}\right].$$

If $\alpha = 0$, then $[u_{\mathscr{F}}]^{0} = cl\{x \in \mathbb{R} : u_{\mathscr{F}}(x) > 0\}$. Here, $cl\{x \in \mathbb{R} : u_{\mathscr{F}}(x) > 0\}$ is the closure of $\{x \in \mathbb{R} : u_{\mathscr{F}}(x) > 0\}$.

A fuzzy set $\mathscr{F}$ in $\mathbb{R}$ is said to be a fuzzy interval, if the following conditions are satisfied [12]:

i. $[u_{\mathscr{F}}]^{\alpha}$ is compact, for any $\alpha \in [0,1]$;

ii. $[u_{\mathscr{F}}]^{\alpha}$ is convex, for any $\alpha \in [0,1]$;

iii. $[u_{\mathscr{F}}]^{\alpha}$ is normal, i.e., there exist $x \in \mathbb{R}$ such that $u_{\mathscr{F}}(x) = 1$.

The set of fuzzy intervals are denoted by $\mathscr{F}(\mathbb{R})$. For any $\mathscr{F} \in \mathscr{F}(\mathbb{R})$ there exist $a,b,c,d \in \mathbb{R}$ and $L : [a,b] \to \mathbb{R}$ non-decreasing and $R : [c,d] \to \mathbb{R}$ non-increasing such that the membership function $u_{\mathscr{F}}$ is given as below [12]:

$$u_{\mathscr{F}}(x) = \begin{cases} L(x), & a \leq x \leq b, \\ 1, & b \leq x \leq c, \\ R(x), & c \leq x \leq d, \\ 0, & \text{otherwise.} \end{cases}$$

For any $\alpha \in [0,1]$ the α-level set of a trapeziodal fuzzy interval of $\mathscr{F}$ with membership function u is given by [12]:

$$\begin{aligned} [u_{\mathscr{F}}]^{\alpha} &= \left[L^{-1}(0)(1-\alpha) + \alpha L^{-1}(1), R^{-1}(0)(1-\alpha) + \alpha R^{-1}(1)\right] \\ &= [a(1-\alpha) + \alpha b, (1-\alpha)d + \alpha c] \\ &= \left[u_{\mathscr{F}}^{-}, u_{\mathscr{F}}^{+}\right]. \end{aligned}$$

Let $\mathscr{F}_1 = (a_1, a_2, a_3, a_4)$ and $\mathscr{F}_2 = (b_1, b_2, b_3, b_4) \in \mathscr{F}_{\mathscr{T}}(\mathbb{R})$ be two trapeziodal fuzzy intervals and $\lambda \in \mathbb{R}^{+}$, then the following conditions hold:

i. $\mathscr{F}_1 + \mathscr{F}_2 = (a_1 + b_1, a_2 + b_2, a_3 + b_3, a_4 + b_4)$,

ii. $\lambda \mathscr{F}_1 = (\lambda a_1, \lambda a_2, \lambda a_3, \lambda a_4)$,

iii. $\mathscr{F}_1 \trianglerighteq \mathscr{F}_2 \Longleftrightarrow a_1 \geq b_1$, $a_2 \geq b_2, a_3 \geq b_3$, and $a_4 \geq b_4$.

Let $\mathscr{F}_1, \mathscr{F}_2 \in \mathscr{F}(\mathbb{R})$. If there exist $\mathscr{F}_3 \in \mathscr{F}(\mathbb{R})$ such that $\mathscr{F}_1 = \mathscr{F}_2 + \mathscr{F}_3$, then $\mathscr{F}_3$ is said to be Hukuhara difference between $\mathscr{F}_1$ and $\mathscr{F}_2$ denoted by $\mathscr{F}_3 = \mathscr{F}_1 -_H \mathscr{F}_2$ [13].

Let $\mathscr{F}_1 = (a_1, a_2, a_3, a_4)$ and $\mathscr{F}_2 = (b_1, b_2, b_3, b_4) \in \mathscr{F}_{\mathscr{T}}(\mathbb{R})$ be two trapeziodal fuzzy intervals. If $a_i - b_i \leq a_j - b_j$ then the Hukuhara difference between $\mathscr{F}_1$ and $\mathscr{F}_2$ is defined by:

$$\mathscr{F}_1 -_H \mathscr{F}_2 = (a_1 - b_1, a_2 - b_2, a_3 - b_3, a_4 - b_4), \text{ for all } i \leq j.$$

In this paper, we use $\mathscr{F}_1 - \mathscr{F}_2$ instead of $\mathscr{F}_1 -_H \mathscr{F}_2$ for $\mathscr{F}_1, \mathscr{F}_2 \in \mathscr{F}(\mathbb{R})$.

Hukuhara difference between two fuzzy inrevals does not always exist. Regarding the existence of the Hukuhara difference, there is an extensive literature described in [13]. Here, we review some existing facts on the Hukuhara difference which are interesting for our paper.

Let $\mathscr{F}_1, \mathscr{F}_2 \in \mathscr{F}(\mathbb{R})$. $\mathscr{F}_1 -_H \mathscr{F}_2$ exists if and only if

$$(\mathscr{F}_1)_\alpha^- - (\mathscr{F}_2)_\alpha^- \leq (\mathscr{F}_1)_\beta^- - (\mathscr{F}_2)_\beta^- \leq (\mathscr{F}_1)_\beta^+ - (\mathscr{F}_2)_\beta^+ \leq (\mathscr{F}_1)_\alpha^+ - (\mathscr{F}_2)_\alpha^+$$

for all $\alpha, \beta \in (0,1]$ and $\beta > \alpha$.

Throughout this paper, we use $\mathscr{F}_1 - \mathscr{F}_2$ instead of $\mathscr{F}_1 -_H \mathscr{F}_2$ for $\mathscr{F}_1, \mathscr{F}_2 \in \mathscr{F}(\mathbb{R})$.

Defuzzification is the conversion of a fuzzy quantity to a precise value. It means that, the process of converting a fuzzified output into a single classical value with respect to a fuzzy quantity. In the literature, various defuzzification techniques exist. In this article, we employ the signed distance method.

The *signed distance* of fuzzy set $\mathscr{F}$ is given as follows [14]:

$$d^*(\mathscr{F}) = \frac{1}{2} \int_0^1 \left[(\mathscr{F})_\alpha^- + (\mathscr{F})_\alpha^+ \right] d\alpha.$$

If $\mathscr{F} \in \mathscr{F}_{\mathscr{T}}(\mathbb{R})$ then the signed distance of $\mathscr{F} = (a,b,c,d)$ as

$$d^*(\mathscr{F}) = \frac{1}{4}(a+b+c+d).$$

For other specific cases, an expansion of the signed distance formula is proposed in [15].

6.3.3 Fuzzy Interval Cooperative Games

A cooperative fuzzy interval game is a pair $< N, \mathscr{U} >$, where $N = \{1, 2, ..., n\}$ is the set of players and $\mathscr{U} : 2^N \to \mathscr{F}(\mathbb{R})$ maps the coalitions $S \in 2^N$ into fuzzy intervals $\mathscr{U}(S) \in \mathscr{F}(\mathbb{R})$ with $v(\emptyset) = 0$, where 0 is a fuzzy interval with membership function given in [16]:

$$u_0(x) = \begin{cases} 1, & x = 0, \\ 0, & x \neq 0. \end{cases}$$

We denote by $\mathscr{F}(\mathbb{R})^N$ the set of all fuzzy payoff vectors, and $\mathscr{F}G^N$ the family of all cooperative fuzzy interval games.

We call a fuzzy interval game $< N, \mathscr{U} >$ superadditive, for any $S, T \in 2^N$ such that $S \cap T = \emptyset$,

$$\mathscr{U}_\alpha (S \cup T) \trianglerighteq \mathscr{U}_\alpha (S) + \mathscr{U}_\alpha (T)$$

for all $\alpha \in (0, 1]$.

Moreover, if for any coalition $T \subseteq S$ and any $\alpha, \beta \in (0, 1]$ such that $\beta > \alpha$, the superadditive game $< N, \mathscr{U} >$ also satisfies:

$$\mathscr{U}_\alpha^- (S) - \mathscr{U}_\alpha^- (T) \leq \mathscr{U}_\beta^- (S) - \mathscr{U}_\beta^- (T) \leq \mathscr{U}_\beta^+ (S) - \mathscr{U}_\beta^+ (T) \leq \mathscr{U}_\alpha^+ (S) - \mathscr{U}_\alpha^+ (T).$$

A fuzzy interval game $< N, \mathscr{U} >$ is size monotonic if $< N, |\mathscr{U}| >$ size monotonic, i.e., $|\mathscr{U}|(T) \geq |\mathscr{U}|(S)$ for all $\alpha \in [0, 1]$, $S, T \in 2^N$ with $S \subset T$ where $|\mathscr{U}(S)| = \mathscr{U}_\alpha^+ (S) - \mathscr{U}_\alpha^- (S)$. We will denote the class of fuzzy interval games that are size monotonic and superadditive by $SM\mathscr{F}G^N$. Here we can say that Hukuhara difference always defined on $SM\mathscr{F}G^N$.

The Shapley value of cooperative games with fuzzy characteristic functions are suggested in [17]. Let $< N, \mathscr{U} >$ be a fuzzy interval game. Then the Shapley value is defined as follows:

$$\phi(\mathscr{U}) = (\phi_1(\mathscr{U}), \phi_2(\mathscr{U}), \ldots, \phi_n(\mathscr{U})),$$

with

$$\phi_i(\mathscr{U}) := \sum_{S \subset N \setminus \{i\}} \frac{|S|!\,(N - 1 - |S|)!}{N!} (\mathscr{U}(S \cup \{i\}) -_H \mathscr{U}(S)).$$

The fuzzy interval CIS-value ($\mathscr{F}CIS$-value) is defined by

$$\mathscr{F}CIS : SM\mathscr{F}G^N \rightarrow \mathscr{F}(\mathbb{R})^N,$$

$$\mathscr{F}CIS_i(\mathscr{U}) = \mathscr{U}(\{i\}) + \frac{1}{|N|} \left(\mathscr{U}(N) - \sum_{j \in N} \mathscr{U}(\{j\}) \right),$$

for all $i \in N$ and $\mathscr{U} \in SM\mathscr{F}G^N$ [18].

The fuzzy $ENSC$-value ($\mathscr{F}ENSC$-value) assigns to every game $\mathscr{U}$ the $\mathscr{F}CIS$-value of its dual game given by [18].

$$\begin{aligned}
\mathscr{F}ENSC &: SM\mathscr{F}G^N \rightarrow \mathscr{F}(\mathbb{R})^N, \\
\mathscr{F}ENSC_i(\mathscr{U}) &= \mathscr{F}CIS_i(\mathscr{U}^*), \\
&= -\mathscr{U}(N \setminus \{i\}) + \frac{1}{|N|} \left(\mathscr{U}(N) + \sum_{j \in N} \mathscr{U}(N \setminus \{j\}) \right).
\end{aligned}$$

6.3.4 An Application

In this section we will use the example from article [9] as an application one of the inherent features of business (and all real strategic situations) is the phenomenon of uncertainty. It is not only a challenge for entrepreneurs and managers who are forced to make decisions in conditions of uncertainty. Mathematically describing a strategic situation taking into account uncertainty and drawing conclusions based on such a description is also a great challenge for scientists dealing with game theory. Nowadays, we know certain ways to deal with this problem. The application of the fuzzy approach in a real-life example of a cooperative strategic game is presented below.

We consider an attempt to create an alliance by grey competing IT (Information Technology) companies from the SME (Small and Medium Enterprises) sector. The owners of these companies met in various circumstances in 2019–2020, while in the first half of 2021 they noticed that in addition to the area of activity in which they compete directly with each other, there are also many areas in which their competencies complement each other, and thanks to the establishment of formal alliance, companies could fill each other's critical shortcomings, as well as increase the likelihood of taking advantage of impending opportunities and better respond to threats arising from the environment. Thanks to this, each of the companies could benefit from strategic cooperation with other companies. Based on this observation, in 2021 the owners of the grey companies started negotiations to create a new company that would be a platform for cooperation between them. Various possible configurations for concluding such an alliance (establishing a new company) were considered. There are strict rules for calling a company small or medium up to 50 people is small. A micro-company is an enterprise with 1 to 9 employees. Turnover and/or total annual balance sheet must not exceed 2 million euros. As reports by the Polish Agency for Enterprise Development show, Poles are very eager to set up microenterprises. There are as many as 1.96 million of them in Poland and they account for as much as 95.0% of all companies nationwide. It is microenterprises that have the largest share of the entire SME sector in the creation of Polish GDP. According to 2021 data, it is about 29.5%. A small business is an enterprise that has between 10 and 50 employees. The company's total annual balance sheet and/or turnover must not exceed 10 million euros. According to the latest data for 2021, small companies account for 4.1% of all enterprises, with 84 353. A medium-sized company is an enterprise that has between 50 and 250 employees. The turnover and/or total annual balance sheet of a medium-sized company does not exceed 50 million euros. There are 14.7 thousand medium-sized companies in Poland (0.7% of all companies) [19]. Consider three IT companies: Company A, experienced (about 10 years on the market), medium company (employs about 50 people) providing an advanced. IT tool for optimizing marketing campaigns (its clients are marketing agencies), and creating a department providing programming services for various entities. Company B, is moderately experienced (about 5 years on the market), a small company (employment of about 10 people) providing software development services for various entities. Company C, little experienced (about 2 years on the market), a micro company (employment about 5 people) dealing with training programmers and creating a department providing programming services for various entities.

Table 6.1: The crisp profit of the *firms*.

Firms	A	B	C	A-B	A-C	B-C	A-B-C
Effectiveness of administration	50	60	50	60	60	60	70
Acquiring new employees	20	30	80	20	80	70	80
Operational efficiency	30	60	50	60	30	70	60
Potential to introduce new products	60	50	70	70	70	70	70
Ability to attract new customers	50	20	40	80	30	40	80
Financial stability of the company	70	30	30	60	60	40	60
Total profit	280	250	320	350	330	350	430

As can see in Table 6.1, we can construct the crisp game given in Table 6.2.

Table 6.2: The crisp total profits of the *firms*.

S	$\emptyset$	$\{1\}$	$\{2\}$	$\{3\}$	$\{1,2\}$	$\{1,3\}$	$\{2,3\}$	$\{1,2,3\}$
$v(S)$	0	280	250	320	350	330	350	430

Let us fuzzify the profits of our three Firms by using fuzzy intervals given in Table 6.3.

Table 6.3: The total fuzzy profits of the *firms*.

S	$U(\{S\})$
$\{\emptyset\}$	$(0, 0, 0, 0)$
$\{1\}$	$(180, 220, 300, 320)$
$\{2\}$	$(210, 240, 280, 300)$
$\{3\}$	$(260, 300, 340, 360)$
$\{1,2\}$	$(320, 340, 360, 380)$
$\{1,3\}$	$(300, 320, 340, 380)$
$\{2,3\}$	$(300, 340, 360, 380)$
$\{1,2,3\}$	$(360, 400, 440, 480)$

We calculate the crisp solutions of the game as follows:

$$\begin{aligned} \phi(v) &= (138.33, 133.33, 158.33), \\ CIS(v) &= (140, 110, 180), \\ ENSC(v) &= (136.67, 156.67, 136.67). \end{aligned}$$

Now we calculate the fuzzy solutions of the game as follows:

$$\begin{aligned} \phi(\mathscr{U}) &= \begin{pmatrix} (150, 153.33, 180, 206.67), \\ (170, 186.67, 186.67, 193.33), \\ (183.33, 206.67, 206.67, 233.33) \end{pmatrix}, \\ \mathscr{F}CIS(\mathscr{U}) &= \begin{pmatrix} (83.333, 100, 140, 153.33), \\ (113.33, 120, 120, 133.33), \\ (163.33, 180, 180, 193.33) \end{pmatrix}, \\ \mathscr{F}ENSC(\mathscr{U}) &= \begin{pmatrix} (126.67, 126.67, 140, 160), \\ (126.67, 146.67, 160, 160), \\ (106.67, 126.67, 140, 160) \end{pmatrix}. \end{aligned}$$

Now we can defuzzify the solutions by using the signed distance formula of fuzzy sets as follows:

$$\begin{aligned} d^*(\phi(\mathscr{U})) &= (172.5, 184.17, 207.5), \\ d^*(\mathscr{F}CIS(\mathscr{U})) &= (119.17, 121.67, 179.17), \\ d^*(\mathscr{F}ENSC) &= (138.34, 148.34, 133.34). \end{aligned}$$

Comparison of the solutions is given in Table 6.4.

Table 6.4: Comparison of the solutions.

	Firm A		Firm B		Firm C	
	Crisp	Fuzzy	Crisp	Fuzzy	Crisp	Fuzzy
Shapley	138.33	172.5	133.33	184.17	158.33	207.5
CIS	140	119.17	110	121.67	180	179.17
ENSC	136.67	138.34	156.67	148.34	136.67	134.34

It can be seen in Table 6.4 that fuzzy Shapley solutions are advantageous for all firms. On the other hand, crisp *CIS* solutions are advantageous for Firm *A* and Firm *C* but fuzzy *CIS* solutions fuzzy solutions are advantageous for Firm *B*. Moreover, fuzzy *ENSC* solution is advantageous for Firm *A*. Based on this results, the owners of the companies decided to establish a new company $(A-B-C)$, where the shareholders in appropriate proportions were all three companies: *A*, *B* and *C*. Moreover, company *A* acquired a minority stake in companies *B* and *C*.

This model deals with cooperative games whose characteristic functions are fuzzy intervals. Fuzzy interval solution concepts, that is, solutions that assign each cooperative fuzzy interval game a set made up of vectors of fuzzy intervals, can be used to solve reward or cost-sharing issues in scenarios depicted by fuzzy interval cooperative games. Such solutions educate the participants about the variety of personal rewards that can result from cooperation within the grand coalition and are also useful after cooperation to distribute among the participants the final cut of the grand coalition's value, i.e., the significant profit realized by the grand coalition. A natural generalization of the model of cooperative interval games using fuzzy intervals is the model of fuzzy interval cooperative games [20–28].

6.4 Specific Applications in Business

6.4.1 Strategic Alliances

Firms often form strategic alliances to share resources, reduce risks, and enhance their competitive positions. Game theory models help assess the stability of such alliances and determine how to distribute benefits equitably.

A real-life example of game theory applied to a strategic alliance in business is the partnership between The Coca-Cola Company and Keurig Dr Pepper (formerly known as Dr Pepper Snapple Group) in the United States.

Background: In the beverage industry, companies like Coca-Cola and Keurig Dr Pepper often compete for market share in various product categories, including carbonated soft drinks, juices, and bottled water. However, in 2018, these two industry giants announced a strategic alliance to collaborate on a specific product category: single-serve, at-home beverage systems, commonly known as "pod" or "capsule" machines [29].

Game Theory in the Alliance:

1. **Cooperation in Market Entry**: Both Coca-Cola and Keurig Dr Pepper recognized the potential of the growing single-serve beverage market. By forming a strategic alliance, they cooperated to gain entry into this market segment, which was largely dominated by brands like Keurig and Nespresso.

2. **Competitive Landscape**: The single-serve beverage market was competitive, with established players and evolving consumer preferences. Game theory principles came into play as Coca-Cola and Keurig Dr Pepper had to strategically position their alliance to compete effectively. They analyzed potential competitive responses and how to differentiate their offerings.

3. **Shared Resources**: The alliance allowed both companies to share their resources and expertise. Keurig Dr Pepper provided its innovative Keurig brewing technology, while Coca-Cola contributed its vast portfolio of beverage brands. Game theory guided the negotiation of terms to ensure that the benefits of this resource-sharing arrangement were fairly distributed.

4. **Pricing and Marketing Strategies**: In a competitive environment, pricing and marketing strategies were crucial. Game theory helped both companies anticipate the pricing decisions of competitors like Keurig and Nespresso and adjust their own pricing and marketing tactics accordingly.

5. **Distribution and Retailer Relationships**: The alliance also involved considerations regarding distribution and relationships with retailers. Game theory guided decisions on how to navigate negotiations with retailers, ensuring that the alliance's products were prominently featured and effectively marketed in stores.

Outcomes and Impact:
The strategic alliance between Coca-Cola and Keurig Dr Pepper resulted in the introduction of Coca-Cola branded pods for Keurig's KOLD and K-Cup systems. This allowed consumers to enjoy Coca-Cola's popular brands, such as Coca-Cola, Diet Coke, and Sprite, in single-serve formats at home.

The game theory-informed alliance provided several benefits:

1. **Market Expansion**: The alliance allowed both companies to expand their product offerings into a growing market segment, strengthening their presence in the at-home beverage category.

2. **Diversification**: By collaborating in the single-serve beverage segment, Coca-Cola and Keurig Dr Pepper diversified their product portfolios beyond traditional bottled and canned beverages.

3. **Resource Optimization**: The alliance allowed for the efficient use of resources and expertise, reducing the risk associated with solo market entry.

4. **Competitive Response**: Game theory considerations helped both companies anticipate competitive responses and make strategic adjustments, enabling them to compete effectively.

In this example, game theory played a pivotal role in shaping the strategic alliance between Coca-Cola and Keurig Dr Pepper, allowing them to cooperate in a specific market segment while navigating the competitive landscape to achieve mutual benefits.

6.4.2 Mergers and Acquisitions

When companies merge or acquire one another, they must consider both cooperative aspects (integration of operations) and competitive aspects (market dominance). Game theory aids in evaluating the impact of these strategic decisions.

A real-life example of game theory applied to mergers and acquisitions (M&A) in business can be observed in the bidding war for the acquisition of 21st Century Fox assets by The Walt Disney Company and Comcast Corporation in 2018 [30].

Background: In December 2017, 21st Century Fox, a major entertainment conglomerate, announced its intention to sell a significant portion of its assets, including film and television studios, cable networks, and international assets. This decision sparked a competitive bidding process, with multiple media and entertainment companies vying to acquire these valuable assets.

Game Theory in the Bidding War:

1. **Competitive Bidding**: The acquisition of 21st Century Fox's assets represented a strategic opportunity for both Disney and Comcast to expand their media and entertainment portfolios. Game theory principles were evident as both companies assessed their competitors' intentions and formulated bidding strategies.

2. **Sequential Decision-Making**: The bidding war followed a sequential game framework. Initially, Disney made an offer to acquire 21st Century Fox's assets. Comcast then entered the game by making a counteroffer, sparking a back-and-forth bidding process.

3. **Anticipating Rival Moves**: Both Disney and Comcast had to anticipate each other's potential moves. Game theory helped them evaluate how much to bid in each round, taking into account factors such as the perceived value of the assets, the willingness of 21st Century Fox's board and shareholders to accept offers, and the risk of overpaying.

4. **Cooperative and Competitive Elements**: While Disney and Comcast were competitors in the bidding process, there were also cooperative elements involved. For example, 21st Century Fox aimed to maximize the sale price of its assets, creating a cooperative relationship with both Disney and Comcast. However, Disney and Comcast were also competing to secure a favorable deal for themselves.

5. **Strategic Alliances**: Throughout the bidding war, Disney and Comcast both sought strategic alliances with other companies to strengthen their bids. Disney, for instance, had the backing of Twenty-First Century Fox's board and had already acquired key regulatory approvals. Comcast, on

the other hand, sought financing and support from potential partners like Verizon.

Outcomes and Impact:
Ultimately, The Walt Disney Company emerged victorious in the bidding war and acquired the majority of 21st Century Fox's assets for approximately $71.3 billion. Disney's strategic victory allowed it to significantly expand its content library, strengthen its position in the streaming media market, and enhance its competitive advantage in the entertainment industry.

This real-life example demonstrates how game theory principles, including competitive bidding, sequential decision-making, and the anticipation of rival moves, played a crucial role in shaping the outcome of a high-stakes merger and acquisition process. Both Disney and Comcast strategically navigated the bidding war to maximize their benefits and secure valuable assets.

6.4.3 Supply Chain Management

In supply chain networks, cooperation among suppliers, manufacturers, and distributors is crucial for efficiency. Game theory helps optimize decisions related to pricing, production, and distribution while considering potential conflicts of interest. A real-life example of game theory applied to supply chain management in business can be found in the collaboration between Procter & Gamble (P&G) and Walmart, one of the world's largest retailers [31]. This partnership illustrates how game theory principles can optimize supply chain operations. Background: P&G, a consumer goods company, supplies a wide range of products to Walmart, which demands a high level of coordination and efficiency due to the scale and complexity of both companies' operations. In such a large and intricate supply chain relationship, game theory plays a role in decision-making and cooperation [32].

Game Theory in Supply Chain Management:

1. **Inventory Management**: Both P&G and Walmart need to manage their inventory efficiently. Game theory helps them make decisions regarding how much inventory to hold, considering factors like lead times, demand variability, and costs. Cooperative game theory principles can be applied to find optimal inventory levels that benefit both parties.

2. **Ordering and Reordering Strategies**: Decisions on when and how much to order are critical in supply chain management. P&G and Walmart engage in sequential games, where one party's ordering decision affects the other's. Game theory helps them anticipate each other's moves and determine the best ordering strategies to minimize costs and stockouts.

3. **Promotions and Pricing**: P&G and Walmart often collaborate on promotions and pricing strategies. These decisions involve elements of cooperation and competition. Game theory helps them negotiate pricing terms,

discounts, and promotional activities to maximize joint profits while considering market dynamics and competitor responses.

4. **Supply Chain Investments**: Game theory guides decisions on investments in supply chain infrastructure, such as distribution centers, transportation, and technology. Both parties need to strategically allocate resources to ensure the smooth flow of products while minimizing costs and lead times.

5. **Collaborative Forecasting**: P&G and Walmart engage in collaborative forecasting to improve demand planning accuracy. Game theory can help optimize the information-sharing process by addressing concerns related to data accuracy, privacy, and strategic behavior.

Outcomes and Impact:
The application of game theory to supply chain management in the P&G-Walmart partnership has several benefits:

1. **Efficient Inventory**: By using game theory to optimize inventory levels and ordering strategies, both P&G and Walmart reduce carrying costs, minimize stockouts, and enhance customer service.

2. **Cost Reduction**: Cooperative game theory aids in cost-sharing decisions, such as distribution center locations and transportation routes, leading to cost reductions for both parties.

3. **Improved Forecasting**: Collaborative forecasting using game theory principles leads to more accurate demand predictions, reducing the bullwhip effect and helping both companies better plan their production and distribution activities.

4. **Enhanced Customer Satisfaction**: The efficient supply chain management resulting from game theory applications ensures that products are readily available to customers, enhancing overall satisfaction.

This real-life example demonstrates how game theory principles can be applied to optimize supply chain operations by balancing cooperation and competition between two major companies. By using game theory to make strategic decisions, P&G and Walmart have created a mutually beneficial partnership that enhances efficiency and effectiveness throughout their supply chain.

6.4.4 Regulatory Strategies

Companies often engage with regulators and government bodies. Game theory assists in understanding how to cooperate or compete effectively in regulatory negotiations, compliance, and public policy advocacy.

A real-life example of game theory applied to regulatory strategies in business can be seen in the context of telecommunications spectrum auctions. These auctions are conducted by government regulatory bodies to allocate portions of the electromagnetic spectrum to telecommunications companies for wireless communication services. An example of such an auction is the Federal Communications Commission (FCC) spectrum auctions in the United States [33].

Background: The electromagnetic spectrum is a finite and valuable resource used for wireless communication, including mobile phones and data services. Regulatory authorities, such as the FCC in the U.S., must allocate spectrum licenses to telecommunications companies to ensure fair competition, efficient use of resources, and optimal service quality for consumers.

Game Theory in Spectrum Auctions:

1. **Simultaneous Multiple-Round Auctions (SMRAs)**: Spectrum auctions often use SMRAs, a format derived from game theory. In SMRAs, participants submit bids in multiple rounds, with the auctioneer revealing the current highest bid after each round. Bidders must strategically adjust their bids based on the revealed information and their assessment of competitors' strategies.

2. **Bid Strategies**: Game theory guides telecom companies' bid strategies. Bidders need to consider how much spectrum they need, the value of specific frequency bands, and their competitors' willingness to pay. Cooperative elements come into play as bidders may form consortiums or alliances to jointly acquire spectrum licenses while competing with other bidders.

3. **Winner's Curse**: The winner's curse is a phenomenon in auctions where the winning bidder often overestimates the value of the won license. Game theory helps bidders assess their own valuations, taking into account the potential for overpayment and the strategic behavior of rivals.

4. **Bidder Collusion**: Game theory principles are applied to detect and prevent collusion among bidders. Regulatory authorities use auction rules and anti-collusion measures to ensure a competitive bidding environment.

Outcomes and Impact:
The application of game theory to spectrum auctions has several outcomes and impacts:

1. **Efficient Spectrum Allocation**: Game theory helps regulators allocate spectrum efficiently by encouraging companies to bid based on their true valuation of spectrum licenses. This leads to more effective use of the electromagnetic spectrum.

2. **Revenue Maximization**: Governments aim to maximize revenue from spectrum auctions while ensuring that licenses go to companies capable of providing high-quality services. Game theory helps achieve this balance by creating competitive pressure on bidders.

3. **Promoting Competition**: Game theory-based auction designs promote competition within the telecommunications industry, which benefits consumers through improved services, lower prices, and increased innovation.

4. **Reducing Collusion**: Regulatory strategies derived from game theory principles aim to reduce collusion and ensure fair competition. These strategies include rules that prevent communication among bidders during the auction.

In summary, spectrum auctions, such as those conducted by the FCC, demonstrate the application of game theory in regulatory strategies for allocating valuable resources in a manner that promotes competition, efficient spectrum use, and revenue maximization while preventing collusion and ensuring fair market access for telecommunications companies.

6.4.5 Market Entry and Expansion

When entering new markets, firms must balance cooperation with local partners and competition with existing players. Game theory models guide market entry strategies, including joint ventures, franchising, or independent expansion.

A real-life example of game theory applied to market entry and expansion in business can be seen in the global expansion strategies of ride-sharing companies, particularly Uber and Lyft. Both companies have engaged in intense competition while strategically making decisions about when, where, and how to enter new markets and expand their services [34, 35].

Background: Uber and Lyft are two of the world's largest ride-sharing platforms, providing transportation services through mobile apps. Their expansion strategies involve complex decision-making influenced by market dynamics, competition, and regulatory environments.

Game Theory in Market Entry and Expansion:

1. **Sequential Game of Entry**: When entering new markets, Uber and Lyft engage in a sequential game. One company typically enters a new city first, and the other follows suit. The second entrant observes the actions and strategies of the first entrant and adjusts its market entry strategy accordingly.

2. **Competitive Pricing**: Pricing is a critical factor in ride-sharing competition. Both Uber and Lyft continuously adjust their pricing strategies to

attract riders and drivers while undercutting each other's fares. This pricing competition can be viewed as a game where each company seeks to maximize its market share and profitability.

3. **Driver and Rider Incentives**: Game theory principles apply to the design of driver and rider incentives. Both companies offer promotions, bonuses, and incentives to attract and retain drivers and riders. The effectiveness of these incentives depends on how they are strategically deployed to outcompete the rival platform.

4. **Regulatory Strategies**: Game theory guides the regulatory strategies of Uber and Lyft. As they expand into new markets, they must navigate varying regulatory landscapes. Game theory helps them decide when to cooperate with regulators, when to challenge regulations, and when to exit markets where regulatory hurdles are too high.

5. **Market Selection**: The decision of which markets to enter is influenced by a variety of factors, including population density, transportation infrastructure, and potential demand. Game theory helps assess the risks and rewards associated with market selection, taking into account competitive responses.

Outcomes and Impact:
The application of game theory to market entry and expansion strategies by Uber and Lyft has several outcomes and impacts:

1. **Consumer Choice**: Competition between Uber and Lyft provides consumers with choices, leading to better service quality, lower prices, and more convenient transportation options.

2. **Innovation**: The competitive environment encourages both companies to innovate, introducing new features, services, and technologies to gain a competitive edge.

3. **Market Growth**: The intense competition between Uber and Lyft has driven the growth of the ride-sharing industry, expanding the overall market for such services.

4. **Regulatory Landscape**: Game theory helps both companies navigate the regulatory landscape by strategically engaging with regulators and adapting to changing regulations, ensuring their continued operations in various markets.

5. **Global Expansion**: Both Uber and Lyft have expanded internationally, using game theory principles to make informed decisions about which countries and cities to enter and how to compete effectively in global markets.

In summary, the market entry and expansion strategies of ride-sharing companies like Uber and Lyft exemplify the application of game theory in the real world of business. Their competitive interactions and strategic decision-making have shaped the dynamics of the ride-sharing industry and influenced how they enter and expand in markets around the world.

6.5 Conclusion

Cooperation and competition are two fundamental forces that shape the landscape of business strategy. Game theory provides the analytical framework necessary to understand, balance, and optimize these dynamics. By effectively leveraging cooperation and competition, businesses can form alliances, make strategic decisions, and navigate complex market environments, ultimately enhancing their competitive advantage and long-term success.

Avinash Dixit and Barry Nalebuff, already in the preface to their *"The Art of Strategies"* [36] devoted to the application of game theory, writes: *"Strategic thinking is the art of (...) finding a way to cooperate, even when the rest of the players are driven by self-interest, not benevolence. (...)"*

We can hope that, guided by reason and logical thinking, we as humanity are able to enter a higher level of civilization. The application of the methods of the exact sciences (especially game theory) to the analysis of empirical social research and solving practical problems of societies seems appropriate to achieve this goal.

Acknowledgment

The research and the publication of their results were carried out and financed under the "Implementation Doctorate" Program of the Polish Ministry of Education and Science implemented in 2022–2026 (Contract No. PP RU00022465 of August 5, 2022).

This research was also funded by project PROMETEO/2021/063 from the Comunidad Valenciana.

References

[1] Morrish, S.C. and Hamilton, R.T. 2002. Airline alliances—who benefits?. *Journal of Air Transport Management*, 8(6): 401–407.

[2] Star Alliance. n.d.. The History of Star Alliance. https://www.staralliance.com/en/history.

[3] Pakora, B. 2024. What a codeshare flight is and why it matters to travelers. *Forbes Advisor*. https://www.forbes.com/advisor/credit-cards/travel-rewards/codeshare-flight/.

[4] Networks. 2016, September 13. Game Theory within the Apple-Samsung Rivalry. Course blog for INFO 2040/CS 2850/Econ 2040/SOC 2090, Cornell University. https://blogs.cornell.edu/info2040/2016/09/13/game-theory-within-the-apple-samsung-rivalry/.

[5] Brandenburger, A.M. and Nalebuff, B.J. 1996. *Co-opetition*. New York: Currency/Doubleday.

[6] Casadesus-Masanell, R. and Ghemawat, P. (2006). Dynamic mixed duopoly: A model motivated by Linux vs. Windows. *Management Science*, 52(7): 1072–1084.

[7] Networks. 2019, September 25. A beautiful duopoly—How economic theory can explain the competition between Uber and Lyft. A Beautiful Duopoly. Course blog for INFO 2040/CS 2850/Econ 2040/SOC 2090. https://blogs.cornell.edu/info2040/2019/09/25/a-beautiful-duopoly/.

[8] Branzei, R., Dimitrov, D. and Tijs, S. 2008. *Models in Cooperative Game Theory* (Vol. 556). Springer Science & Business Media.

[9] Driessen, T.S.H. and Funaki, Y. 1991. Coincidence of and collinearity between game theoretic solutions. *Operations-Research-Spektrum*, 13(1): 15–30.

[10] Van Den Brink, R. and Funaki, Y. 2009. Axiomatizations of a class of equal surplus sharing solutions for TU-games. *Theory and Decision*, 67(3): 303–340.

[11] Zadeh, L.A. 1965. Fuzzy sets. *Information and Control*, 8(3): 338–353.

[12] Dubois, D.J. 1980. *Fuzzy Sets and Systems: Theory and Applications* (Vol. 144). Academic Press.

[13] Dubois, D., Kerre, E., Mesiar, R. and Prade, H. 2000. Fuzzy interval analysis. In *Fundamentals of Fuzzy Sets* (pp. 483–581). Springer, Boston, MA.

[14] Yao, J.S. and Wu, K. 2000. Ranking fuzzy numbers based on decomposition principle and signed distance. *Fuzzy Sets and Systems*, 116(2): 275–288.

[15] Özcan, İ. and Aytar, S. 2022. Application of polygonal fuzzy numbers to the basic economic order quantity model. *Opsearch*, 59: 482–493.

[16] Mallozzi, L., Scalzo, V. and Tijs, S. 2011. Fuzzy interval cooperative games. *Fuzzy Sets and Systems*, 165(1): 98–105.

[17] Yu, X. and Zhang, Q. 2010. An extension of cooperative fuzzy games. *Fuzzy Sets and Systems*, 161(11): 1614–1634.

[18] Özcan, İ. and Alparslan Gök, S.Z. 2021. On the fuzzy interval equal surplus sharing solutions. *Kybernetes*, 51(9): 2753–2767.

[19] EU Commission et al. 2015. User guide to the sme definition.

[20] Alparslan Gök, S.Z. and Özcan, İ. 2023. On big boss fuzzy interval games. *European Journal of Operational Research*, 306(3): 1040–1046.

[21] Alparslan Gök, S.Z., Ergün, S., Kırlar, B.B., Özcan, İ and Tayman, A. 2023. Fuzzy perspective of online games by using cryptography and co-operative game theory. *International Journal of Uncertainty, Fuzziness and Knowledge-Based Systems*, 31(06): 891–915.

[22] Özcan, İ. and Alparslan Gök, S.Z. 2021. On cooperative fuzzy bubbly games. *Journal of Dynamics & Games*, 8(3): 267–275.

[23] Özcan, I. and Alparslan Gök, S.Z. 2021. On the fuzzy interval equal surplus sharing solutions. *Kybernetes*, 51(9): 2753–2767.

[24] Özcan, İ., Sledzinski, J.D., Alparslan Gök, S.Z., Butlewski, M. and Weber, G.W. 2023. Mathematical encouragement of companies to cooperate by using cooperative games with fuzzy approach. *Journal of Industrial and Management Optimization*, 19(10): 7180–7195.

[25] Özcan, İ., Alparslan Gök, S.Z. and Weber, G.W. 2024. Peer group situations and games with fuzzy uncertainty. *Journal of Industrial and Management Optimization*, 20(1): 428–438.

[26] Özcan, İ. and Alparslan Gök, S.Z. 2024. On big boss grey games. *Journal of Dynamics and Games*.

[27] Özcan, İ., Alparslan Gök, S.Z. and Weber, G.W. 2023. On sequencing fuzzy interval games. *Management and Production Engineering Review*, 14(4): 10–18.

[28] Özcan, İ. and Alparslan Gök, S.Z. 2022. Airport situations and games with fuzzy uncertainty. In *International Conference on Science, Engineering Management and Information Technology*, pp. 50–60. Cham: Springer Nature Switzerland.

[29] Khosla, S. 2018, March 23. Keurig and Dr. Pepper Merger. https://www.whartonugpevc.com/articles/2018/3/29/keurig-and-dr-pepper-merger.

[30] Lee, E. 2018, June 20. Disney ups its bid for 21st century fox to $71.3 billion. Business and Policy. The New York Times. https://www.nytimes.com/2018/06/20/business/dealbook/disney-comcast-21st-century-fox.html.

[31] Trebilcock, B. 2023, July 18. How P&G stays on top of its game. Supply Chain Management Review. https://www.scmr.com/article/insiders-view-of-proctor-and-gamble-supply-chain-success.

[32] Grean, M. and Shaw, M.J. 2002. Supply-chain partnership between P&G and Wal-Mart. *In*: Shaw, M.J. (ed.). E-Busine.

III

OPERATION RESEARCH IN SIGNAL PROCESSING

Chapter 7

Pixel-based Snow Cover Classification over Complex Mountainous Regions by Random Forest Algorithm

Cansu Aksu,[1] *Semih Kuter*[2,*] and *Zuhal Akyurek*[3]

7.1 Introduction

With the recent technological improvements in all kinds of data recording instruments, the era of *big data* has revolutionized the Earth Sciences from the *data-poor* field to the *data-rich* field [22]. This blast in the volume of data has resulted in an urgent call from the *Earth System Sciences Community* on how to improve accuracy, enhance the level of model intelligence, scale up opera-

[1]Middle East Technical University, Graduate School of Natural and Applied Sciences, Department of Geodetic and Geographic Information Technologies, 06800 Ankara, Turkey.
[2]Cankiri Karatekin University, Faculty of Forestry, Department of Forest Engineering, 18200 Cankiri, Turkey.
[3]Middle East Technical University, Faculty of Engineering, Department of Civil Engineering, 06800 Ankara, Turkey.
* Corresponding author: semihkuter@karatekin.edu.tr

tions, and reduce cost against the continuous and exponentially growing size of the retrieved datasets [44]. The parallel developments in the storage capacity and computational power offer immense potential for the implementation of *Artificial Intelligence* (AI) and *Machine Learning* (ML) methods in the various fields of Earth Sciences [22, 44].

As in all branches of science and engineering, *earth observation* and *remote sensing* areas have also witnessed an increasing trend in the use of AI and ML during the past decade [13]. The Earth is a dynamic system where numerous physical, chemical, and biological processes occur, and these processes typically exhibit high degrees of interconnectivity [42]. Despite these interconnected characteristics, these processes and their interactions often display complex, non-linear relationships, and can even include extreme anomalies, shifts, and rapid changes that surpass our current understanding [26, 27].

The Northern Hemisphere harbors an overwhelming majority of the planet's snow cover (i.e., 98%), rendering its presence critical to the survival of human and animal populations alike [49]. The snowmelt-induced runoff is a crucial source of water in high-altitude regions across North America, Europe, and Asia, playing a key role in sustaining hydrological cycles and ecosystems. The extent and depth of snow cover are intrinsically linked to snowfall and temperature patterns, which directly influence the snowmelt rate. Therefore, fluctuations in snow precipitation and temperature serve as pivotal indicators of weather and climate dynamics, with far-reaching implications for environmental processes and human societies [41].

The amount and extent of snow cover exerts a multifaceted influence on local climate regimes, beyond its impact on precipitation amounts. The presence of fresh snow cover can augment surface albedo, characterized by high reflectivity, while the snow layer functions as an effective thermal insulator, owing to its high thermal emissivity. Moreover, snowmelt may act as a heat sink [45], and the duration of the snowmelt season bears a direct effect on carbon flux, soil respiration, and the thickness of the active layer of the soil in regions characterized by permafrost [7]. Hence, precise data on snow cover is a fundamental prerequisite for undertaking hydrological, climatic, and greenhouse gas cycle studies in the mid and upper latitudes of the Northern Hemisphere [45].

Observations spanning various time periods corroborate a diminishing trend in the snow cover area in the Northern Hemisphere in recent years [49]. This finding carries profound implications for climate research and water resources management, as snow properties play a pivotal role in investigating climatic dynamics and ascertaining the maximum possible runoff [45]. In particular, snowmelt represents a major contribution to river flow and groundwater recharge in mountainous regions, providing water resources to billions of humans. Mountain regions are highly sensitive to climate change. Consequently, ensuring precise and continuous monitoring of snow cover is crucial. However, in areas featuring complex topography, the data collected from *in-situ* observation stations within the study site may suffer from limitations, potentially compromising the accuracy of the estimates [1].

Satellite-based methods have been employed in snow cover monitoring for nearly six decades, enabling the generation of snow cover maps with high temporal resolution, even daily, using advanced multispectral sensors such as MODIS [38], and with sufficient spatial resolution via satellites such as Landsat and Sentinel 2 [12]. The AVHRR sensor operated by NOAA is also a valuable asset in this context [30].

Snow-covered area (SCA), an essential hydrological parameter, has been employed for streamflow forecasting since the 1940s, initially relying on aerial photography, and subsequently utilizing satellite data by incorporating classification algorithms [34]. Optical remote sensing data can furnish two distinct features of SCA: binary snow cover maps (indicating the presence or absence of snow, i.e., snow/no snow) and *fractional snow-covered area* (fSCA) acquired by subpixel snow cover mapping. In this study, we utilize a variant of binary snow cover maps where the areas with no snow are further delineated based on additional classification criteria.

The mapping of snow cover using remote sensing data has been revolutionized by the application of ML algorithms, which have demonstrated superior performance over other traditional methods such as statistical regression analysis and spectral unmixing in a range of studies. Specifically, *Random Forest* (RF), *Support Vector Machine* (SVM), *Artificial Neural Network* (ANN) and *Multivariate Adaptive Regression Splines* (MARS) algorithms have emerged as particularly effective approaches for this task [18, 24, 25, 29].

Even though snow cover monitoring at optical wavelengths using either traditional model-based remote sensing techniques or ML algorithms is well-studied, as briefly mentioned above, there still exist several challenges and limitations such as cloud and canopy obscuration, cloud-snow confusion at land-snow boundary over snow-covered areas, reduced reflectance at optical-infrared wavelengths due to topography/relief- or cloud-casted shadows, etc.

In this study, we utilize the RF algorithm to classify snow cover, clouds, water bodies, and land in three mountainous regions, namely, European Alps, Tatra Mountains, and Kackar Mountains using Sentinel-2 imagery. The classification incorporates three distinct morphological phases of snow, namely fresh snow, fully-formed snowpack, and melting snow. The primary objective is to evaluate the RF algorithm's performance under different input scenarios and to address the common challenges faced in snow cover mapping using optical remote sensing data, including the erroneous classification of shadows and clouds as snow, particularly over complex mountainous terrains.

7.2 Dataset and Methodology

7.2.1 Satellite Imagery and Study Area

Remotely-sensed snow cover products are typically offered in two fundamental formats, which utilize spectral channels in the visible (0.4–0.7 μm) and near-infrared (1–4 μm) ranges. These formats consist of (i) binary snow cover maps; and (ii) fSCA maps [15, 36]. The binary snow cover maps label each pixel as either containing snow or not, while the fSCA maps reveal the area fractions of different land cover types (e.g., soil, rock, vegetation, snow, water, etc.) within a low or moderate spatial resolution pixel. To generate binary snow cover maps, a thresholding approach based on the *normalized difference snow index* (NDSI) is often employed in the traditional approach. This approach utilizes the snow's high reflectance in *visible* (VIS) bands and low reflectance in *short wave-infrared* (SWIR) part of the electromagnetic spectrum [10, 16].

This research conducts a comprehensive evaluation of the RF algorithm's performance in detecting four distinct classes, namely, cloud, land, snow, and water using Sentinel-2 imagery from diverse mountainous regions. When viewed within the binary snow cover mapping perspective as discussed above, the no snow class in our RF-based classification sceheme further comprises all other elements that do not belong to snow class by assigning cloud, land, and water labels to the associated pixels.

Sentinel-2 is a high-resolution multispectral imaging mission operated by the *European Space Agency* (ESA), which is designed for monitoring land, ocean, and atmospheric conditions. This polar-orbiting mission is intended to meet the coverage and revisit requirements and provide reliable datasets. To achieve these objectives, the Sentinel-2 constellation includes two identical satellites: Sentinel-2A and Sentinel-2B, which were launched on 23 June 2015 and 7 March 2017, respectively. Because the twin satellites are positioned in the same sun-synchronous orbit with a phase delay of 180 degrees, they are capable of revisiting the same location every 5 days at the equator and every 2/3 days over mid-latitudes, while providing a swath width of 290 km [11]. The spectral band designation of Sentinel-2 is given in Table 7.1.

Digital elevation model (DEM) is a terrain representation that is critical for Earth sciences, hydrological applications, and remote sensing analysis. DEM can be acquired through various techniques such as photogrammetry, interferometry, and ground/laser surveying [6]. The production of DEMs dates back to the 1960s and involves using two different angles in the visible or near-infrared regions of the electromagnetic spectrum [46]. In this study, DEM data obtained from the *shuttle radar topography mission* (SRTM) is utilized to improve the precision of snow cover classification by considering the effects of elevation on snow cover distribution. The DEM data are acquired for the three study regions and then resampled to 20 m resolution using the cubic spline method.

Table 7.1: Spectral bands of Sentinel-2.

Spectral Band	*Resolution (m)*	*2A Central Wavelength (nm)*	*2B Central Wavelength (nm)*
Band 1	60	442.7	442.2
Band 2	10	492.4	492.1
Band 3	10	559.8	559.0
Band 4	10	664.6	664.9
Band 5	20	704.1	703.8
Band 6	20	740.5	739.1
Band 7	20	782.8	779.7
Band 8	10	832.8	832.9
Band 8a	20	864.7	864.0
Band 9	60	945.1	943.2
Band 10	60	1373.5	1376.9
Band 11	20	1613.7	1610.4
Band 12	20	2202.4	2185.7

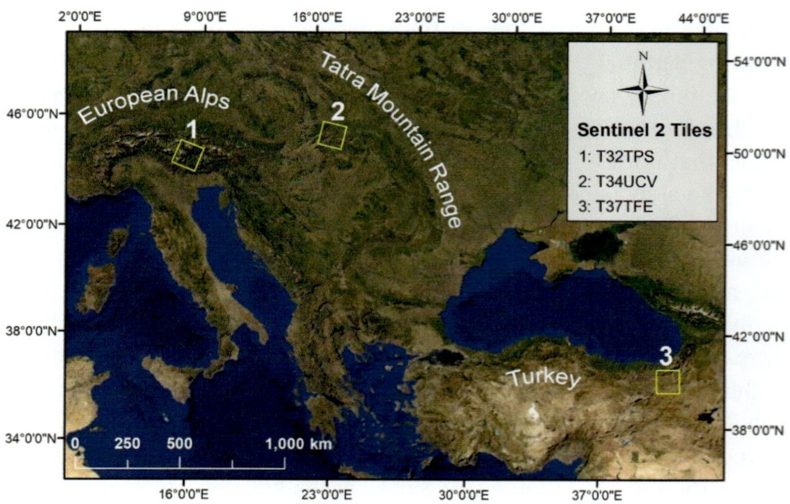

Figure 7.1: The locations of the selected Sentinel-2 tiles.

This study focuses on three distinct mountainous regions: the European Alps, the Tatra Mountains, and the Kackar Mountains in Turkey. For each region, a single Sentinel-2 tile is selected. Figure 7.1 illustrates the selected tile locations.

These three mountainous regions are chosen to observe the behavior of the RF algorithm under different climates, snow duration, and depths. In each study area, a total of three Sentinel-2 images are used; one for the period when snow accumulation begins (from November to December), one for the period when snow is at its peak (from January to March), and one for the period when snow begins to melt (from April to June). The cloud contamination is also taken into consideration when selecting these images and the maximum scene cloud coverage is set as $\leq 30\%$.

The details on the selected Sentinel-2 images are given in Table 7.2.

7.2.2 Atmospheric and Topographic Correction of Sentinel-2 Images

Satellite-detected object reflectance can be distorted by atmospheric absorption and scattering, thereby reducing the accuracy of remotely sensed images. As such, atmospheric correction is advisable for change detection and classification operations, as it can produce superior outcomes compared to non-atmospherically corrected data [43]. Topographic correction pertains to adjusting the solar reflectance values of uneven terrain surfaces to address the differences in reflectance that occur. Such differences can result in a broad spectrum of reflectance values for similar vegetation types, leading to darker-than-expected shaded areas and brighter-than-expected illuminated areas. To attain more precise reflectance values, particularly in areas with rough terrain, correcting for terrain incident angles and shadows via topographic correction is indispensable [39]. Performing atmospheric and topographic corrections on images can enhance the accuracy of classification tasks. Consequently, it is often recommended to conduct both types of corrections prior to any land cover classification effort [48].

In this study, before generating the various predictor variable combinations as input to the RF classification scheme, the Sentinel-2 image dataset (cf. Table 7.2) is subjected to atmospheric and topographic correction using Sen2Cor. Sen2Cor utilizes the *atmospheric/topographic correction for satellite imagery* (ATCOR) algorithm [40] to apply the atmospheric and topographic correction. For topographic correction, SRTM DEM data with a resolution of 90 m and an empirical *bidirectional reflectance distribution function* (BRDF) method with three adjustable variables are used.

Table 7.2: Information on the selected Sentinel-2 tiles.

Region	*Sentinel-2 Tile No*	*Acquisition Date*
1-Alps	T32TPS	5 December 2018 24 January 2019 13 June 2019
2-Tatra	T34UCV	29 November 2018 23 January 2020 8 April 2019
3-Kackar	T37TFE	19 December 2019 19 March 2018 8 April 2018

To obtain *bottom-of-atmosphere* (BOA) reflectances, Sen2Cor performs Level-2A processing on Sentinel Level-1C data. This processing encompasses cloud detection, scene classification, retrieval of *aerosol optical thickness* (AOT) and *water vapour* (WV), and conversion of *top-of-atmosphere* (TOA) reflectance to BOA reflectance. Optional procedures, such as cirrus correction, terrain correction, adjacency correction, and empirical BRDF corrections, are also available. The main processing phases of Sen2Cor are depicted in Figure 7.2.

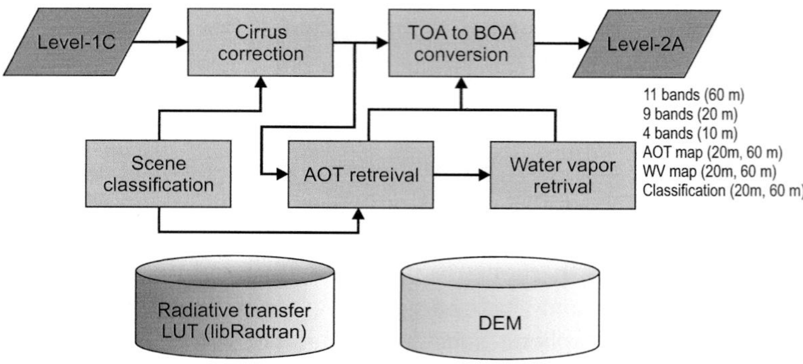

Figure 7.2: Sen2Cor main processing steps (adopted from [32]).

7.2.3 Principal Component Analysis

Principal component analysis (PCA) is a method employed to rearrange the information within a multispectral dataset into a set of principal components, which can consist of multiple images. PCA is a widely used technique in remote sensing for reducing the dimensionality of multispectral image data. Multispectral remote sensing images often contain a large number of bands (i.e., measurements at different wavelengths), which can make it difficult to identify patterns and relationships in the data [21].

PCA works by transforming the original data into a new set of variables called principal components, which are linear combinations of the original bands. The first principal component accounts for the most variation in the data, and each subsequent principal component accounts for as much of the remaining variation as possible [21]. Principal components are statistically independent (orthogonal) of one another and arranged in a decreasing sequence of variability, with the initial component having the highest variance. By retaining only the principal components that capture the most important information in the data, PCA can help to reduce the dimensionality of the image while preserving the most important information [33].

PCA has many applications in remote sensing, including image compression, feature extraction, and classification. It can be particularly useful in situations where there are correlations between different bands in the image, as it can help to separate out these correlations and identify the underlying patterns in the data [21].

Thus, PCA is applied on the atmospherically and topographically corrected Sentinel-2 spectral bands using the sci-kit learn library in Python, and the first three principal components are utilized as predictor variables to observe whether they lead to improved accuracy in the RF-based classification.

7.2.4 Predictor Variable Combinations

This study utilizes Sentinel-2 bands 2-7, 8a, 11, and 12 at 20 m resolution (cf. Table 7.2), three normalized band indices derived from these bands, and the elevation values from the DEM data. Bands 2-4, whose native spatial resolutions are 10 m, are resampled to 20 m by using ESA's Sen2Cor scene processing module [32].

In this study, to distinguish among the four classes - cloud, land, snow, and water - three different indices are used as inputs for different predictor variable combinations in the RF classification scheme. These indices are NDSI, *normalized difference vegetation index* (NDVI), and *normalized difference water index* (NDWI).

NDSI is an index that uses VIS and SWIR bands to distinguish between clouds and snow cover. NDSI-based approaches have been widely used for snow cover mapping across the globe and have consistently produced accurate results [31]. Moreover, NDSI is often employed as an input variable for RF and other machine-learning algorithms in snow cover mapping. NDSI is given by the following equation:

$$NDSI = \frac{(Green - SWIR)}{(Green + SWIR)}, \tag{7.1}$$

where the green and SWIR bands of Sentinel-2 are band 3 and band 11, respectively.

NDVI, calculated from the near-infrared (NIR) and red bands, is a widely used index for distinguishing vegetation from other land cover types [20]. When combined with NDSI, NDVI is often used to map snow cover in areas with dense vegetation, shadows, and low illumination [23]. In addition, NDVI is frequently employed to differentiate clouds from other land cover types since cloud pixel values are zero for both NDVI and NDSI [14]. The NDVI is realized by the following equation:

$$NDVI = \frac{(NIR - Red)}{(NIR + Red)}, \tag{7.2}$$

where the red and NIR bands correspond to band 4 and band 8 of Sentinel-2, respectively.

NDWI is a widely used index for mapping and monitoring water bodies. It measures the difference between the reflectance values of NIR and green light to highlight areas with high water content [35]. NDWI is an effective tool for detecting water in diverse landscapes, including wetlands, rivers, lakes, and oceans. Its high accuracy and sensitivity make it a valuable resource for studying water resources and their dynamics over time. The NDWI is obtained through the following calculation:

$$NDWI = \frac{(Green - NIR)}{(Green + NIR))}, \tag{7.3}$$

where green and NIR bands are band 3 and band 8 for Sentinel-2, respectively.

This study utilizes four different predictor variable combinations for the RF classification, each including different types of data: (i) the *Sb* combination includes Sentinel-2 bands at 20 m resolution and three normalized band indices (i.e., NDSI, NDVI, and NDWI), and for this input combination, no atmospheric/topographic correction is applied; (ii) the *AT* combination includes atmospherically and topographically corrected Sentinel-2 bands, as well as NDSI, NDVI, and NDWI; (iii) the third combination, *Dem*, includes the same data as *AT* combination with the addition of elevation data from the DEM; (iv) and the fourth combination, *PCb*, employes the first three principal components obtained from the atmospherically and topographically corrected Sentinel-2 bands together with NDSI, NDVI and NDWI. It should be noted that atmospherically and topographically corrected Sentinel-2 bands are used in all combinations, except for *Sb*. These four predictor variable combinations are given in Table 7.3.

Table 7.3: Predictor variable combinations used in the RF classification.

Combination name	*Content of the combination*
Sb	Bands 2-7, 8a, 11, 12, NDSI, NDVI, and NDWI
AT	Bands 2-7, 8a, 11, 12, NDSI, NDVI, and NDWI
Dem	Bands 2-7, 8a, 11, 12, NDSI, NDVI, and NDWI, elevation data
PCb	The first three PCA bands (derived from Bands 2-7, 8a, 11, 12), NDSI, NDVI, and NDWI

7.2.5 Selection of Training and Test Datasets

There are multiple approaches used to gather training and testing datasets for remote sensing applications, such as acquiring in-situ data, generating data as polygons utilizing a *Geographic Information Systems* (GIS) software, and seeding of training data [19]. For this particular study, polygon shapefiles are used to select the training and testing datasets from all nine images. The data are evaluated based on expert visual analysis of the atmospherically and topographically corrected Sentinel-2 images.

A dedicated Python code is employed to extract the input data to the manually drawn polygons and arrange the training and test datasets for the RF model. Two sets of training datasets are prepared with 300 and 1000 samples for each category (i.e., cloud, land, snow, and water) by simple random sampling. The number of training samples per category is determined based on earlier research [47].

The reason for choosing 300 samples for each class simply comes from the traditional 30*p* rule [47], which is a widely used heuristic in remote sensing for determining the minimum number of training samples necessary to achieve accurate classification of an image. This rule recommends using at least 30*p* pixels per class for training a classifier, where *p* stands for the number of spectral bands included, as having too few training samples can lead to poor generalization and decreased accuracy in the classification of new, unseen data. It is important to note that while the 30*p* rule provides a useful starting point, the optimal number of training samples may vary depending on a variety of factors, including the complexity of the scene being classified, the spatial resolution of the imagery, and the classification algorithm is used. The main motivation behind creating two different training datasets with two different sizes is to observe whether increasing the size of the usual training sample size per class (i.e., 300) by about three times (i.e., 1000) significantly affects the accuracy of RF models or not.

To ensure an unbiased accuracy assessment, it's important to use a reserved test dataset that is not employed during the training and not biased towards any particular class. While the ideal test dataset would be ground reference points or polygons [19], this is not always possible. In this study, the test data points for all classes are randomly obtained from the Sentinel-2 images using ArcMap 10.7 software. The sample sizes for the test data are determined based on the following multinomial distribution formula often used in sample size (N) calculation [19, 37]:

$$N = \frac{B\prod_i(1-\prod_i)}{b_i^2}, \tag{7.4}$$

where $\prod_i$ is the proportion of a population in the ith class out of k classes that has the proportion closest to 50%, b_i indicates the desired precision (e.g., 5%) for this class, B is the upper $(\alpha/k) \times 100$th percentile of the chi square (χ^2) distribution with 1 degree of freedom, and k denotes the number of classes. In this study, a minimum of 100 samples per class is employed to populate the error matrix and evaluate the accuracy of RF classification.

7.2.6 Random Forest

RF is an algorithm that uses a group of *classification and regression trees* (CARTs) to perform supervised classification and regression tasks [17]. RF is based on the bagging concept of Breiman [4], which uses a bootstrap aggregation technique to randomly select a subset of training data for each tree. The

RF algorithm accumulates (averages) these trees to obtain the "bagged" estimation. RF has an *out-of-bag* (OOB) error, which is used as an acceptable estimator of the error expected for unseen data [5]. RF has several features that make it attractive in engineering and scientific studies, including its ability to handle multi-class classification and regression problems, model interactions between predictor variables, and train on subsets of data, making it suitable for high-dimensional complex datasets. The algorithmic structure of RF can be briefly described as [9]:

The Random Forest Algorithm

Let $D = \{(x_1, y_1), \cdots, ((x_N, y_N)\}$ represents the training data, with $x_i = (x_{i,1}, \cdots, x_{i,p})^T$. For j = 1 to J:

1. Take a bootstrap sample D_j of size N from D.

2. Fit a binary recursive partitioning tree using the training data D_j, which is the bootstrap sample:

1. (a) Commence the construction of the binary recursive partitioning tree by initializing a single node that contains all the observations.

2. (b) Iteratively execute the subsequent steps for each unpartitioned node, in a recursive manner, until the predefined stopping criterion is satisfied:

1. (a) i. Randomly choose a subset of m predictors from the p available predictors.

2. (b) ii. Determine the optimal binary partition among all possible binary partitions of the m predictors identified in Step (i).

3. (c) iii. Partition the node into two descendant nodes utilizing the split obtained in Step (ii).

In order to generate a prediction for a new observation x,

* $\hat{f}(x) = \frac{1}{J}\sum_{j=1}^{J} \hat{h}_j(x)$ for regression,

* $\hat{f}(x) = \arg\max_y \sum_{j=1}^{J} I(\hat{h}_j(x) = y)$ for classification,

where $\hat{h}_j(x)$ denotes the predicted output of the response variable at a given value of x, which is estimated using the jth tree in the model.

The RF algorithm involves adjusting three fundamental parameters to optimize its performance. These parameters include the number of trees in the forest (N_{tree}), the number of predictor variables to be randomly selected at each split (M_{try}), and the minimum number of observations per tree leaf (L_{size}) [5, 9].

The literature commonly adopts $\sqrt{p}$ and $\lceil p \rceil$ as the default values of M_{try} for regression and classification, respectively, with p representing the dimension of the input parameter space (i.e., the total number of predictor variables) and $\lceil \bullet \rceil$ representing the ceiling operator. These default values of M_{try} are frequently utilized in remote sensing applications [2].

According to prior studies [3, 9], the literature commonly assumes a default value of L_{size} equal to 1 for classification and 5 for regression. In the field of remote sensing, these default values are generally accepted for classification and regression applications [24].

Increasing the number of trees (N_{tree}) in an RF model leads to a decrease in its variance. As demonstrated in [5], as N_{tree} increases, the generalization error of RF eventually reaches a limit, indicating that larger N_{tree} values do not result in over-fitting [3]. In the remote sensing literature, a wide range of N_{tree} values have been applied in different studies, and based on stability in the generalization error, a reasonable choice for N_{tree} is typically considered to be 500 [24].

The above-mentioned default values for M_{try}, L_{size}, and N_{tree} are also preferred in our RF-based classification scheme. Python's sci-kit learn library's *RandomForestClassifier* function is utilized to build the RF models with various input combinations (cf. Table 7.3).

7.2.7 Accuracy Assessment

Thematic maps created through the classification of remote sensing images may contain errors caused by various factors, such as atmospheric conditions, data acquisition, and subjective human interpretation. Despite efforts to correct these errors, some may still persist. As a result, it is crucial to conduct a thorough accuracy assessment of the classified image [19]. This involves comparing the classified data with ground reference data and using statistical metrics to evaluate its accuracy level.

To assess the accuracy of the classification results in relation to the ground reference data (test data), researchers commonly employ a confusion matrix (i.e., error matrix). The confusion matrix provides a two-dimensional table that compares the predicted classification labels with the actual labels of the test data. One dimension of the matrix represents the true values of the categories, while the other represents the predicted values [8]. By presenting a summary of the number of correctly and incorrectly classified instances, the confusion matrix allows for a detailed assessment of the performance of the classification model.

In this study, we utilize the confusion matrix as a means of evaluating the accuracy of the RF-based classification, in conjunction with the *overall accuracy*

(OA) and *Cohen's Kappa coefficient* (κ). The overall performance of the classifier can be evaluated using the OA, which indicates the percentage of correctly classified pixels out of the total pixels [8]. On the other hand, κ offers a measure of agreement between the predicted and actual classification labels, taking into account the degree of agreement that can be expected by chance, and it is given by the following expression [19]:

$$\kappa = \frac{N\sum_{i=1}^{k} x_{ii} - \sum_{i=1}^{k}(x_{i+} \times x_{+j})}{N^2 - \sum_{i=1}^{k}(x_{i+} \times x_{+j})}, \tag{7.5}$$

where k denotes the number of rows (i.e., land cover classes) in the confusion matrix, x_{ii} represents the number of observations in row i, and column i, x_{i+} and x_{+j} are the marginal totals for row i and column j, respectively, and finally, N stands for the total number of observations.

The coefficient can range from -1 to 1, with values closer to 1 indicating a higher level of agreement between the predicted and actual labels beyond what could be expected by chance. On the other hand, values closer to 0 suggest that the agreement is no better than what could be achieved through random guessing, while negative values indicate systematic disagreement between predicted and actual labels [8].

7.3 Results and Discussions

7.3.1 Quantitative Results from the Error Matrix

In this study, 9 Sentinel-2 images are classified using RF by attaining cloud, land, snow, and water labels to each pixel in accordance with the input combinations explained in Table 7.3. The accuracy metrics obtained from the associated error matrices in terms of OA and κ on each image for each input combination are given in Table 7.4.

In general, the most accurate results are obtained for Alps on 5 December 2018 with both Sb_{300} and Sb_{1000} input combinations (OA = 0.94, κ = 0.89) and the Tatra Mountains on 8 April 2019 with PCb_{1000} input combination (OA = 0.93, κ = 0.87). On the other hand, the least accurate performance is observed for the Kackar scene on 19 December 2019 with Dem_{300} input combination (OA = 0.66, κ = 0.50).

The results of the accuracy assessment reveal that including the elevation data in the training process (i.e., *Dem* input combination) does not yield a significant improvement in the classification. Additionally, OA and κ metrics indicate that the RF models trained with *PCb* input combination generally exhibit improved performance on the scenes.

Table 7.4: OA and κ metrics on the Sentinel-2 tiles (please note that subscripts 300 and 1000 indicate the training data size of the associated input combination (cf. Table 7.3)).

Region	*Date*	*Metric*	Sb_{300}	Sb_{1000}	AT_{300}	AT_{1000}	Dem_{300}	Dem_{1000}	PCb_{300}	PCb_{1000}
	5 Dec 2018	OA	0.94	0.94	0.90	0.93	0.84	0.86	0.90	0.90
		κ	0.89	0.89	0.83	0.87	0.70	0.75	0.83	0.83
Alps	24 Jan 2019	OA	0.79	0.78	0.80	0.81	0.80	0.80	0.78	0.79
		κ	0.66	0.65	0.69	0.70	0.70	0.70	0.67	0.68
	13 Jun 2019	OA	0.82	0.82	0.83	0.82	0.82	0.82	0.84	0.84
		κ	0.73	0.72	0.73	0.72	0.72	0.72	0.75	0.76
	29 Nov 2018	OA	0.75	0.73	0.69	0.74	0.74	0.74	0.78	0.78
		κ	0.66	0.64	0.59	0.66	0.66	0.65	0.70	0.71
Tatra	23 Jan 2020	OA	0.78	0.77	0.79	0.80	0.72	0.73	0.72	0.71
		κ	0.69	0.67	0.70	0.72	0.62	0.63	0.60	0.59
	8 Apr 2019	OA	0.91	0.90	0.91	0.91	0.91	0.92	0.92	0.93
		κ	0.85	0.82	0.83	0.83	0.84	0.85	0.86	0.87
	19 Dec 2019	OA	0.81	0.87	0.78	0.79	0.66	0.77	0.70	0.78
		κ	0.71	0.80	0.67	0.68	0.50	0.66	0.54	0.66
Kackar	19 Mar 2018	OA	0.84	0.83	0.84	0.84	0.81	0.83	0.81	0.84
		κ	0.70	0.66	0.69	0.70	0.62	0.67	0.64	0.69
	8 Apr 2018	OA	0.74	0.72	0.75	0.75	0.74	0.75	0.77	0.75
		κ	0.62	0.60	0.63	0.64	0.62	0.64	0.68	0.64

The decision of using 300 samples for each category in this research is solely based on the established practice of selecting from 10x to 30x the number of spectral bands for remotely sensed images (i.e., 30p rule), as indicated in [47]. Our study also examines a sample size that is three times larger than the commonly selected size and discovers that a training size of 1000 per category produces slightly superior outcomes than 300 per category for all input combinations over nearly all scenes. A study conducted by [28] that assessed classification accuracy on Landsat images with varying training sample sizes (ranging from 20 to 240 per category) for different algorithms, determined that the RF algorithm was significantly influenced by the training sample sizes and the accuracy increased with the size of the training samples, which is also in line with the findings of our research.

The false color RGB composite of the Alps scene on 13 June 2019 and its RF-classified image using PCb_{1000} input combination are given in Figure 7.3. This false-color composite in Figure 7.3 is created utilizing the Sentinel-2 band combination of Red = 11, Green = 8a, and Blue = 3. This particular band combination renders a visual representation whereby ice and snow are displayed in a bright blue hue, while water bodies exhibit a near-black appearance. Furthermore, saturated soil is visually detectable in blue, and clouds are observed as white.

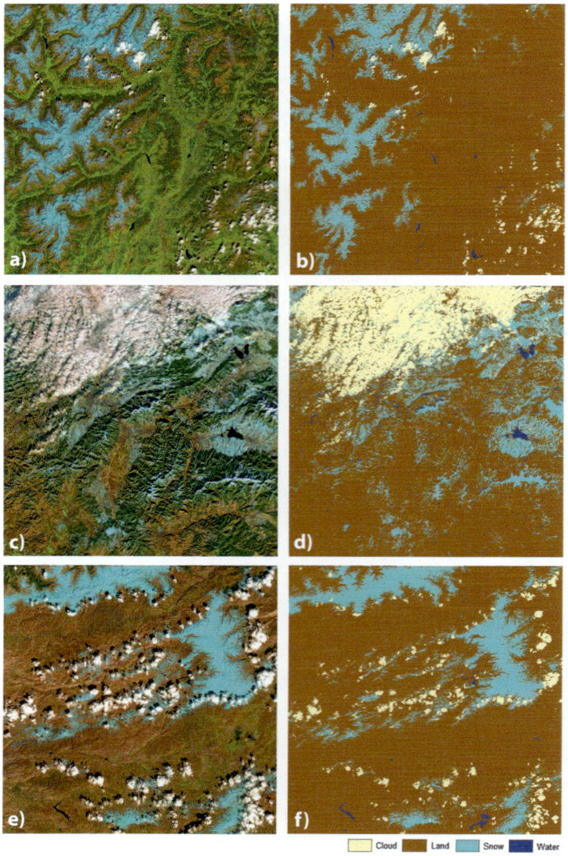

Figure 7.3: a) RGB false color composite and b) RF-classified images of the Alps scene on 13 June 2019; c) RGB false color composite and d) RF-classified images of the Tatra scene on 8 April 2019; e) RGB false color composite and f) RF-classified images of Kackar scene on 8 April 2018 (please note that all RF-classified images use the PCb_{1000} input combination, and a full sentinel-2 tile at 20 m resolution is an image of 5490 rows by 5490 columns, roughly composed of 30.14 M pixels).

7.3.2 Visual Analysis on the Classified Image Subsets

In addition to the quantitative accuracy metrics presented in Table 7.4, we also conduct a thorough qualitative assessment of the classified Sentinel-2 images by visually examining various close-up image subsets. This allows us to observe the performance of the RF algorithm in greater detail. In the visual analysis, the employed Sentinel-2 RGB false color composites use the same band combination as described in Figure 7.3.

The first close-up image subset is from the Alps acquired on 24 January 2019, given in Figure 7.4, where snow cover over a slope-casted shadow surface is

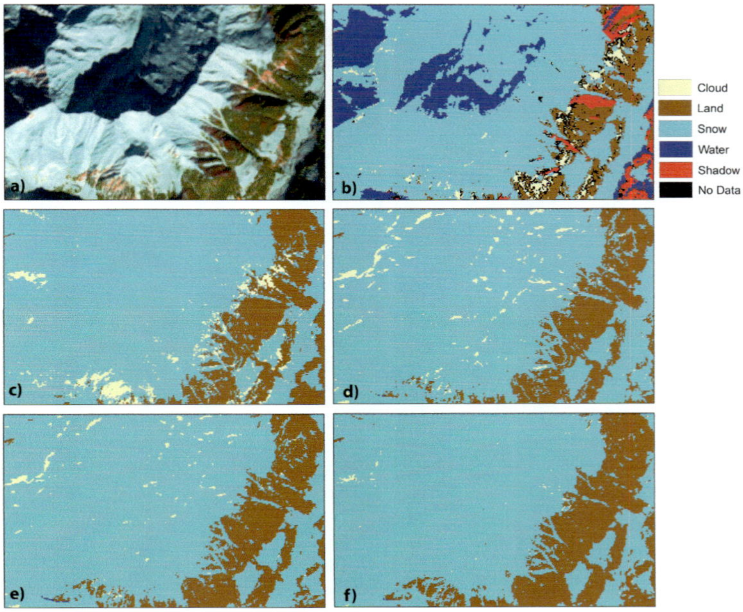

Figure 7.4: Close-up scenes from the Alps subset on 24 January 2019; a) RGB false color composite, b) Sen2Cor scene classification image, c) RF-classified image with Sb_{300}, d) RF-classified image with AT_{300}, e) RF-classified image with AT_{1000}, and f) RF-classified image with PCb_{300}. (please note that *shadow* and *no data* labels in the legend only apply for the Sen2Cor scene classification image).

clearly observed. Additionally, there is no apparent cloud contamination in the scene (cf. Figure 7.4a). Figure 7.4b shows the image generated by the Sen2Cor processor. The snow cover over the slope-casted shadow areas are mislabelled as water. Additionally, at the fringes of the slopes where land and snow meet, there are pixels misclassified as cloud. Although the RF trained with Sb_{300} input combination (cf. 7.4c) seems to surpass the problem of mislabelling of snow over the shadow areas, it still suffers from the issue of mislabelling pixels as a cloud at the land-snow boundaries. Even though increasing the size of the training data appears to alleviate the issue, its effect is limited, and mislabelled cloud pixels are still persistent in the image (cf. 7.4d).

On the other hand, the use of atmospherically and topographically corrected input combination with large training data size option (i.e., AT_{1000}) significantly reduces the number of pixels mislabelled as cloud at the land-snow boundary (cf. 7.4e). And finally, the input combination that utilizes the PC bands, even with small-sized training data (i.e., PCb_{300}), completely resolves the issue, as shown in Figure 7.4f.

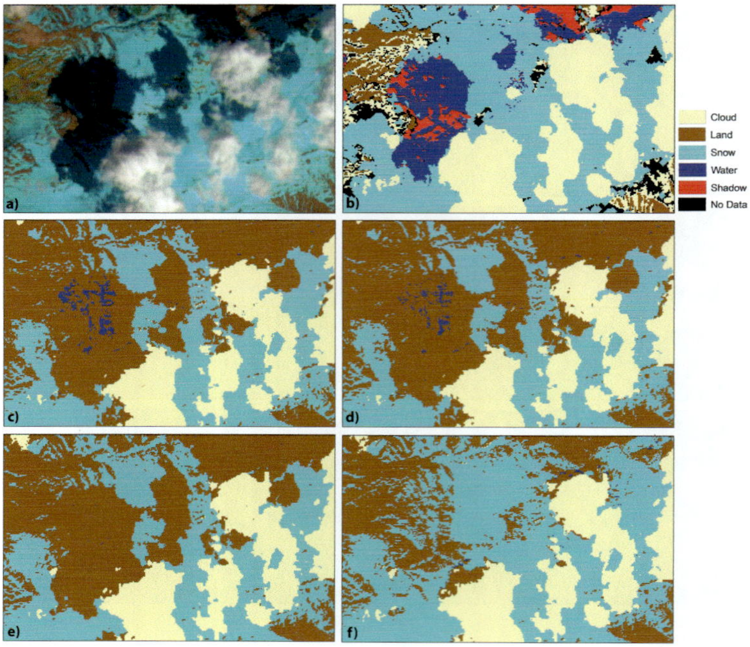

Figure 7.5: Close-up scenes from the Kackar subset on 8 April 2018; a) RGB false color composite, b) Sen2Cor scene classification image, c) RF-classified image with Sb_{300}, d) RF-classified image with Sb_{1000}, e) RF-classified image with Dem_{1000}, and f) RF-classified image with PCb_{300}. (please note that *shadow* and *no data* labels in the legend only apply for the Sen2Cor scene classification image).

The second close-up visual assessment is performed on a spatial subset from Kackar taken on 8 April 2018 (cf. Figure 7.5). In this specific subset, the cloud contamination is apparent, and cloud-casted shadows are observed over land and snow-covered areas (cf. Figure 7.5a). Most of the land and snow-covered areas under the cloud shadows are identified as water and shadow in the Sen2Cor-generated image (cf. Figure 7.5b). Mislabelling of pixels at the land-snow boundary as cloud is also observed, as in Figure 7.4b. For the Sb_{300} input combination (cf. Figure 7.5c), it is seen that the rate of misclassified land and snow pixels as water is notably reduced, and increase in the size of the training data further improves the performance (cf. Figure 7.5d). However, it should also be noted that the snow-covered areas under the cloud shadows are misclassified as land for Sb_{300} and Sb_{1000} input combinations. Even though Dem_{1000} input parameter set seems to remove the water labeling of land and snow pixels (cf. Figure 7.5e), the underestimation of snow under cloud shadows still remains a problem. When it comes to the PCb_{300} combination (cf. Figure 7.5f), it successfully labels both land and snow pixels under the cloud shadows.

7.4 Conclusion

Snow cover is a critical component of the Earth's climate system that plays a crucial role in regulating the planet's energy balance and hydrological cycle. Snow reflects incoming solar radiation, which helps to cool the surface, and acts as a reservoir for water that is gradually released as it melts. Changes in the spatiotemporal extent of snow cover have significant impacts on both regional and global scales, affecting water resources, agricultural productivity, and natural disasters.

Overall, snow cover is a critical component of the Earth's climate system, with far-reaching implications for both natural and human systems. As climate change continues to intensify, it is increasingly essential to understand the complex interactions between snow cover and other components of the climate system, to predict and mitigate the impacts of climate change. Detecting snow cover using remote sensing techniques is crucial as it provides a cost-effective and convenient alternative to other methods currently used for mapping snow cover at regional and global scales.

Satellite observations have become an indispensable tool for monitoring seasonal snow in mountainous regions worldwide, owing to the significant spatial and temporal variations of snow cover. These observations provide comprehensive and detailed information, which complements in-situ measurements and model simulations. Although classical snow mapping methods and ML techniques have their pros and cons, the latter usually offers higher precision and adaptability. However, the implementation of ML-based approaches demands more computational resources and data.

The present study is designed to evaluate the efficacy of the RF algorithm in detecting snow cover by leveraging various input combinations of Sentinel-2 bands, normalized band indices, principal component bands, and DEM data while varying the number of training samples per class. The assessment is conducted across three distinct regions with disparate climatic characteristics, encompassing three different phases of snow cover spanning from formation, peak, and melting phases. The primary objective is to gauge the performance of the RF algorithm in detecting snow cover over heterogeneous and complex mountainous terrains.

The investigation reveals that a larger training sample size of 1000 yields superior accuracy levels as compared to a smaller sample size of 300, across all input combinations. These findings suggest that the performance of the RF algorithm is subject to variations in the number of training samples employed for the classification of remote sensing data. The observed sensitivity of the RF algorithm to the size of training samples underscores the importance of optimal sample selection to ensure the reliability of the classification outcome.

The input combination obtained by PCA, encompassing the first three principal components generated from the Sentinel-2 bands combined with the three normalized band indices, emerges as the most accurate input combination overall. The PCA input combination demonstrates remarkable efficacy in detecting snow cover in regions with shadow cast from topography and on the fringes of slopes. Moreover, the PCA input combination stands out by requiring fewer input parameters than alternative combinations. The reduced input requirement of the PCA combination greatly benefits future RF classification exercises, as it decreases the RF algorithm's runtime.

Accurate snow mapping over forested areas is particularly important, as forests can significantly affect the snowpack by intercepting snowfall and reducing the amount of snow that reaches the ground. On the other hand, snow cover detection over forested areas using optical remote sensing techniques presents several challenges. Vegetation cover, canopy shadowing, and mixed pixels with overlapping signals from snow, vegetation, and soil, all pose significant obstacles in accurately identifying snow cover over forested regions. Moreover, the complexity of forested landscapes and their diverse topographic features over the areas with rough topography further exacerbate the difficulties in snow cover detection.

Therefore, it is important to continue our research efforts in snow mapping over complex forested mountainous areas to improve the accuracy and reliability of the satellite-derived snow cover products. This can be achieved through improved data sources, more advanced modeling techniques, and a better understanding of the interaction between snow, trees, and the ground. The development of ML algorithms and AI can also help overcome these challenges associated with the detection of snow-covered forest areas, especially during periods of cloud cover. In addition, interdisciplinary collaborations between remote sensing experts, snow hydrologists, and forest ecologists can help advance our knowledge in this field and lead to more accurate and reliable snow cover products.

References

[1] Basang, D., Barthel, K. and Olseth, J.A. 2017. Satellite and ground observations of snow cover in Tibet during 2001–2015. *Remote Sensing*, 9(11): 1201.

[2] Belgiu, M. and Drăguţ, L. 2016. Random forest in remote sensing: A review of applications and future directions. *ISPRS Journal of Photogrammetry and Remote Sensing*, 114: 24–31.

[3] Biau, G. and Scornet, E. 2016. A random forest guided tour. *Test*, 25(2): 197–227.

[4] Breiman, L. 1996. Bagging predictors. *Machine Learning*, 24: 123–140.

[5] Breiman, L. 2001. Random forests. *Machine Learning*, 45: 5–32.

[6] Bühler, Y., Christen, M., Kowalski, J. and Bartelt, P. 2011. Sensitivity of snow avalanche simulations to digital elevation model quality and resolution. *Annals of Glaciology*, 52(58): 72–80.

[7] Cohen, J. and Rind, D. 1991. The effect of snow cover on the climate. *Journal of Climate*, 4(7): 689–706.

[8] Congalton, R.G. and Green, K. 2009. *Assessing the Accuracy of Remotely Sensed Data: Principles and Practices*. CRC Press, Boca Raton, FL, 2nd Edition.

[9] Cutler, A., Cutler, D.R. and Stevens, J.R. 2012. *Random Forests*, pp. 157–175. Springer US, Boston, MA.

[10] Dozier, J. 1989. Spectral signature of alpine snow cover from the Landsat thematic mapper. *Remote Sensing of Environment*, 28: 9–22.

[11] Drusch, M., Del Bello, U., Carlier, S., Colin, O., Fernandez, V., Gascon, F., Hoersch, B., Isola, C., Laberinti, P., Martimort, P., Meygret, A., Spoto, F., Sy, O., Marchese, F. and Bargellini, P. 2012. Sentinel-2: ESA's optical high-resolution mission for GMES operational services. *Remote Sensing of Environment*, 120: 25–36.

[12] Gascoin, S., Grizonnet, M., Bouchet, M., Salgues, G. and Hagolle, O. 2019. Theia Snow collection: High-resolution operational snow cover maps from Sentinel-2 and Landsat-8 data. *Earth System Science Data*, 11(2): 493–514.

[13] Gevaert, C.M. 2022. Explainable AI for earth observation: A review including societal and regulatory perspectives. *International Journal of Applied Earth Observation and Geoinformation*, 112: 102869.

[14] Ghasemian, N. and Akhoondzadeh, M. 2018. Introducing two Random Forest based methods for cloud detection in remote sensing images. *Advances in Space Research*, 62(2): 288–303.

[15] Hall, D.K., Foster, J.L., Salomonson, V.V., Klein, A.G. and Chien, J.Y.L. 2001. Development of a technique to assess snow-cover mapping errors from space. *IEEE Transactions on Geoscience and Remote Sensing*, 39(2): 432–438.

[16] Hall, D.K., Riggs, G.A. and Salomonson, V.V. 1995. Development of methods for mapping global snow cover using moderate resolution imaging spectroradiometer data. *Remote Sensing of Environment*, 54: 127–140.

[17] Hastie, T., Tibshirani, R. and Friedman, J. 2009. *The Elements of Statistical Learning: Data Mining, Inference, and Prediction*. Springer Series in Statistics. Springer, NY, USA, 2nd edition.

[18] Hou, J., Huang, C., Zhang, Y. and Guo, J. 2020. On the value of available MODIS and Landsat8 OLI image pairs for MODIS fractional snow cover mapping based on an artificial neural network. *IEEE Transactions on Geoscience and Remote Sensing*, 58(6): 4319–4334.

[19] Jensen, J.R. 1996. *Introductory Digital Image Processing: A Remote Sensing Perspective*. Number Ed. 2. Prentice-Hall Inc.

[20] Jin, Y., Liu, X., Chen, Y. and Liang, X. 2018. Land-cover mapping using Random Forest classification and incorporating NDVI time-series and texture: A case study of central Shandong. *International Journal of Remote Sensing*, 39: 8703–8723.

[21] Jolliffe, I.T. 2002. *Principal Component Analysis*. Springer, New York.

[22] Karpatne, A., Ebert-Uphoff, I., Ravela, S., Babaie, H.A. and Kumar, V. 2019. Machine learning for the geosciences: Challenges and Opportunities. *IEEE Transactions on Knowledge and Data Engineering*, 31(8): 1544–1554.

[23] Klein, A.G., Hall, D.K. and Riggs, G.A. 1998. Improving snow cover mapping in forests through the use of a canopy reflectance model. *Hydrological Processes*, 12(10-11): 1723–1744.

[24] Kuter, S. 2021. Completing the machine learning saga in fractional snow cover estimation from MODIS Terra reflectance data: Random forests versus support vector regression. *Remote Sensing of Environment*, 255: 112294.

[25] Kuter, S., Akyurek, Z. and Weber, G.-W. 2018. Retrieval of fractional snow covered area from MODIS data by multivariate adaptive regression splines. *Remote Sensing of Environment*, 205: 236–252.

[26] Kuter, S., Bolat, K. and Akyurek, Z. 2022. A machine learning-based accuracy enhancement on EUMETSAT H-SAF H35 effective snow-covered area product. *Remote Sensing of Environment*, 272: 112947.

[27] Lary, D.J., Alavi, A.H., Gandomi, A.H. and Walker, A.L. 2016. Machine learning in geosciences and remote sensing. *Geoscience Frontiers*, 7(1): 3–10.

[28] Li, C., Wang, J., Wang, L., Hu, L. and Gong, P. 2014. Comparison of classification algorithms and training sample sizes in urban land classification with Landsat thematic mapper imagery. *Remote Sensing*, 6(2): 964–983.

[29] Liu, C., Huang, X., Li, X. and Liang, T. 2020. MODIS fractional snow cover mapping using machine learning technology in amountainous area. *Remote Sensing*, 12(6): 962.

[30] Liu, Y., Peters-Lidard, C.D., Kumar, S., Foster, J.L., Shaw, M., Tian, Y. and Fall, G.M. 2013 . Assimilating satellite-based snow depth and snow cover products for improving snow predictions in Alaska. *Advances in Water Resources*, 54: 208–227.

[31] Luo, J., Dong, C., Lin, K., Chen, X., Zhao, L. and Menzel, L. 2022. Mapping snow cover in forests using optical remote sensing, machine learning and timelapse photography. *Remote Sensing of Environment*, 275: 113017.

[32] Main-Knorn, M., Pflug, B., Louis, J., Debaecker, V., Müller-Wilm, U. and Gascon, F. 2017. Sen2Cor for sentinel-2. *In Image and Signal Processing for Remote Sensing XXIII*, volume 10427, pp. 37–48. SPIE.

[33] Mather, P.M. and Koch, M. 2011. *Computer Processing of Remotely-Sensed Images: An Introduction*. John Wiley and Sons, Ltd., West Sussex, UK.

[34] Maurer, E.P., Rhoads, J.D., Dubayah, R.O. and Lettenmaier, D.P. 2003. Evaluation of the snow-covered area data product from MODIS. *Hydrological Processes*, 17(1): 59–71.

[35] McFeeters, S.K. 1996. The use of the Normalized Difference Water Index (NDWI) in the delineation of open water features. *International Journal of Remote Sensing*, 17(7): 1425–1432.

[36] Metsämäki, S., Pulliainen, J., Salminen, M., Luojus, K., Wiesmann, A., Solberg, R., Böttcher, K., Hiltunen, M. and Ripper, E. 2015. Introduction to GlobSnow Snow Extent products with considerations for accuracy assessment. *Remote Sensing of Environment*, 156: 96–108.

[37] Ozdarici Ok, A. and Akyurek, Z. 2012. A segment-based approach to classify agricultural lands by using multi-temporal optical and microwave data. *International Journal of Remote Sensing*, 33(22): 7184–7204.

[38] Parajka, J. and Blöschl, G. 2006. Validation of MODIS snow cover images over Austria. *Hydrology and Earth System Sciences*, 10(5): 679–689.

[39] Riaño, D., Chuvieco, E., Salas, J. and Aguado, I. 2003. Assessment of different topographic corrections in Landsat-TM data for mapping vegetation types (2003). *IEEE Transactions on Geoscience and Remote Sensing*, 41(5): 1056–1061.

[40] Richter, R. and Schläpfer, D. 2005. Atmospheric/topographic correction for satellite imagery. *DLR report DLR-IB*, 565.

[41] Robinson, D.A. 1993. Monitoring northern hemisphere snow cover. *Glaciological Data*, pp. 1–25.

[42] Salcedo-Sanz, S., Ghamisi, P., Piles, M., Werner, M., Cuadra, L., Moreno-Martínez, A., Izquierdo-Verdiguier, E., Muñoz-Marí, J., Mosavi, A. and Camps-Valls, G. 2020. Machine learning information fusion in Earth observation: A comprehensive review of methods, applications and data sources. *Information Fusion*, 63: 256–272.

[43] Song, C., Woodcock, C., Seto, K., Lenney, M. and Macomber, S. 2000. Classification and change detection using Landsat TM Data: When and how to correct atmospheric effects? *Remote Sensing of Environment*, 75: 230–244, 11.

[44] Sun, Z., Sandoval, L., Crystal-Ornelas, R., Mousavi, S.M., Wang, J., Lin, C., Cristea, N., Tong, D., Carande, W.H., Ma, X., Rao, Y., Bednar, J.A., Tan, A., Wang, J., Purushotham, S., Gill, T.E., Chastang, J., Howard, D., Holt, B., Gangodagamage, C., Zhao, P., Rivas, P., Chester, Z., Orduz, J. and John, A. 2022. A review of Earth Artificial Intelligence. *Computers and Geosciences*, 159: 105034.

[45] Takala, M., Luojus, K., Pulliainen, J., Derksen, C., Lemmetyinen, J., Kärnä, J.-P., Koskinen, J. and Bojkov, B. 2011. Estimating northern hemisphere snow water equivalent for climate research through assimilation of space-borne radiometer data and ground-based measurements. *Remote Sensing of Environment*, 115(12): 3517–3529.

[46] Tsai, Y.S., Dietz, A., Oppelt, N. and Kuenzer, C. 2019. Wet and dry snow detection using Sentinel-1 SAR data for mountainous areas with a machine learning technique. *Remote Sensing*, 11(8): 895.

[47] Van Niel, T.G., McVicar, T.R. and Datt, B. 2005. On the relationship between training sample size and data dimensionality: Monte Carlo analysis of broadband multi-temporal classification. *Remote sensing of Environment*, 98(4): 468–480.

[48] Vanonckelen, S., Lhermitte, S. and Van Rompaey, A. 2013. The effect of atmospheric and topographic correction methods on land cover classification accuracy. *International Journal of Applied Earth Observation and Geoinformation*, 24: 9–21.

[49] Wang, X., Wu, C., Peng, D., Gonsamo, A. and Liu, Z. 2018. Snow cover phenology affects alpine vegetation growth dynamics on the Tibetan Plateau: Satellite observed evidence, impacts of different biomes, and climate drivers. *Agricultural and Forest Meteorology*, 256-257: 61–74.

Chapter 8

Construction of a Circular Graphical Network via an RNA-seq Data Set

Vilda Purutçuoğlu[1,*] and *Elif Doğan Dar*[2]

8.1 Introduction

Circular data contains directional information which represents a recurrent phenomenon that repeats in fixed time intervals. Such features can be observed in various fields such as geology ([76]), ecology ([71]), biology ([61]), and psychology ([49]). Since the traditional statistical theory does not take into account the circular nature of such data, all statistical theory has to be built from the ground up according to the topology of the circular data. Our aim in this work is to build network models for such circular data. Hence, in this study we are interested in interactions among circular variables which can be seen as a network structure. Nodes show variables and edges denote interactions among them. Indeed, networks are a big part of scientific literature in various fields where interactions

[1]Middle East Technical University, (METU, Turkey).

[2]Heidelberg University Hospital, Heidelberg, Germany.

* Corresponding author: vpurutcu@metu.edu.tr

among different entities are examined. However, the construction of networks by circular random variables has been worked on in only a few laboratories. Therefore, we aim to build a novel non-parametric circular graphical network based on circular regression.

Relations between circular variables are examined earlier via different approaches. Circular correlation coefficients are used to determine whether there is a relation/interaction between variables or not ([40, 24, 35, 25]). Some of these measures consist of rank-based coefficients as suggested by Fisher (1993) [25], and Fisher and Lee (1982) [24], correlations related to sine and cosine components of the variables as proposed by Jupp and Mardia (1981) [40], and a correlation coefficient is used by Jammalamadaka and Sarma (1988) [35] which is an analog of the Pearson's correlation in the circular setting. Unlike our model, these approaches take only the pairwise interactions into account, instead of looking at the cumulative effect of all variables at once. While considering networks, there needs to be a more holistic view where combined effects are taken into account. Unlike these previous studies, here, the effects of all variables in a network are inferred simultaneously by using a model-based method.

To investigate interactions between circular variables, multivariate circular distributions are also proposed. In these methods, the data are fitted to various distributions, and the independence is tested via parametric tests such as the likelihood ratio test and the Wald test ([60, 44]), whereas, there are certain drawbacks of these approaches. First of all, many of these distributions are bi-variate or tri-variate, and therefore, are not applicable to larger systems. For example, Singh et al. (2002) [85] propose a bi-variate circular distribution to explore the relation between dihedral angles of peptides. Similarly, Shieh and Johnson (2005) [84] apply the bivariate von Mises distribution to model the relation between wind directions at two different times of the day at the same place. Here, we can only investigate the relationship between two circular entities. In rare cases, multivariate distributions with a higher number of variables are utilized. For example, Mardia et al. (2008) [61] construct a multivariate von Mises distribution and apply it to dihedral angles of gamma turns on peptides. Their model can also be performed for models with more than two variables. However, strict parametric assumptions are made, which can be sub-optimal. Unlike these methods, we apply a non-parametric estimation method without any distributional assumptions, rendering the model more flexible.

We also see various circular regression models for the investigation of the interactions among circular variables. Earlier models consist of mixed models, i.e., regression of circular variables on linear variables and vice versa. Gould (1969) [30] regresses circular variables on linear variables and uses the von Mises distribution for the errors with a maximum likelihood approach. Laycock (1975) [51] implements trigonometric regression in such a way that a linear variable will be regressed on a circular variable. Mardia and Sutton (1978) [60] define an angular-linear joint distribution and apply the conditional distribution of a linear variable on the circular variable for the regression. Johnson and Wehrly (1978) [38] also define various angular-linear distributions and use conditional distributions and maximum likelihood estimation (MLE) for regression. They have also built some parametric tests for these models. However, they regress only one circular variable on one linear variable or they regress a linear variable on other linear and angular variables. Lund (1999) [58] applies the least circular distance (LCD) which is a circular alternative to the Euclidean least squares for the regression of a circular variable on linear variables and one circular variable. He also shows that LCD estimation gives the same results as the maximum likelihood estimation by accepting von Mises errors for the regression. He assumes the additive effect of linear covariates with a link function that maps the real line to the circle and Fourier series approximation for the circular regressor. He uses Akaike Information Criteria (AIC) for model selection and numerical methods for MLE estimation. He also calculates the square of a circular correlation measure which is applied to predicted and observed values resulting in a circular analog of the correlation coefficient R^2. When specific distributions are assumed for the errors, goodness of fit tests have to be applied to verify that the assumption is valid. Some goodness of fit statistics for von Mises distribution can be found in Lund (1999) [58] and Lockhart and Stephens (1985) [56]. Fisher and Lee (1982) [24] generalize an earlier model by using a link function that sends the real line to the circle in a monotone way and they solve the identifiability problem of earlier models.

There are also models where both dependent and independent variables are circular. Sarma and Jammamaladaka (1993) [79] build a circular regression model with one dependent and one independent circular variable, using polynomial models for the sine and cosine of the dependent variable. Since they estimate sine and cosine separately, they use the usual least squares estimation. Rivest (1997) [75] builds a regression of a circular variable on another but his model only takes some rotation into account. Lund (2002) [59] proposes a tree-based method for circular-circular regression with one regressor using binary trees. Downs and Mardia (2008) [61] regress one circular variable on another by mapping centralized half angles to the real line by tangent function, multiplying by a coefficient, and pulling it back to the circle using arc-tangent. They use a maximum likelihood estimation approach assuming errors which follow von Mises distribution. Rueda et al. (2016) [78] generalize this model by constructing a circular piece-wise regression model where each piece is established by using the model of Downs and Mardia. They apply their method to cell-cycle biology where modeling of multiple phases of the cell cycle cannot be done by one single form. They also construct a generalized Akaike Information Criteria (GAIC) which can be seen as an analog of AIC on the linear case. Kim et al. (2016) [45] also generalize the Downs and Mardia (2002) [17] model by adding an intercept parameter into the link function. They used asymmetric generalized von Mises distribution to model the error which is a more flexible distribution compared to earlier distributions used. Furthermore, they suggest three new distance-based methods where no parametric assumptions must be made. They define a new criterion, relative circular prediction bias (RCPB), to select the best estimator among the suggested ones. Kato et al. (2008) [41] use Möbius transformations to model circular-circular regression, Möbius transformation is mapping from the complex plane to itself in such a way that with some restrictions it can map the unit circle to itself in a one-to-one fashion. Their model resembles the Downs and Mardia model, however, they assume the wrapped Cauchy distribution for the error instead of the von Mises distribution which results in some desired properties. Kato and Jones (2010) [42] generalize their method by introducing a family of distributions which corresponds to the Möbius transformations on the circle. They use the Bayesian information criterion (BIC) as well as AIC for the model selection. Di Marzio et al. (2013) [16], however, apply non-parametric methods for the regression. Moreover, they perform local smoothing and avoid forcing a specific shape as a general link function. Alonso-Penaa (2021) [1] suggest non-parametric tests for many models in the literature. However, all these aforementioned models consider only one regressor. Therefore, they cannot be used for large systems.

Additionally, there are multiple circular regression models. The model that we consider using in this study is a multiple circular-circular regression model by Kim and SenGupta (2016) [44]. In this model, they generalize the model in SenGupta's earlier model (SenGupta, 2013 [80]) by including multiple regressors. They implement the circular version of the mean square error where they minimize the mean of the circular distances between estimated values and observations. Then, they prove the asymptotic normality of the suggested estimators for inference purposes.

Experimental methods are also used for network building ([93, 3]). However, these methods are time-consuming and require high amount of resources. Furthermore, false positive and false negative rates are high for many of them. Cheaper, holistic, and computational methods might guide experimental work for further validation and result in increased efficiency. For example, McDonald and Roshbash (2001) [62] used microarray data for this purpose. Yet, their method only accounts for fold change instead of a model-based approach. Accordingly, our model is a novel approach that constructs this circular network by using a model-based approach utilizing its circular nature. We will apply a circular version of the Gaussian Graphical Network (GGM) model [15]. In GGM, the nodes are assumed to be following a Gaussian distribution for the inference (Edwards, 2000), and zeros in the precision matrix correspond to the conditional independence of two nodes given others [20]. This problem can equivalently be formulated with all nodes regressed on others one by one, and coefficients of regressors give the strength of the conditional dependence between nodes. Many methods are established for the underlying inference problem including the coordinate descent for the maximization of the L1-penalized log-likelihood [97], lasso [63], blockwise coordinate descent [4] and graphical lasso [26]. However, all these models are built for linear data.

There are also some network models built for circular data. In a recent paper, Gottard and Panzera (2021) [29] build three separate graphical models based on different multivariate circular distributions. Their approach works under strict distributional assumptions. Similarly, Razavian, Kamisetty, and Langmead (2011) [74] build a graphical model where their network structure is assumed to follow multivariate von Mises distribution. In Boomsma et al. (2008) [6], authors build mixed (linear and directional data) probabilistic Hidden Markov Models (HMMs). In their model, the underlying mechanism should be chain-like since they use HMMs. Moreover, Leguey et al. (2019) [52] build a tree-structured Bayesian network model cleverly utilizing bi-variate wrapped Cauchy distributions. Here, the network needs to have a tree shape and all node couples should follow bivariate-wrapped Cauchy distribution. Unlike these models, our model does not assume any underlying distribution or network shape. Therefore, it is applicable to a wider range of problems. But, in this study, we show its application in gene expression data.

Hereby, in the following section, we initially introduce the interactions of circular variables and circadian gene networks. Then in Section 8.2 and 8.3, we describe our model and inference strategy. Finally, in Section 8.4, we present our application via a real data set, and in Section 8.5, we conclude our findings.

8.2 Circadian Gene Networks

There are different methods used by researchers to reveal interactions among circadian genes. Earlier work on this subject mostly consisted of experimental methods. Wallach et al. (2013) [93] use a yeast-two-hybrid approach where the interaction of each pair of proteins is tested in a yeast medium. However, this approach results in high quantities of false negatives and false positives. Two genes interacting in yeast medium might not interact in human cells or two proteins artificially put together might not encounter in the cell due to spatial limitations. Therefore, they try to validate co-immunoprecipitation experiments in human cells as well as the enrichment of the network from the literature. Baggs et al. (2009) [3] use a perturbation-based model where they knock down certain genes and analyze the expression of clock genes after the knockdown. But this method is also not immune to false negatives, because, usually in biological networks there are alternative paths that result in undetected interactions. These methods are computationally demanding and require a high amount of resources. False positive and false negative rates are high for many of them.

Therefore, computational methods and high-throughput data sets are also suggested in the literature to make a grid search smaller for experimental validation. Since experimental approaches are expensive ad computationally costly, there should be alternative approaches for building interactions by using high-throughput datasets. For example, McDonald and Roshbash (2001) [62] use microarray data for this purpose. They fit cosine curves and apply the cross-correlation coefficients and fold changes for the analysis. Yet, their method only accounts for the fold change instead of a model-based approach. Moreover, they treat data as linear although data is inherently circular in nature. We propose transformations of this linear data to circular data to utilize the circular nature of the data. To detect circadian genes and reveal their periodic patterns, sine waves have been used (Hickey et al. 1984 [33]), however, circadian patterns do not always follow sinusoidal waveforms. Luan and Li (2004) [57] used cubic B-splines instead, which give a more flexible fit compared to the sine waves. However, they need guide genes which are known as circadian for the implementation of their method. For non-sinusoidal form detection, also, Lomb–Scargle periodograms are used by Glynn et al. (2006) [28], the Laplace periodogram is applied by Liang et al. (2009) [55] and Bayesian methods have been suggested by Chudova et al. (2009) [12]. Unfortunately, it is not quite obvious how to transform RNA-seq data to angular data by using these methods, unlike traditional sinewave forms. Therefore, we proceed with the sinewave forms.

8.3 Circular Network Model and Inference

In this chapter, we use a novel non-parametric circular network model. We explain the estimation and inference steps in detail. Then, we apply our model to circadian gene interaction networks which are circular in nature.

8.3.1 Model

In this work, we use the multiple circular regression model defined by Kim and SenGupta (2016) [45] to describe the interactions. For each node, we build a separate regression model where that node, which is a gene or protein in the biological networks, is the dependent variable and the remaining nodes are independent variables. In this network, the regression coefficients which are statistically zero indicate that there is no edge among associated nodes. That is there is no interaction among the related genes or proteins. Other coefficients represent an edge between the dependent variable and the corresponding independent variable, i.e., support the existence of functional/physical interactions.

Considering that $\phi_1, ..., \phi_{p-1}$ are independent variables and θ is the dependent variable, our model can be represented as below:

$$\theta_i = \mu_\theta + 2\arctan\left(\alpha + \beta_1 \sin\left(\frac{(\phi_{1,i} - \mu_{\phi_1})}{2}\right) + \dots\right.$$
$$\left. + \beta_{p-1} \sin\left(\frac{\left(\phi_{p-1,i} - \mu_{\phi_{p-1}}\right)}{2}\right)\right) + \varepsilon_i$$

in which ε_i follows a circular distribution with a mean 0, $\alpha \in \mathbb{R}$ and, $\beta_1, ...\beta_{p-1} \in \mathbb{R}$ denote the intercept and slope parameters, respectively. Moreover, μ_ϕ denotes the centering constant of ϕ for $i = 1, 2, ..., n$ and μ_θ refers to the centering constant of θ. Additionally, $arctan(.)$ and $sin(.)$ represent the arctangent and sinus of the given values, in order. Lastly, n and p are the number of data points and the number of nodes in our system, respectively. Here, the values of circular random variables $\phi_1, ...\phi_{p-1}$ and θ lie within the range of $[-\pi, \pi)$. As a result, by dividing values by 2, we get unique values for sin. α in the model is put because otherwise, whenever $\phi_1 = \mu_{\phi_1}, ..., \phi_k = \mu_{\phi_k}$, we would get $\theta = \mu_\theta$. Using this regression model, we build p regressions where the dependent variable is one of the nodes at each time. In this expression, $\beta_1, ..., \beta_{p-1}$ give the strength of the interactions among the corresponding node θ and the remaining nodes $\phi_1, ...\phi_{p-1}$ in the network.

8.3.2 Estimation and Inference of Model Parameters

For the estimation of these parameters, the following circular mean-square error (CMSE), which is the circular version of the traditional nonparametric least squares estimation, is performed.

$$CMSE = E\left(1 - \cos\left(\widehat{\mu_0} - \mu_0\right)\right), \tag{8.1}$$

where μ_0 and $\widehat{\mu_0}$ are the circular parameter and its estimation, respectively. Moreover, $E(.)$ and $cos(.)$ show the expectation and cosine of the given function, in order. Here, CMSE$= 0$ when $\widehat{\mu_0} - \mu = 0$ and CMSE$= 2$ when $\widehat{\mu_0} - \mu = \pm\pi$. By minimizing sample CMSE via

$$Q\left(\alpha, \beta_{1,...,}\beta_{p-1}, \mu_\theta, \mu_{\phi_{1,...,}}, \mu_{\phi_{p-1}}\right) = \frac{1}{n}\sum_{i=1}^{n}\left[1 - \cos\left(\theta_i - \hat{\theta}_i\right)\right],$$

under

$$\hat{\theta}_i = \mu_\theta + 2\arctan\left(\alpha + \beta_1 \sin\left(\frac{(\phi_{1,i} - \mu_{\phi_1})}{2}\right) + \dots + \beta_{p-1} \sin\left(\frac{\left(\phi_{p-1,i} - \mu_{\phi_{p-1}}\right)}{2}\right)\right).$$

these resulting estimators are called the least circular mean square estimators (LCMSE) and for large n, they are asymptotically normal (Kim and Sengupta 2016). However since CMSE does not have a closed-form, the numerical solution of Equation 8.1 is found by an iterative algorithm like the Newton-Ruphson algorithm. For the inference, the asymptotic normality feature is suggested by Kim and SenGupta [44]. Thus, we implement the bootstrap method to augment data and then to estimate the variance-covariance matrix as well as regression coefficient.

In order to decide whether the correlation coefficients are significantly different than 0 or not, while regressing the node i on other nodes where $i = 1, 2, ..., p$, we test the null hypothesis $H_0 : \beta_j = 0$ where $j \in \{1, 2, \ldots, p\} - \{i\}$. As LCMSE are asymptotically normal, we have $\frac{\widehat{\beta}_j}{\mathrm{se}\left(\widehat{\beta}_j\right)} \sim N(0,1)$ while $se()$ shows the standard error of the given random variable. Then, by computing the bootstrap standard deviation as the standard error of the coefficients, we apply the z-test under the significance level $\alpha = 0.05$. Later, we calculate the Wald-test to test the null hypothesis $H_0 : \beta_x = \beta_j = 0$ where $j \in \{1, 2, \ldots, p\} - \{i, x\}$ given all other coefficients significant. Here, since $\left[\widehat{\beta}_x, \widehat{\beta}_j\right] [\mathrm{Cov}(\hat{\beta})]^{-1} \left[\widehat{\beta}_x, \widehat{\beta}_j\right]' \sim \chi_2^2$ in which $\mathrm{Cov}(.)$ and $(.)'$ indicate the variance-covariance matrix of the coefficients and the transpose of the given vector respectively. We can compare the calculated p-value with α. By repeating the same process for each node separately, and representing their significant regression coefficient as 1, accordingly, denoting a non-significant coefficient as 0, a binary $(0,1)$ adjacency matrix is generated. Since the matrix shows the neighborhood matrix of the network, it has to be symmetric and square. In order to get such a matrix, the regression coefficient constructed for each node is checked for significance. If the regression coefficient β_{ij} for the ith response gene and the jth predictor gene as well as β_{ji} for the jth response and ith predictor gene are significant, the ijth and the jith entries of the adjacency matrix will be both denoted as 1, and 0 otherwise.

8.4 Application by Circadian Gene Interactions

Here, we implement the proposed method on real data which is circadian gene interaction networks. Circadian genes are expressed in the cells in a recurrent manner with a period of approximately 24 hours ([13, 89]). Therefore, interactions among these genes naturally produce a circular network. Hence, in our study, the proposed method is used after the transformation of the linear RNA-seq data to the circular data. Accordingly, the circadian nature of the data is exploited.

8.4.1 Data Description

We use RNA-seq data of the gene expressions taken on a 24 or more hours time course with 2–3 hours intervals. The full network of circadian genes is large. For this reason, we pick subsets of the circadian genes and explore interactions within these sub-networks. We estimate the network structure by using this data with the proposed method and compare it with the real network from the literature.

There are many public databases including GEO (Edgar et al. 2002 [19]) from NCBI (National Center for Biotechnology Information) or ArrayExpress (Athar et al. 2019 [2]) from EBI (European Bioinformatics Institute) which store RNA-seq data from different experiments. The organ or tissue this data is obtained from has crucial importance on the upcoming steps of the investigations since circadian rhythms are highly organ-specific. We only model a subset of genes that are circadian in nature for the organ/tissue under investigation. We used CGDB (Circadian Gene DataBase) to detect these genes. CGDB (Li et al. 2017 [54]) is an online source containing circadian genes from different organisms including humans with the information of the organism, tissue/cell, and PubMed ID as well as the source of the evidence whether it is experimentally identified or predicted. We gather circadian genes by choosing the specific tissue/cell type the data is coming from and experimentally identified evidence sources. Since all variables are circular in our model, we choose only circadian genes of corresponding tissue/cell type from microarray data to begin with.

After determining the nodes of the network, we need to extract the interactions among them too. However, there are no databases that specifically give interactions among circadian genes only. For that reason, we use the STRING (Szklarczyk 2019 [90]) database which is a database of known and predicted protein interactions. We look only at interaction sources of curated databases, co-occurrence, and co-expression and use this network as the true network to compare with our model results.

Although circadian gene expressions are circular, microarray data is linear. To reveal circular structure, we transform these periodic expressions into circular data by using sinusoidal waves: expression levels taken at 2–3 hours time intervals can be fit by a sine wave (Hickey 1984 [33]) with a period of 24 hours and angles in the sine can be used as the corresponding angle at this point. After doing this for each gene and sample, we slice a time point and use these angles as our circular data.

Since circadian genes are organ-specific, data should come from one organ/tissue. To capture the circadian nature, we need consecutive time points, i.e., measurements every 2 hours and we need data for at least one day. We searched databases GEO from NCBI and ArrayExpress from EBI for this purpose. We find data set "GSE113883" from GEO-NCBI from work titled "TimeSignature: A Universal Method for Robust Detection of Circadian State from Gene Expression" (Braun et al. 2018 [7]). Whole blood transcriptional profiles of 11 individuals with intermediate circadian phenotype, were generated by RNA sequencing, using Illumina NextGen 500. Whole blood was collected every 2 hours over 28 hours (15 time-points), yielding a total of 165 samples, of which 153 passed the quality assurance and underwent further analysis. This data contains the whole profile but we need only circadian genes expressed in that specific cell. Hence, afterward, we gather known circadian genes from the literature. We gather circadian genes from CGDB by choosing blood tissue/cell type from humans with experimentally identified evidence. There are 55 genes in total in this database for this search. After gathering circadian gene information, we need to gather interaction information. For this purpose, we use the STRING (Search Tool for Recurring Instances of Neighbouring Genes) database [90]. We ended up with a network of 55 genes shown in Figure 8.1.

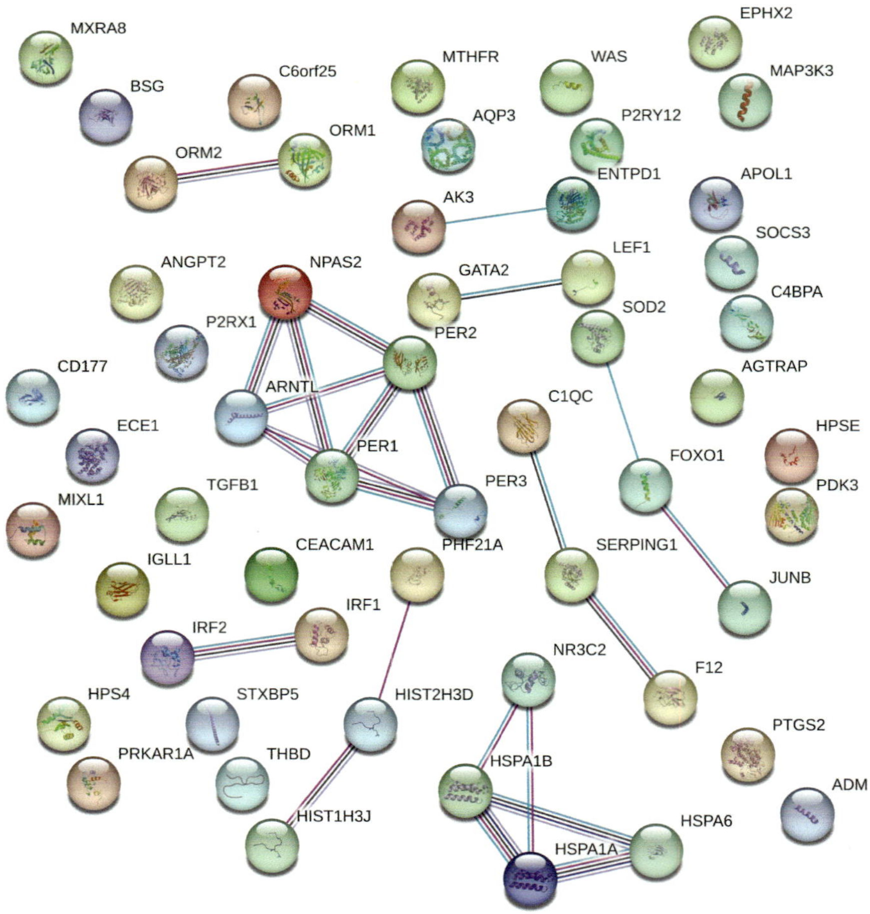

Figure 8.1: Network structure with all circadian proteins from the STRING database.

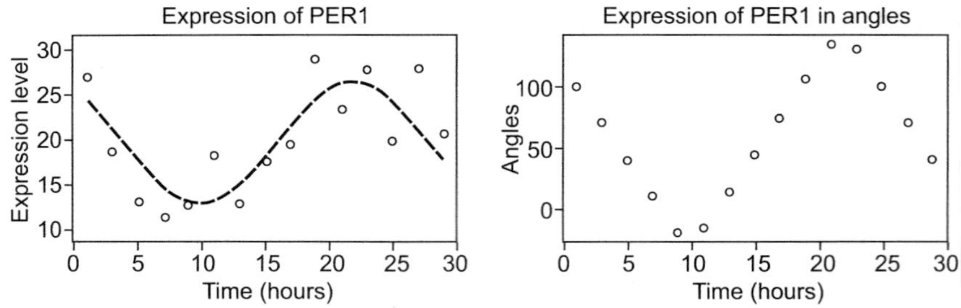

Figure 8.2: Sine wave fitting for the gene PER1.

We have sinusoidal fits for each gene and we slice data at a given time point. Then after pulling points to the sine wave, we use arcsin of the values to transfer them to the unit circle. An example of this transforming procedure can be seen in Figure 8.2 and an example of circular gene expression values after transformation can be seen in Table 8.1. We check the p-values for the fits. Furthermore, from the tabulated values, it is observed that some of the genes have multiple repeated entries. This stems from the fact that expressions of these genes are tightly controlled. Some of them are having a peak at a similar time of the day. Therefore, when we make this circular transformation, we end up with some repeated measures. An example of this can be seen in Figure 8.3.

So, we end up with genes with reasonable sine wave fits. Also, these genes have varying expression patterns, hence, our method is applicable. As can be seen in Figure 8.4, our algorithm caught the existent edge while mistakenly claiming one additional edge.

Table 8.1: Circular gene expressions after transformation.

	sample1	sample2	sample3	sample5	sample7	sample8	sample9	sample10	sample11
F12	43.178	51.128	13.007	65.247	1.828	-68.116	74.816	-72.993	-16.830
SERPING1	37.146	75.000	75.000	75.000	75.000	75.000	75.000	75.000	70.157
ORM1	-25.924	75.000	34.881	75.000	-5.284	75.000	-105.000	53.123	75.000
ORM2	51.277	-55.572	4.142	44.096	23.785	88.484	36.998	47.176	33.505
HPSE	75.000	-3.368	-47.186	56.621	13.157	-5.631	-105.000	75.000	13.009
TGFB1	-4.703	75.000	75.000	75.000	75.000	75.000	75.000	75.000	75.000
C1QC	-60.627	75.000	75.000	75.000	25.213	66.438	-9.486	-38.032	-36.225
PER1	-105.000	-105.000	32.872	-24.309	-105.000	-105.000	55.373	-105.000	3.439
PER2	-34.371	72.032	14.747	75.000	6.711	-22.940	75.000	12.928	-26.055
PER3	40.602	-42.646	-56.661	-97.019	-64.938	-1.422	81.331	-10.272	-61.368
ARNTL	75.000	61.619	51.283	70.619	30.103	75.000	75.000	-43.379	75.000
NPAS2	74.496	-105.000	45.167	75.000	74.048	75.000	75.000	74.979	75.000

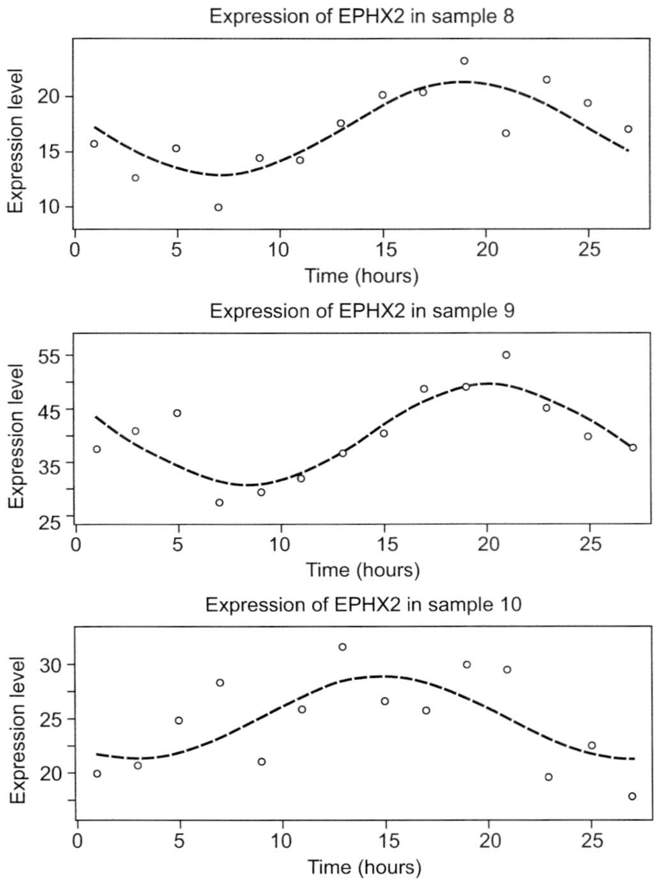

Figure 8.3: Sine wave fitting of the gene EPHX2 for samples 8, 9 and 10.

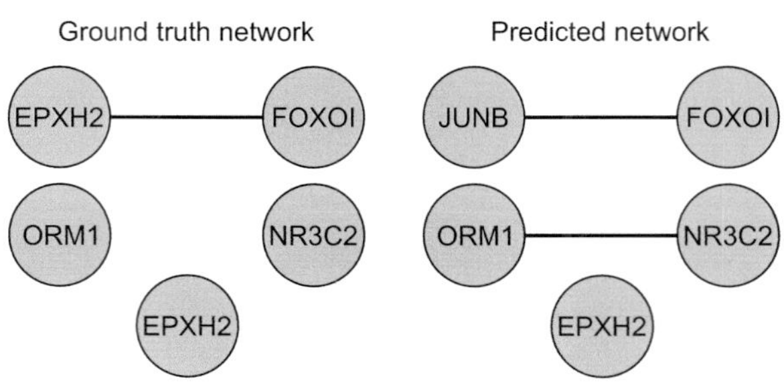

Figure 8.4: An example of circadian sub-network. Our method detected the existent edge correctly while mistakenly claiming one additional edge.

8.5 Conclusion

In this study, we constructed a novel non-parametric circular graphical network model. Inference of the model parameters were done via least circular mean square estimation. To utilize the asymptotic normality assumptions for the large sample size [44], we have initially augmented the data by bootstrapping to estimate the standard error of the coefficients, and deduced the significance of the regression coefficients for the graphical model. We have applied the constructed model to a gene expression data set to evaluate the performance of the proposed approach. The results have shown success in accuracy. However, the model currently only works for smaller systems. Additionally, strictly controlled circadian genes, which follows the same form each day, causes multicollinearity problem in the regressions occasionally. As an extension of this study, we think of improving the searching time of the optimal regression coefficients, specifically for the high-dimensional networks. Also, utilizing the heights of the waves in addition to the current transformation will be considered to solve the occasional multicollinearity problem. Furthermore, we consider developing a parametric inference, rather than circular least square estimators for fitting a graphical network model, since in the current inference, we cannot use the distributional knowledge of data unless the sample size is augmented extensively by bootstrapping.

Acknowledgement

Vilda Purutçuoğlu would like to thank Huawei Turkey R&D center for their support for circular data analysis and the METU research project (No:11201) for financial support.

References

[1] Alonso-Pena, M., Ameijeiras-Alonso, J. and Crujeiras, R.M. 2021. Non-parametric tests for circular regression. *Journal of Statistical Computation and Simulation*, 91(3): 477–500.

[2] Athar, A., Füllgrabe, A., George, N., Iqbal, H., Huerta, L., Ali, A., Snow, C., Fonseca, N.A., Petryszak, R., Papatheodorou, I., Sarkans, U. and Brazma, A. 2019. ArrayExpress update – from bulk to single-cell expression data. *Nucleic Acids Research*, 47(D1): D711–D715.

[3] Baggs, J.E., Price, T.S., DiTacchio, L., Satchidananda, P., FitzGerald, G.A. and Hogenesch, J.B. 2009. Network features of the Mammalian Circadian clock. *PLoS Biol.*, 7(3): 1–13.

[4] Banerjee, S., Li, Y., Wang, Z. and Sarkar, F.H. 2008. Multi-targeted therapy of cancer by genistein. *Cancer Letters*, 269(2): 226–242.

[5] Bezdek, J. 1981. *Pattern Recognition with Fuzzy Objective Function Algorithms*. Plenum Press, New York.

[6] Boomsma, W., Mardia, K.V., Taylor, C.C., Ferkinghoff-Borg, J., Krogh, A. and Hamelryck, T. 2008. A generative, probabilistic model of local protein structure. *Proceedings of the National Academy of Sciences of the United States of America*, 105(26): 8932–8937.

[7] Braun, R., Kath, W.L., Iwanaszko, M., Kula-Eversole, E., Abbott, S.M., Reid, K.J., Zee, P.C. and Allada, R. 2018. Universal method for robust detection of circadian state from gene expression. *Proceedings of the National Academy of Sciences of the United States of America*, 115(39): E9247–E9256.

[8] Cai, Y.D. and Chou, K.C. 1998. Artificial neural network model for predicting HIV protease cleavage sites in protein. *Adv. Eng. Softw.*, 29(2): 119–128.

[9] Cai, Y., Liu, X., Xu, X. and Chou, K. 2002. Support vector machines for predicting HIV protease cleavage sites in protein. *J. Comput. Chem.*, 23: 267–274.

[10] Chen, B., Ding, Y. and Wild, D.J. 2012. Assessing drug target association using semantic linked data. *PLoS Comput. Biol.*, 8(7).

[11] Chormungea, S. and Jenab, S. 2018. Correlation based feature selection with clustering for high dimensional data. *J. Electr. Syst. Inf. Technol.*

[12] Chudova, D., Ihler, A., Lin, K.K., Andersen, B. and Smyth, P. 2009. Bayesian detection of non-sinusoidal periodic patterns in circadian expression data. *Bioinformatics*, 25(23): 3114–3120.

[13] Czeisler, C.A., Duffy, J.F., Shanahan, T.L., Brown, E.N., Mitchell, J.F., Rimmer, D.W., Ronda, J., Silva, E.J., Allan, J.S., Emens, J.S., Dijk, D.R. and Kronauer, R.E. 1999. Stability, precision, and near-24-hour period of the human circadian pacemaker. *Science*, 284: 2177–2181.

[14] Dar, E.D., Purutçuoğlu, V. and Purutçuoğlu, E. 2020. Detection of HIV-1 protease cleavage sites via hidden markov model and physicochemical properties of amino acids. pp. 171–193. *In*: Machado, J., Özdemir, N. and Baleanu, D. (eds.). *Numerical Solutions of Realistic Nonlinear Phenomena. Nonlinear Systems and Complexity*, vol 31. Springer, Cham.

[15] Dempster, A.P. 1972. Covariance selection. *Biometrics*, 157–175.

[16] Di Marzio, M., Panzera, A. and Taylor, C.C. 2013. Non-parametric regression for circular responses. *Scandinavian Journal of Statistics*, 40(2): 238–255.

[17] Downs, T.D. and Mardia, K.V. 2002. Circular regression. *Biometrica*, 89: 683–697.

[18] Durbin, R., Eddy, S., Krogh, A. and Mitchison, G. 1998. *Biological Sequence Analysis*. Cambridge, UK.

[19] Edgar, R., Domrachev, M. and Lash, A.E. 2002. Gene expression Omnibus: NCBI gene expression and hybridization array data repository. *Nucleic Acids Research*, 30(1): 207–210.

[20] Edwards, T. 2000. *Contradictions of Consumption: Concepts, Practices, and Politics in Consumer Society*. Buckingham: Open University Press, 179.

[21] Eisner, J. 2002. An interactive spreadsheet for teaching the forward-backward algorithm. *Proc. of the ACL Workshop on Effective Tools and Methodologies for Teaching NLP and CL*, 10–18.

[22] Faber, J. and Fonseca, L. 2014. How sample size influences research outcomes. *Dental Press Journal of Orthodontics*, 19(4): 27–29.

[23] Fan, J., Feng Yang, and Wu, Y. 2009. Network exploration via the adaptive LASSO and SCAD penalties. *Ann. Appl. Stat.*, 3(2): 521–541.

[24] Fisher, N.I. and Lee, A.J. 1982. Nonparametric measures of angular-angular association. *Biometrika*, 69: 315–321.

[25] Fisher, N.I. 1993. *Statistical Analysis of Circular Data*. Cambridge University Press.

[26] Friedman, J., Hastie, T. and Tibshirani, R. 2008. Sparse inverse covariance estimation with the graphical lasso. *Biostatistics Jul.*, 9(3): 432–441.

[27] Gallo, R.C., Salahuddin, S.Z., Popovic, M., Shearer, G.M., Kaplan, M., Haynes, B.F., Palker, T.J., Redfield, R., Oleske, J., Safai, B., White, Cl., Foster, P. and Markham, P.D. 1984. Frequent detect on and isolation of cytopathic retroviruses (HTLV-III) from patients with AIDS and at risk for AIDS. *Science*, 224(4648): 500–503.

[28] Glynn, E.F., Chen, J. and Mushegian, A.R. 2006. Detecting periodic patterns in unevenly spaced gene expression time series using Lomb-Scargle periodograms. *Bioinformatics*, 22(3): 310–316.

[29] Gottard, A. and Panzera, A. 2021. Graphical models for circular variables.

[30] Gould, A.L. 1969. A regression technique for angular variates. *Biometrics*, 25(4): 683–700.

[31] Gustafson, D.E. and Kessel, W.C. 1978. Fuzzy clustering with a fuzzy covariance matrix. *Proc. IEEE CDC*, 761–766.

[32] Hartigan, J.A. and Wong, M.A. 1979. Algorithm AS 136: A K-means clustering algorithm. *J. Royal Stat. Soc. Series C (Applied Statistics)*, 28: 100–108.

[33] Hickey, D.S., Kirkland, J.L., Lucas, S.B. and Lye, M. 1984. Analysis of circadian rhythms by fitting a least squares sine curve. *Comput. Biol. Med.*, 14(2): 217–223.

[34] Jaeger, S. and Chen, S.-S. 2010. Information fusion for biological prediction. *J. Data Sci.*, 8: 269–288.

[35] Jammalamadaka, S.R. and Sarma, Y.R. 1988. A correlation coefficient for angular variables. *Stat. Theory Data Anal. II (Ed. Matusita)*, 115(39): 349–364.

[36] Jayavardhana Rama, G.L. and Palaniswami, M. 2005. Cleavage knowledge extraction in HIV-1 protease using hidden Markov model. *Proc. 2nd International Conference on Intelligent Sensing and Information Processing*, 469–473.

[37] Jianying, H., Brown, M.K. and Turin, W. 1996. HMM based online handwriting recognition. *IEEE Trans. Pattern Anal. Mach. Intell.*, 18: 1039–1045.

[38] Johnson, R.A. andWehrly, T.E. 1978. Some angular-linear distributions and related regression models. *Journal of the American Statistical Association*, 73(363): 602–606.

[39] Juang, B. and Rabiner, L. 1991. Hidden Markov models for speech recognition. *Technometrics*, 33(3): 251–272.

[40] Jupp, P.E. and Mardia, K.V. 1981. A general correlation coefficient for directional data and related regression problems. *Biometrika*, 68: 738.

[41] Kato, S., Shimizu, K. and Shieh, G. 2008. A circular-circular regression model. *Statistica Sinica*, 18: 633–645.

[42] Kato, S. and Jones, M.C. 2010. A family of distributions on the circle with links to, and applications arising from, Möbius transformation. *Journal of the American Statistical Association*, 105(489): 249–262.

[43] Kim, G., Kim, Y., Lim, H. and Kim, H. 2010. An MLP-based feature subset selection for HIV-1 protease cleavage site analysis. *Artif. Intell. Med.*, 48: 83–89.

[44] Kim, S., SenGupta, A. and Arnold, B. 2016. A multivariate circular distribution with applications to the protein structure prediction problem. *Journal of Multivariate Analysis*, 143.

[45] Kim, S. and Sengupta, A. 2016. Multivariate-multiple circular regression. *Journal of Statistical Computation and Simulation*, 87: 1–15.

[46] Kohl, N.E., Emini, E.A., Schlief, W.A., Davis, L.J., Heimbach, J., Dixon, R.A.F., Scolnik, E.M. and Sigal, I.S. 1988. Active human immunodeficiency virus protease is required for viral infectivity. *Proc. Nutl. Sci. USA.*, 85(15): 4686–4690.

[47] Kouemou, G.L. 2011. History and theoretical basics of Hidden Markov Models. *In: (Ed.). Hidden Markov Models Theory and Applications, IntechOpen.*

[48] Krishnapuram, R., Joshi, A., Nasraoui, O. and Yi, L. 2001. Low-complexity fuzzy relational clustering algorithms for Web mining. *IEEE Trans. Fuzzy Syst.*, 9(4): 595–607.

[49] Kubiak, T. and Jonas, C. 2007. Applying circular statistics to the analysis of monitoring data: Patterns of social interactions and mood. *European Journal of Psychological Assessment*, 23: 227–237.

[50] Kurz, G., Gilitschenski, I. and Hanebeck, U. 2014. Efficient evaluation of the probability density function of a wrapped normal distribution. *Proceedings of IEEE ISIFWorkshop on Sensor Data Fusion: Trends, Solutions, Applications (SDF 2014).*

[51] Laycock, P.J. 1975. Optimal design: Regression models for directions. *Biometrika*, 62(2): 305–311.

[52] Leguey, I., Larrañaga, P., Bielza, C. and Kato, S. 2019. A circular-linear dependence measure under Johnson-Wehrly distributions and its application in Bayesian networks. *Information Sciences*, 486: 240–253.

[53] Li, H. and Gui, J. 2006. Gradient directed regularization for sparse Gaussian concentration graphs, with applications to inference of genetic networks. *Biostatistics*, 7(2): 302–317.

[54] Li, S., Shui, K., Zhang, Y., Lv, Y., Deng, W., Ullah, S., Zhang, L. and Xue, Y. 2017. CGDB: A database of circadian genes in eukaryotes. *Nucleic Acids Research*, 45(D1): D397–D403.

[55] Liang, K.C., Wang, X. and Li, T.H. 2009. Robust discovery of periodically expressed genes using the Laplace periodogram. *BMC Bioinformatics*, 10(15).

[56] Lockhart, R.A. and Stephens, M.A. 1985. Tests of fit for the von Mises distribution. *Biometrika*, 72.3: 647–652.

[57] Luan, Y. and Li, H. 2004. Model-based methods for identifying periodically expressed genes based on time course microarray gene expression data. *Bioinformatics*, 20(3): 332–339.

[58] Lund, U. 1999. Least circular distance regression for directional data. *Journal of Applied Statistics*, 26(6): 723–733.

[59] Lund, U.J. 2002. Tree-based regression for a circular response. *Communications in Statistics-Theory and Methods*, 31: 1549–1560.

[60] Mardia, K. and Sutton, T. 1978. A model for cylindrical variables with applications. *Journal of the Royal Statistical Society: Series B (Methodological)*, 40: 229–233.

[61] Mardia, K.V., Hughes, G., Taylor, C.C. and Singh, H. 2008. A multivariate von Mises distribution with applications to bioinformatics. *Canadian Journal of Statistics*, 36: 99–109.

[62] McDonald, A.S. 2001. The prevalence and effects of test anxiety in school children. *Educational Psychology*, 21(1): 89–101.

[63] Meinshausen, N. and Bühlmann, P. 2006. High-dimensional graphs and variable selection with the Lasso. *The Annals of Statistics*, 34(3): 1436–1462.

[64] Miller, M., Schneider, J., Sathyanarayana, B.K., Toth, M.V., Marshall, G.R., Clawson, L., Selk, L., Kent, S.B.H. and Wlodawer, A. 1989. Structure of complex of synthetic HIV-l protease with a substrate-based inhibitor at 2.3 A resolution. *Science*, 246: 1149–1152.

[65] Mitra, P., Murthy, C.A. and Pal, S.K. 2002. Unsupervised feature selection using feature similarity. *IEEE Trans. Pattern Anal. Mach. Intell.*, 24(3): 301–312.

[66] Munch, K. and Krogh, A. 2006. Automatic generation of gene finders for eukaryotic species. *BMC Bioinform.*, 7: 263.

[67] Murtagh, F. 1985. *Multidimensional Clustering Algorithms*. Physica-Verlag.

[68] Nakai, K., Kidera, A. and Kanehisa, M. 1988. Cluster analysis of amino acid indices for prediction of protein structure and function. *Protein Eng.*, 2: 93–100.

[69] Nanni, L. 2006. Comparison among feature extraction methods for HIV-1 protease cleavage site prediction. *Pattern Recognit.*, 39(4): 711–713.

[70] Niu, B., Yuan, X.C., Roeper, P., Su, Q., Peng, C.R., Yin, J.Y., Ding, J., Li, H. and Lu, W.C. 2013. HIV-1 protease cleavage site prediction based on two-stage feature selection method. *Protein Pept. Lett.*, 20: 290–298.

[71] Otieno, S. and Anderson-Cook, C.M. 2006. Measures of preferred direction for environmental and ecological circular data. *Environmental and Ecological Statistics*, 13: 311–324.

[72] Pachter, L., Alexandersson, M. and Cawley, S. 2002. Applications of generalized pair hidden Markov models to alignment and gene finding problems. *J. Comput. Biol.*, 9: 389–399.

[73] Park, H. and Jun, C. 2009. A simple and fast algorithm for k-medoids clustering. *Expert Syst. Appl.*, 36: 3336–3341.

[74] Razavian, N.S., Kamisetty, H. and Langmead, C.J. 2011. *The von mises Graphical Model: Structure Learning*. Technical Report CMU-CS-11-108, Carnegie Mellon University.

[75] Rivest, L.-P. 1997. A decentred predictor for ciruclar-circular regression. *Biometrika*, 84(3): 717–726.

[76] Roberts, N.M., Tikoff, B., Davis, J.R. and Stetson-Lee, T. 2019. The utility of statistical analysis in structural geology. *Journal of Structural Geology*, 125: 64–73.

[77] Rögnvaldsson, T., You, L. and Garwicz, D. 2015. State of the art prediction of HIV-1 protease cleavage sites. *Bioinformatics*, 31: 1204–1210.

[78] Rueda, C., Fernández, M.lA., Barragán, S., Mardia, K.V. and Peddada, S.D. 2016. Circular piecewise regression with applications to cell-cycle data. *Biometrics Dec.*, 72(4): 1266–1274.

[79] Sarma, Y. and Jammalamadaka, S. 1993. Circular regression. Statistical science and data analysis. *Proceedings of the Third Pacific Area Statistical Conference*, 109–128.

[80] Sengupta, P. 2013. The laboratory rat: Relating its age with human's. *International Journal of Preventive Medicine*, 4(6): 624.

[81] Schäfer, J. and Strimmer, K. 2005. An empirical Bayes approach to inferring large-scale gene association networks. *Bioinformatics*, 21(6): 754–764.

[82] Schroff, R.W., Gottlieb, M.S., Prince, H.E., Chai, L.L. and Fahey, J.L. 1983. Immunological studies of homosexual men with immunodeficiency and Kaposi's sarcoma. *Clin. Immunol. Immunopathol.*, 27(3): 300–314.

[83] Sesane, M. and Geyer, S. 2017. The perceptions of community members regarding the role of social workers in enhancing social capital in metropolitan areas to manage HIV and AIDS. *Social Work*, 53(1): 1–26.

[84] Shieh, G. and Johnson, R. 2005. Inferences based on a bivariate distribution with von Mises marginals. *Annals of the Institute of Statistical Mathematics*, 57: 789–802.

[85] Singh, H., Hnizdo, V. and Demchuk, E. 2002. Probabilistic model for two dependent circular variables. *Biometrika*, 89: 719–723.

[86] Starner, T. and Pentland, A. 1995. Real-time American Sign Language recognition from video using Hidden Markov models. *Proc. of International Symposium on Computer Vision - ISCV*, 265–270.

[87] Strug, D.L., Grube, B.A. and Beckerman, N.L. 2008. Challenges and changing roles in HIV/AIDS. *Social Work in Health Care*, 35(4): 1–19.

[88] Stultz, C.M. 1993. Structural analysis based on state-space modeling. *Protein Sci.*, 2: 305–314.

[89] Sukumaran, S., Almon, R.R., DuBois, D.C. and Jusko, W.J. 2010. Circadian rhythms in gene expression: Relationship to physiology, disease, drug disposition and drug action. *Advanced Drug Delivery Reviews*, 62(9-10): 904–917.

[90] Szklarczyk, D., Gable, A.L., Lyon, D., Junge, A.,Wyder, S., Huerta-Cepas, J., Simonovic, M., Doncheva, N.T., Morris, J.H., Bork, P., Jensen, L.J. and Mering, C.V. 2019. STRING v11: Protein-protein association networks with increased coverage, supporting functional discovery in genomewide experimental datasets. *Nucleic Acids Research*, 47(D1): D607–D613.

[91] Thompson, T.B., Chou, K.C. and Zheng, C. 1995. Neural network prediction of the HIV-1 protease cleavage sites. *J. Theor. Biol.*, 177(4): 369–379.

[92] Turhal, U., Gök, M. and Durgut, A. 2015. Comparison among feature encoding techniques for HIV-1 protease cleavage specificity. *Int. J. Intell. Syst. Appl. Eng.*, 3(2): 62–66.

[93] Wallach, T., Schellenberg, K., Maier, B., Kalathur, R.K.R., Porras, P., Wanker, E.E., Futschik, M.E. and Kramer, A. 2013. Dynamic circadian protein-protein interaction networks predict temporal organization of cellular functions. *PLoS Genetics*, Mar;9.

[94] Wang, M., Zhou, Z., Khan, M.J., Gao, J. and Loor, J.J. 2015. Clock circadian regulator (CLOCK) gene network expression patterns in bovine adipose, liver, and mammary gland at 3 time points during the transition from pregnancy into lactation. *J. Dairy Sci.*, 98(7): 4601–4612.

[95] Wermuth, N. 1976. Analogies between multiplicative models in contingency tables and covariance selection. *Biometrics*, 95–108.

[96] White, J.V. 1994. Protein classification by stochastic modeling and optimal filtering of amino-acid sequences. *Math. Biosci.*, 119: 35–75 .

[97] Yuan, M. and Lin, Y. 2007. Model selection and estimation in the Gaussian graphical model. *Biometrika*, 94(1): 19-35.

[98] Zhang, C., Bickis, M.G., Wu, F.X. and Kusalik, A.J. 2006. Optimally connected hidden markov models for predicting MHC-binding peptides. *J. Bioinform. Comput. Biol.*, 4(5): 959–980.

[99] Zhao, T., Liu, H., Roeder, K., Lafferty, J.D. andWasserman, L.A. 2012. The huge package for high-dimensional undirected graph estimation in R. *Journal of Machine Learning Research: JMLR*, 13: 1059–1062.

Chapter 9

Imputation of Missing Value in EMG Signaling Data

Fatemeh Sarasir and *Vilda Purutçuoğlu**

9.1 Introduction

There are many collections of information that show changes over time. These are found in different real-life situations, especially, in healthcare and life sciences. As modern technology improves, patient information like electromyograms, vital signs, and lab results are stored via complex time series using monitoring and electronic health record systems. Additionally, not having enough data is a common issue that impacts many different areas. There are many reasons why data might not be there. Someone may have intentionally excluded it or unintentionally overlooked its inclusion. It could also be because the records were lost, even though predetermined time intervals can result in certain variables not being measured during certain periods, leading to missing data in Electronic Health Records (EHRs). Moreover, when collecting signals, a few things can cause data to be lost, such as things that interfere with the emission of the signal or problems with the connections between the electrodes and the body [20], or if the wireless transmission doesn't work properly [9]. Also, when using multiple

Middle East Technical University.
* Corresponding author: vpurutcu@metu.edu.tr

methods, a problem can occur if someone has tried to collect the data. There are numerous other factors that can result in missing data as well.

The absence of information can have adverse effects on the accuracy of statistical data and pose challenges in analyzing the data, such as making predictions for diagnoses. Conversely, it has been observed that absent values often possess implicit meanings [22]; for instance, the absence of data can reveal the choices adopted by healthcare providers while handling electronic health record information [12]. Hence, it holds valuable theoretical and practical implications to approximate absent values [39].

Our study is about Electromyography (EMG) signaling data which is one of the common data types in many biomedical and biostatistical analyses. This data helps researchers and doctors diagnose muscle problems by measuring the electrical impulses produced when muscles contract. When muscle cells get a signal, they create electricity or energy. Accordingly, there are two main types of EMG observations. These are the surface EMG (sEMG), which records signals from the muscles on the skin's surface, and intramuscular EMG (IEMG), which involves putting a small needle electrode directly into the muscle to catch the signals [1].

The applications of EMG signals are diverse and encompass areas such as upper-limb prostheses [40, 41] and [27], electric wheelchair control [21], and muscle-computer interaction [24]. In these contexts, it is crucial to have complete and uninterrupted EMG signals to ensure effective and successful integration. Nevertheless, in practice, the acquisition of EMG data is not devoid of information loss. The frequency of measurements differs between patients and across variables in the data stored in the EHRs. Consequently, discretizing clinical data into a sampling rate of sEMG signals might need to be reduced to match other signals from the body. This could lead to not having enough data [42].

The presence of missing values within the EMG signal can lead to a decline in the overall performance of healthcare applications, such as myoelectric pattern recognition, which aims to anticipate motor intentions based on sEMG signals [20]. Additionally, missing information not only impacts recognizing patterns in muscle movements but also decreases the precision in classifying movements for controlling prosthetic devices. When the data is incomplete and the proportion of missing values is substantial, it significantly degrades the performance of classification methods and reduces their statistical power. This emphasizes the significance of obtaining a complete dataset. To address this issue, appropriate imputation methods must be employed to effectively estimate the missing data.

In general, there are two ways to deal with missing data: (i) removal of missing data [16] or (ii) filling in the missing information. Sometimes, it is suitable for skipping steps when there is a lot of information, and the observations are not related to a specific time. However, if the lost data is carelessly deleted, it can also lead to significant damage to the original data. Because the accuracy and the reliability of the analytical results may be compromised, leading to a significant waste of data resources. Hence, imputation methods work by replacing missing information with educated guesses and continuing with the analysis as if no data was missing [17].

Thereby, a multitude of optimization techniques have been suggested to address the issue of absent information. In 1977 Dempster et al. [8] released their initial research on limited information. Their approach relies on probability models that can be categorized as a type of data augmentation (DA) technique. In this approach, DA utilizes Bayesian statistical methods to calculate the posterior probability density of a parameter or set of parameters, based on prior knowledge related to the entire data set. The focus then shifts to sampling from the resulting distribution [33]. Additional research exploring various approaches for effectively managing missing data has been conducted and documented by other scholars [13, 25, 18, 6]. A significant contemporary strategy for managing incomplete data is through the implementation of MI (missing information) category techniques. Several experts strongly endorse the use of single and MI techniques as substitutes for probability-based methods, as recent advancements have made them more effective [6, 14, 2, 26]. So, when studying biomedical signal data, researchers mainly use machine learning methods and nonparametric sampling methods, which have been talked about in relevant literature. These methods are often used and work well with important programming languages, but, they don't take into account the specific traits of the data. These traits have a big impact on the accuracy of statistical explanations.

Therefore, in this study, our goal is to overview the well-known MI methods specifically used in EMG data and supported by MATLAB® programming languages, namely, linear imputation, cubic spline imputation, K-Nearest Neighbor (KNN) algorithm, and random forest method that are all dependent on operation research algorithms. Then, we include common imputation approaches that are implemented in almost all data types, namely, mean, median, and moving average approaches. Later, we extend these analyses by covering alternative approaches which have not been applied yet for EMG data types, namely, the normal ratio method and normal ratio weighted with correlation method which is, indeed, designed for climate time series data sets. We select these two approaches since both climate and EMG data share some inherited properties of time series data such as high correlation and nonlinear structure. Finally, we cover the expectation-maximization (EM) approach in our analyses. The EM method is a well-known parametric (i.e., probability distribution-based) statistical MI technique that is used for medical analyses since the missing imputation is one of the challenges in these sorts of analyses due to the sparse observations of patients. In our representation, we classify these ten methods based on two main classes: (i) univariate time series imputation and (ii) multivariate time series imputation. The first group considers filling the missingness for one measurement over time and covers the mean, median, moving average, linear, cubic spline, normal ratio, and normal ratio weighted with correlation imputation methods. The second group assumes that there are multiple measurements over time and KNN, random forest, and EM imputation can be used under this branch. In the implementation, we use the underlying ten methods by comparing their performance in terms of accuracy and computational demand by using three benchmark EMG data sets.

Consequently, the forthcoming segments of the chapter are systematically organized as follows. Section 9.2 presents the missingness mechanism in general and possible pattern in missing data. Section 9.3 expounds upon the importance attributed to Electromyograms (EMG) and the utilization of EMG data within the scope of this study. Sections 9.4 and 9.5 sequentially introduce distinct methodologies for addressing missing data and delineate the metrics employed for evaluating their efficacy, respectively. Moving forward, Section 9.6 elucidates the outcomes attained through the data imputation processes, considering both the precision of imputed values and the associated computational time. Ultimately, Section 9.7 encapsulates the culminating remarks and deductions drawn from the study's findings.

9.2 Missingness Mechanism

Finding a good and useful method to estimate data that is missing is related to the way in which data becomes incomplete [14]. It is believed that utilizing the information obtained from existing observations is highly valuable in the estimation of missing data. If we agree that the necessary information can be obtained from related observations, the process of filling in missing data will become easier. In this situation, it is important to figure out why the data is not complete and what type of data is missing. We need to handle these issues carefully.

In simple terms, researchers agree that there are two ideas about the missingness mechanism. The first idea is that it can be ignored, and the other is that it cannot be ignored. If we ignore the mechanism, we will confidently overlook it when analyzing missing data. But if we don't ignore it, we must include it in our estimation of parameters.

The classification of how missing data occurs is split into three categories. First, data will be classified as missing-completely-at-random (MCAR) if the missing values happen without any relation to the observed or unobserved values. Furthermore, data is considered missing-at-random (MAR) when the reason for missing values is only related to the information that has been already observed. In simpler terms, missing data is called missing-not-at-random (MNAR) when the missing values are related to both the information which we have and the information which we don't have, and this missingness cannot be fully explained by the data that we do have. MCAR and MAR belong to a class of missingness mechanisms where missing data can be ignored, but, MNAR belongs to a class of mechanisms where missing data cannot be ignored. In this study, we aim to collect information even when there is missing data without any pattern.

To understand how the likelihood of missing data is connected to the variables which we measure, we provide thorough explanations of the mechanisms behind missing data.

Let's call the whole set of data the *complete data matrix* as $Y_{com} = (Y_{ij}), (i = 1,2,\ldots,p; j = 1.2,\ldots,p)$ and assign a size of $(N \times p)$- dimensional matrix to it. We mean that the data is fully gathered and includes two parts, namely, Y_{obs} and Y_{mis}, respectively. These parts are related to each other. Thus, $Y_{com} = (Y_{obs}, Y_{mis})$ and $f(Y \mid \theta) \equiv f(Y_{obs}, Y_{mis} \mid \theta)$ show how crowded or spread out the Y_{com} is [16].

Accordingly, the marginal density of Y_{obs} is found by

$$f(Y_{obs} \mid \theta) = \int f(Y_{obs}, Y_{mis} \mid \theta) dY_{mis}.$$

If the likelihood of certain factors being related to observed values, denoted by $L(\theta \mid Y_{obs})$, is determined without considering that the missingness mechanism is proportional to $f(Y_{obs} \mid \theta)$, then in this situation;

$$L(\theta \mid Y_{obs}) \propto f(Y_{obs} \mid \theta).$$

To find out the mechanisms of missing and how it relates to other variables, Rubin [22] uses a matrix with indicators to analyze the statistical mechanism. Here, let $M_com = (M_{ij})$, $(i = 1,2,\ldots,N; j = 1,2,\ldots,p)$ with a size of $(N \times p)-$dimensional matrix as used beforehand, be a grid of numbers (matrix) with only zeros and ones via

$$\begin{cases} 1 & \text{if } Y_{ij} \text{ missing,} \\ 0 & \text{otherwise,} \end{cases}$$

and can be demonstrated as $M_com = (M_{obs}, M_{mis})$. The joint distribution of Y_{com} and M_{com} is represented by

$$f(Y, M \mid \theta, \psi) = f(Y \mid \theta) f(M \mid Y, \psi). \tag{9.1}$$

In Equation 9.1, by integrating out the Y_{mis}, the joint distribution of the data which we observe and the indicator matrix can be achieved as follows:

$$f(Y_{obs}, M \mid \theta, \psi) = \int f(Y_{obs}, Y_{mis} \mid \theta) f(M \mid Y_{obs}, Y_{mis}, \psi) dY_{mis} \tag{9.2}$$

in which $f(M \mid Y_{obs}, Y_{mis},)\psi)$ indicates the missingness mechanism based on a mathematical calculation, and the likelihood of θ and ψ is related to

$$L(\theta, \psi \mid Y_{obs}, M) \propto f(Y_{obs}, M \mid \theta, \psi). \tag{9.3}$$

9.2.1 Patterns of Missing Data

The literature exhibits diverse patterns when it comes to missing information. When discussing a missing data pattern, we describe the way in which we discern both the available and missing values in a dataset. It helps us understand

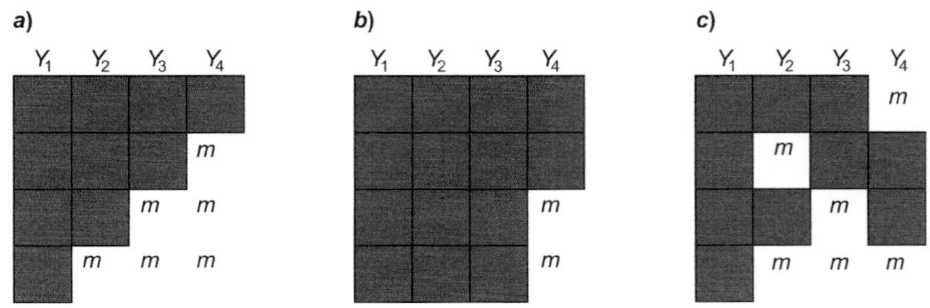

Figure 9.1: Three prototypical missing data patterns: (a) monotone missingness, (b) univariate missingness, and (c) arbitrary missingness.

how the data is organized and to decide how to handle it properly. However, the reason behind the missing data is not disclosed. Missingness can manifest itself in diverse manners [13]. Figure 9.1 shows a few different ways that data can be missing. One common pattern is called monotone missingness, which often happens in studies where measurements are taken repeatedly over time. Sometimes, a dataset that doesn't have a monotone pattern can be changed so that it can by rearranging the data. Rearrangement of the missing pattern to a simple pattern makes the math easier for maximum likelihood and multiple imputation methods, as discussed in Chapter 3. On the other side, univariate missing pattern (b) is not common, but, the arbitrary missing pattern (c) is often observed in research. The information analyzed in this study showed that there were mostly single-variable and randomly missing parts.

9.3 Data Description

9.3.1 *Electromyogram (EMG) Signals and their Significance*

Skeletal muscles are the main muscles in our body and makeup about 40% of our total body weight. They control and adjust the skeleton, and they are often connected to bones by tough collagen tendons. The beginning of a muscle is the part that is connected to the trunk or to the bone that doesn't move much. Muscle cells called muscle fibers form a skeletal muscle, similar to how neurons form a nerve. The shape of each muscle cell resembles that of a cylinder, long and elongated. It has many nuclei near its surface. The formation of skeletal muscle fibers involves the fusion of numerous small muscle cells, resulting in the creation of large cells [30].

The muscle fibers are arranged in a consecutive manner. A certain type of tissue envelops every muscle fiber in our body, keeping them united. Bundles known as fascicles are formed by grouping together these muscle fibers. Between the bundles, one can find collagen, elastic fibers, nerves, and blood vessels. The

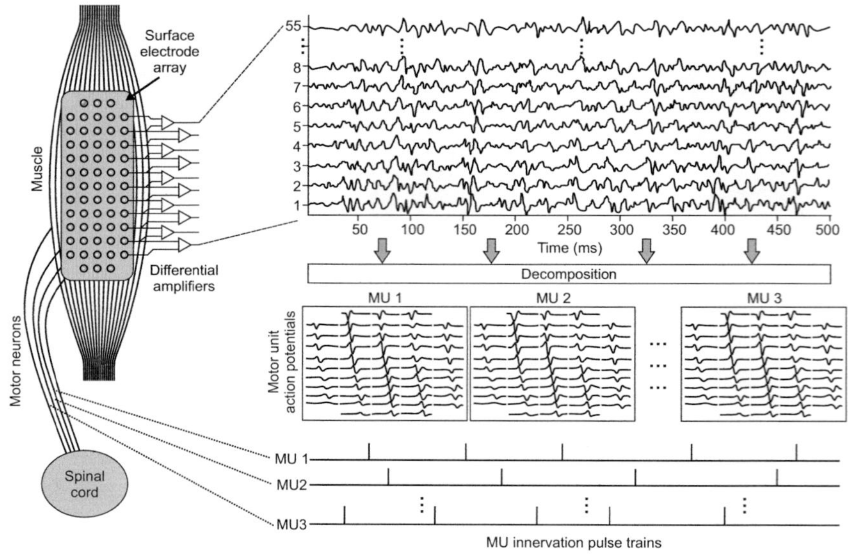

Figure 9.2: Diagram showing how SEMG signals are created (left), captured (top), and separated into different components (bottom right).

whole muscle is covered by a layer of tissue that is connected to the tissue around the muscle fibers and bundles, as well as the tendons that attach the muscle to the bones underneath [4].

Many small fibers constitute the composition of our body's muscles. When these fibers are given a small electrical signal, they squeeze and make a small amount of electricity. We call this electricity action potential. Action potentials (APs) start in the middle of the muscle and move along the fibers in both directions toward the tendons. The neural system sends electrical signals from the brain to the muscles, which makes them move. Involuntary muscle contractions happen because of a series of electrical pulses called innervation pulse trains (IPT) that are sent by the same motoneuron. All fibers coordinating with a specific nerve synchronize their contractions. The collaboration of muscles results in the formation of a motor unit (MU) [30], [4]. However, reconstructed MUAPs and IPTs are crucial measurements for understanding how the muscles and nerves are affected by disease or injury and how well rehabilitation is working. These machines are used in serious studies and tests about different muscle and nerve problems, problems caused by work, and signs of muscle tiredness. The process of making, getting, and breaking down multichannel surface EMG signals is shown in Figure 9.2 [4].

9.3.2 EMG Data Description

This investigation is centered on the scrutiny of three unique manifestations of authentic electromyography (EMG) data, originating from three distinct databases. The core objective is to thoroughly evaluate the effectiveness of assorted imputation techniques with respect to their precision and computational efficiency. The inaugural data set subjected to examination corresponds to the Roessingh Research and Development-MyLeg database designated for activity prediction (MyPredict). This particular repository serves the purpose of fostering research endeavors concerning data-driven strategies for intent recognition and activity prediction within the context of the lower limb. This objective is achieved through the utilization of electromyography. The measurement sessions encompassed a series of trials encompassing various activities including sitting, standing, walking on level ground, ascending and descending stairs, ascending and descending ramps, traversing uneven terrain, and moving within simulated confined spaces. The participant was subjected to measurement utilizing a set of eight inertial measurement units, which were complemented by distinct variants of surface electromyography (sEMG). Therefore, the data set is characterized by a subject seamlessly transitioning between different activities. A comprehensive exposition of this dataset can be found in the study of Schulte et al. (2023) [28].

The subsequent data set is relevant to the Nearlab hand movement intention detection based on the surface electromyography signals data set. This compilation encompasses the electrical signals originating from the hand muscles of a subject, as captured by surface electrodes, during the execution of eight fundamental hand movements and six composite movements. In the Nearlab data set, the arrangement of electrodes is situated both on the dorsal aspect (depicted in the upper segment of Figure 9.3) and the volar aspect (illustrated in the lower segment of Figure 9.3) of the dominant hand of the participant. In its entirety, the data set utilizes a configuration of ten distinct channels, comprising six proximate to the forearm region and an additional set of four channels positioned 3 cm distal to the aforementioned ones. Notably, electrodes designated as 6 and 10 are not discernible within the provided visual representation [32].

The dataset entitled Tai Chi, Physiological Complexity, and Healthy Aging - Gait constitutes the third corpus of data under contemplation sourced from PhysioNet. This specific data set encompasses both gait and electromyography (EMG) data procured within the framework of a hybrid investigation. Within the context of this study, the randomized controlled trial (RCT) segment was conducted with the intention of scrutinizing the immediate effects of a Tai Chi intervention on a cohort of previously unexposed, healthy individuals. One among these participants, who had engaged in regular Tai Chi practice for a span of 3 months, was selected. The study protocol encompassed the concurrent acquisition of both gait and electromyography (EMG) data, acquired during walking sessions spanning a duration of 10 minutes, executed under the single-task con-

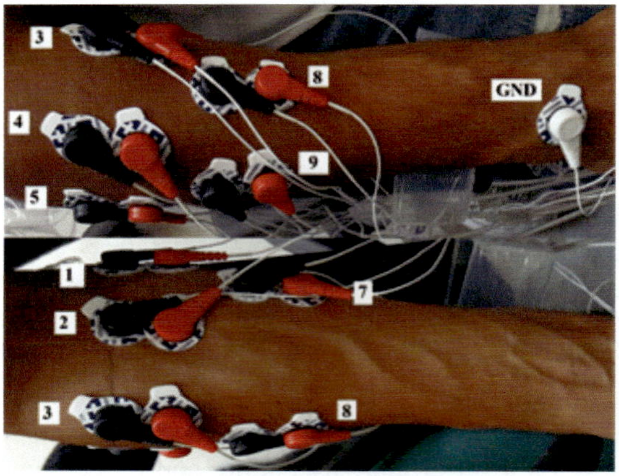

Figure 9.3: The configuration of electrode placement on the dominant hand of a right-handed participant is delineated within the Nearlab dataset.

dition and at the participant's self-selected preferred walking velocity. The overarching objective of this investigation centered on the utilization of the acquired gait and EMG data to augment the comprehension of alterations in mobility attributed to Tai Chi training [34].

9.4 Methods

9.4.1 Deleting Schemes

To handle missing data, deleting schemes, which are alternatively called complete case analysis or listwise deletion, and involve completely removing observations that have missing values from the dataset. This technique is successful in cases where there is a limited amount of missing data and the missing data follows a random pattern. If the missing data does not follow a completely random pattern (MCAR), eliminating portions of the data can lead to erroneous findings and diminish the precision of our analysis. One of the main problems with deleting schemes is that it can make the group being studied smaller. This can make it harder to find important patterns or relationships and increase the chance of not noticing important effects that are there. Moreover, unfairness can arise when certain sections of the data are eliminated, and the missing information is related to the outcome or other variables under investigation. This bias happens when the data that is not available is connected to something about the people being studied, like their wealth or their health, and this is also connected to the result.

In general, deleting certain parts of information can be a simple and direct way to deal with missing data. However, it is important to be careful and only use this method when the missing data is completely random. Utilizing advanced techniques such as multiple imputations or maximum likelihood estimation could be more appropriate when the missing data pertains to the subject of our study or other related events.

9.4.2 Missing Data Imputation Methods

Imputation refers to the act of making educated guesses or providing information to complete missing details. Many ways have been invented to solve the problem of missing information. Here, we will give a short explanation of the techniques used to fill in missing data and the ways in which we measure them in this research. We divide them into two categories: filling in missing data for one measurement over time, also known as univariate time series imputation, and filling in missing data for multiple observations over time, also named multivariate time series imputation.

9.4.3 Univariate Time Series Imputation

Univariate time series imputation offers a viable approach for addressing missing values within a time series dataset containing only one variable. This technique serves to maintain the integrity of temporal relationships and patterns inherent in the data. Univariate time series imputation includes mean imputation, median imputation, moving average imputation, linear interpolation, and spline interpolation methods.

9.4.3.1 Mean Imputation

Traditionally, in EMG applications, the approach to handling missing data involved substituting it with the mean values of neighboring data points [13]. This method calculates the average of the numbers in a column without any missing values. Then, it fills in the missing values in each column separately without being influenced by the other columns. The equation for this calculation can be written like this: Let α_m be a value that is missing, and $\alpha_1, \alpha_2, \alpha_3, \ldots, \alpha_N$ be a group of variables that we can see at the same time. Hereby, the mean imputation is derived via the following calculation where N is the total number of observations.

$$\alpha_m = \frac{1}{N}(\Sigma_{i=1}^{N} \alpha_i). \tag{9.4}$$

Despite some advantages of the mean imputation method like being simple and quick, and working effectively with small number sets, this method has some drawbacks in which it does not take into account the relationships between different characteristics, and it only works on one column at a time. So, using this method on encoded categorical features cannot yield good outcomes.

Overall, instead of offering supplementary details, this technique merely amplifies the sample size and improves the precision of the errors, and it has been shown to be highly suboptimal and ineffective.

9.4.3.2 Median Imputation

An alternative method is a median imputation, where the missing values in a column are substituted with the median value. When the numbers are unevenly distributed, this method uses the middle value instead of the missing values. The median imputation is calculated by the following equation.

$$\alpha_m = \begin{cases} \alpha_m\left[\frac{N+1}{2}\right] & \text{if N is even} \\ \frac{\alpha_m\left[\frac{N}{2}\right]+\alpha_m\left[\frac{N}{2}+1\right]}{2} & \text{if N is odd} \end{cases} \tag{9.5}$$

9.4.3.3 Moving Average Imputation

When employing the moving average imputation method, we fill in missing values by taking the weighted average of surrounding values. In order to determine the moving average approach, it is necessary to consider an equal number of observations prior to and following the unavailable data points. One method of demonstrating the functionality of the process is by utilizing observations, namely, $(\alpha_{mK-2}, \alpha_{mK-1}, \alpha_{mK+1}, and\,\alpha_{mK+2})$ that are in close proximity. We specifically examine the values preceding and succeeding a particular point (represented as K) and utilize them to determine the mean of four values (two on each side of K) where Y is used to represent the observed values. If all the observations in the current window are missing (NAN), the window size will be enlarged until at least two values that are not missing are found. Consequently, the entire equation is displayed in the following manner.

$$\alpha_{mK} = \frac{1}{2N}(\alpha_{mK-N} + \ldots + \alpha_{mK-1} + \alpha_{mK+1} + \ldots + \alpha_{mK+N}) \tag{9.6}$$

9.4.3.4 Linear Interpolation

The process of interpolation entails modifying your data to suit a specific function through mathematical techniques. By utilizing this function, one can infer what the missing data could potentially be. The most straightforward type of interpolation is known as linear interpolation. By using the relationship between adjacent observed values, the linear interpolation methodology can be employed to estimate missing values. The fundamental idea supporting this assumption is that the variable's values exhibit a linear pattern, moving in either an upward or downward direction, either gradually over time or when comparing various data points [7]. The linear imputation method is highly proficient in accurately replacing missing values as by examining the values that were recorded immediately before and after, the closest values to the missing data point are identified. An analogy for linear interpolation between two data points of coordinates

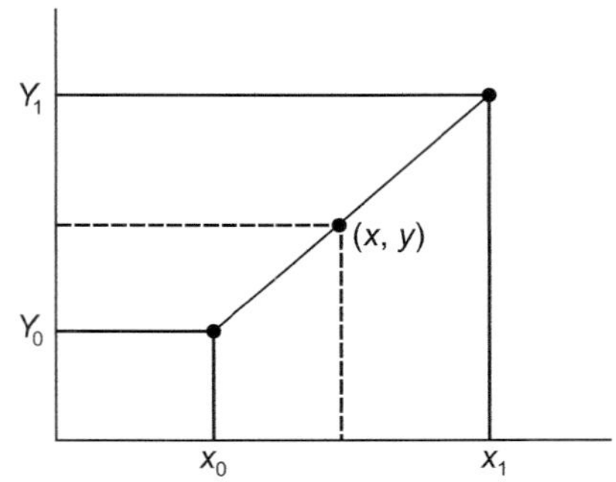

Figure 9.4: The existence of two black points allows us to locate the gray line joining them. This line can be employed to ascertain the value of y corresponding to any given x.

(x_0, y_0) and (x_1, y_1) involves envisioning a straight line that connects these provided points as shown in Figure 9.4. The slope equation enables the identification of the corresponding value y on a straight line for any given value of x in the interval (x_0, x_1).

$$\frac{y = y + 0}{x - x_0} = \frac{y_1 - y_0}{x_1 - x_0}.$$

By examining the shape in Figure 9.4, we can discover the solution geometrically. This particular scenario involves fitting a polynomial based on just one data point $(n = 1)$. In order to calculate y in this equation when x is not given, the following equation is used.

$$y = \frac{y_0(x_1 - x)y_1(x = x_0)}{x_1 - x_0}.$$

The above expression enables the computation of values within the interval of x_0 and x_1 by employing a straight line. The formula used for extrapolation is identical when extending linearly beyond this interval. Linear interpolation on a set of data points $(x_0, y_0), (x_1, y_1), \ldots, (x_n, y_n)$ means connecting the dots in a straight line between each pair of points. Generally, this leads to a line that is uneven and lacks a steady incline as seen in Figure 9.5.

9.4.3.5 Cubic Spline Interpolation

The utilization of spline imputation is prevalent when it comes to replacing missing data points in a given data set. The process entails constructing a curve using the given points and then using this curve to make educated guesses for the missing values in the data set. The technique is highly adaptable and powerful as it effortlessly identifies intricate patterns and correlations within the dataset [5], [31]. The technique of employing math enables the smooth connection of disparate sections within a curve, known as splines. It is made up of segments, or small sections that are described by polynomial equations.

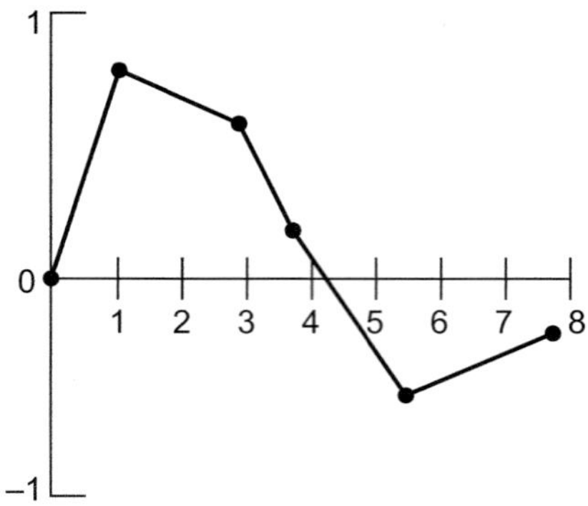

Figure 9.5: Linear interpolation on a data set (gray points) consists of pieces of linear interpolants (gray lines).

Cubic spline interpolation, which is extensively employed, addresses the issue of Runge's phenomenon by using interpolation. The occurrence of Runge's phenomenon arises when we attempt to approximate values within a specified range by applying high-degree polynomials and evenly spaced points. The consequence of this is the emergence of oscillation or wiggles at the range boundaries. In 1901, Runge uncovered this occurrence while studying the errors by resulting from polynomial interpolation in estimating selected functions [23]. By employing this technique, one can obtain a polynomial that is smoother in nature and showcases a smaller margin of error in contrast to alternative interpolating polynomials such as the Lagrange polynomial and Newton polynomial.

Let's say we have a group of $(n+1)$ points, from $(x_0, y_0) to (x_n, y_n)$. There will be a curved line $q_i(x) = y$ connecting each pair of points, and each curve will be in the shape of a cubic polynomial between each successive pair of knots $(x_(i-1), y_(i-1))$ and $(x_i, y_i)(i = 1, 2, \ldots, n)$ connecting to both of them. So, there will be a certain number (n) of polynomials in such a way that the first polynomial will start at a point called (x_0, y_0), and the last polynomial will end at a point called (x_n, y_n). Accordingly, the curvature of the spline interpolation which is a measure of how much a curve is bent or curved is described as the following formula.

$$K = \frac{y''}{(1 + y'^2)^{3/2}}, \tag{9.7}$$

where y' and y'' denote the first and second order to derivative (i.e., rates of change of $y(x)$) with respect to x, in order. By maintaining constant values for both y' and y'' at all points, including the knots, we can minimize the degree of bending in the spline curve [10].

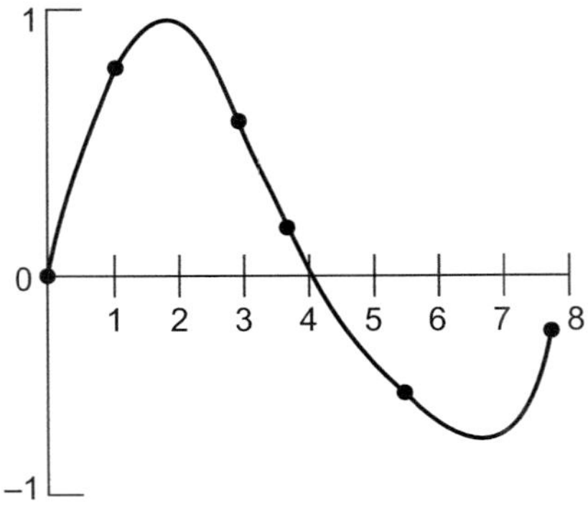

Figure 9.6: Spline interpolation on a data set (gray points) consists of pieces of cubic interpolants (gray lines).

9.4.3.6 Normal Ratio Method

The normal ratio (NR) model incorporates weights w to fill in missing data. The computation of these weights is based on the comparison of nearby stations. At times, we observe that the proportion of variation between various measurements remains consistent when evaluated concurrently [35]. Let's suppose we have a series of totals labeled as T_1, T_2 which we get from the nearby sensors, whereas the T_m is the total amount of a specific interested variable at the target sensor. By performing a calculation, the NR estimate for a missing value can be obtained as follows:

$$\alpha_m = \frac{1}{N}\left[\left(\frac{T_m}{T_1}\right)\alpha_1 + \left(\frac{T_m}{T_2}\right)\alpha_2 + \cdots + \left(\frac{T_m}{T_N}\right)\alpha_N\right] \tag{9.8}$$

For obtaining the weight, we let ρ_i signify the relationship between the target and i-th neighbor sensor and n_i indicate the number of observations whose correlation is calculated based on the weight, which is indicating the Normal Ratio Weighted with Correlations (NRWC) Method. In order to determine the weight w_i ($i = 1, \ldots, N$), the following formula is used:

$$w_i = \frac{r_i^2(n_i - 2)}{1 - r_i^2} \tag{9.9}$$

where r_i and n_i denote the Pearson correlation coefficient and associated sample size, respectively, for the i-th sensor. By prioritizing segments that are more similar to the missing data via weighting their correlation r_i ($i = 1, \ldots, N$), the method can produce a more accurate estimation of the missing data for N observations by improving the (9.8) formula to the following equation:

$$\alpha_m = \frac{1}{\sum w_i}\left[w_1\alpha_1 + w_2\alpha_2 + \cdots + w_N\alpha_N\right] \tag{9.10}$$

9.4.4 Multivariate Time Series Imputation

The purpose of implementing this method is to substitute any missing data while maintaining the integrity of the relationships and connections among the variables. The KNN algorithm, the random forest algorithm, and the EM algorithm can be considered under this group.

9.4.4.1 K-Nearest Neighbor Imputation Method

Nearest neighbor (NN) is a method that uses a type of learning algorithm where a computer program is trained to make predictions based on given examples and labeled data. Through supervised learning, an attempt is made to identify novel patterns in the data by associating established data patterns with unfamiliar data. The KNN technique constitutes a progression of the nearest neighbor approach, wherein its fundamental aim is to discern the K-most proximate neighbors from a collection of N specimens, encompassing the specimen under examination. Addressing the issue of absent data through the employment of the KNN technique commences with the selection of a designated quantity of immediate neighbors or proximate instances denoted as K. Subsequently, this selected cohort of K instances serves as a basis for approximating the absent data points [15, 38].

In the context of accounting for missing values during distance computation, numerous distance metrics, including the Minkowski distance, Manhattan distance, Cosine distance, Jaccard distance, Hamming distance, and Euclidean distance, are applicable for KNN imputation. Among these options, empirical evidence suggests that the Euclidean distance metric demonstrates notable efficacy and efficacy, as attested by prior studies [3]. Consequently, the Euclidean distance measure stands as the preeminent choice for distance computation, being extensively favored within the context of KNN imputation procedures.

In a scenario where the space is characterized by p dimensions, the distance linking two instances, denoted as x_0 and x_i, can be ascertained through the utilization of the subsequent Euclidean distance equation. In this formulation, x_{ip} signifies the observation corresponding to the ith instance within the pth spatial dimension.

$$d(x_0,x_i) = \sqrt{\mid x_{01} - x_{i1} \mid^2 + \mid x_{02} - x_{i2} \mid^2 + \ldots + \mid x_{0p} - x_{ip} \mid^2}. \tag{9.11}$$

The protocol for addressing absent data through the utilization of the Euclidean distance metric unfolds as delineated below:

1. In the realm of absent data instances, a paramount step involves the computation of distances between instances. This entails the application of the prescribed distance formula to each exemplar that encompasses such deficient information.

2. The computed distances are subsequently organized in a sequential manner, thereby, enabling the identification of the K smallest distances among the set.

3. In scenarios where the data adheres to a continuous nature, akin to the scope of our investigation, the procedure for filling in the absent values necessitates the determination of an appropriate imputation value. This value is ascertained by resorting to either the median or the arithmetic mean, derived from the collection of values that correspond to the K distances, thus, identified [38].

It is noteworthy to emphasize that the selection of the parameter K holds significant importance within the framework of KNN imputation. Opting for a smaller value of K can result in a more confined imputation based on local information, whereas, opting for a larger value of K can encompass a wider range of diverse information, albeit potentially introducing a greater degree of noise.

Nevertheless, certain limitations accompany this method. One pivotal assumption is predicated upon the notion that instances possessing akin characteristics also share resemblances in their feature attributes—a proposition that might not invariably hold true within intricate datasets. Additionally, the process of KNN imputation is susceptible to computational inefficiency, particularly, when confronted with voluminous datasets, thereby, constituting a notable drawback.

9.4.4.2 Random Forest Imputation Method

The random forest (RF) algorithm encompasses an ensemble of numerous decision trees, each characterized by uniform nodes, yet drawing upon distinct datasets to generate varied terminal nodes. By amalgamating the decisions emanating from these individual decision trees, the algorithm generates an outcome that encapsulates the aggregate mean of the outcomes produced by these constituent trees.

Shortly, the RF construction commences with an initial node, subsequently giving rise to branching to further nodes. This iterative progression persists until a terminal point, termed a leaf, is attained. Each node within this structure functions as an inquiry point to aid in the classification of data. The branches emerging from a node symbolize distinct potential pathways that the associated data point could traverse. Ultimately, a leaf denotes the culmination of a decision tree, signifying a terminal point bereft of further branching possibilities. The algorithm of RF leverages the terminal nodes, or ultimate determinations, from each node within individual trees to shape its own verdict. This methodology contributes to an elevation in model accuracy, as it aggregates outcomes from a multitude of distinct decision trees, subsequently calculating an average that culminates in an augmented degree of accuracy [29, 19].

When executing this method, it is pertinent to acknowledge that the Gini index is frequently employed. The Gini index serves as the pivotal criterion to determine the branching of nodes within a decision tree. This index encapsulates the expression that guides the decision-making process regarding the partitioning of nodes in the tree structure. As seen in Equation 9.12, the symbol "pi" denotes the proportional frequency of the particular class under observation within the dataset, while c signifies the count of distinct classes present [11].

$$Gini = \Sigma_{i=1}^{c}(p_i)^2. \tag{9.12}$$

Equation 9.12 utilizes both the class labels and their associated probabilities to ascertain the Gini impurity of each branch originating from a node. Consequently, it facilitates the identification of the branch that is more predisposed to manifest.

Entropy can also serve as a criterion to guide the branching of nodes within a decision tree, which is calculated by the following expression.

$$Entropy = \Sigma_{i=1}^{c}(-p_i) \times log_2(p_i). \tag{9.13}$$

In contrast to the Gini index, entropy hinges on the utilization of outcome probabilities to govern the branching process at nodes. This entails making decisions based on the likelihood of specific outcomes. Notably, the computation of entropy involves a higher degree of mathematical complexity as it necessitates the employment of logarithmic functions in its calculation [19].

9.4.4.3 Expectation-Maximization Algorithm

The central aim of Expectation-Maximization (EM) integration within the algorithm is not solely concerned with imputing missing data to attain data completeness. Instead, EM operates by incorporating the presence of missing data as indicative of proportionate information, thereby, facilitating the estimation of the pertinent parameter through the utilization of conditional expectations relating to the absent data points. This approach aligns with the principles outlined by Little and Rubin [13], underscoring the integral role of missing data in parameter estimation.

In our computation employing the expectation-maximization (EM) algorithm, we adopt the presumption of a multivariate normal distribution to model the actual magnitude of the EMG data. The method articulated by the researchers encompasses a sequential computational process characterized by two iterative phases, conventionally designated as prediction and estimation.

- *Prediction step*: In this step, an estimated value denoted by $\tilde{\theta}$ of the latent parameters are employed to anticipate the influence of unavailable observations on the statistics that sufficiently capture the entirety of the data.

- *Estimation step:* Subsequently, in the estimation step, the anticipated sufficient statistics are leveraged to recalibrate and determine a refined estimate for the underlying parameters [19].

The EM algorithm leverages the interdependent relationships between the variables Y_{miss} and θ. Specifically, Y_{miss} encapsulates information pertinent to θ, while θ encodes information about Y_{miss}, delineating the likelihood of Y_{miss}'s occurrences. As a procedural approach, the algorithm iterates through a sequence of steps, wherein it begins by enhancing the observed data set Y_{obs} through augmentation. This is coupled with the utilization of an initial estimation of the latent variables θ. Subsequently, the algorithm iteratively refines the estimation of θ by incorporating insights derived from the augmented data set $Y_{com} = (Y_{obs}, Y_{mis})$ [25, 37]. This iterative process continues to alternate between enhancement and re-estimation, contributing to the convergence of the algorithm. Hence, the complete data distribution function, denoted as Y_{com}, can be factorized in the following manner;

$$f(Y_{com} \mid \theta) = f(Y_{obs} \mid \theta).f(Y_{mis} \mid Y_{obs}, \theta). \tag{9.14}$$

In the left-hand side of Equation 9.18, term $f(Y_{com} \mid \theta)$ represents the probability distribution of empirically collected data, while $f(Y_{mis} \mid Y_{obs}, \theta)$ denotes the predictive distribution conditioned upon variables Y_{mis} and θ, hereby, preserving the interrelations inherent in these variables. The logarithmically transformed likelihood corresponding to entity $Y_{com}$is expressed as follows:

$$l(\theta \mid Y_{com}) = l(\theta \mid Y_{com}, Y_{mis}) = l(\theta \mid Y_{obs}) + logf(Y_{mis} \mid Y_{obs}, \theta). \tag{9.15}$$

Nonetheless, due to the unavailability of the precise value of Y_{mis}, it becomes infeasible to directly compute the log-likelihood of the comprehensive dataset, $(\theta \mid Y_{com})$. Therefore, it is necessary to simplify Equation 9.16 in a subsequent manner as

$$l(\theta \mid Y_{obs}) = l(\theta \mid Y_{com}) - logf(Y_{mis} \mid Y_{obs}, \theta). \tag{9.16}$$

Subsequently, by computing the expectations of both sides of Equation 9.16 with respect to the conditional predictive distribution of variable Y_{mis} given variable Y_{obs}, alongside a present estimation denoted as $\theta^{(t)}, f(Y_{mis} \mid Y_(obs, \theta^{(t)})$, the resulting equation can be formulated as follows.

$$l(\theta \mid Y_{obs}) = Q(\theta mid \theta^{(t)}) - H(\theta \mid \theta^{(t)}) \tag{9.17}$$

where

$$Q(\theta \mid \theta^{(t)}) = \int l(\theta \mid Y_{obs}, Y_{mis}) f(Y_{mis} \mid Y_{obs}, \theta^{(t)}) dY_{mis},$$

$$H(\theta \mid \theta^{(t)}) = \int logf(Y_{mis} \mid Y_{obs}, \theta) f(Y_{mis} \mid Y_{obs}, \theta^{(t)}) dY_{mis}.$$

As per the findings of the conducted research, it can be deduced that estimating $\theta^{(t+1)}$ outperforms estimate $\theta^{(t)}$. This assertion is substantiated by the observed pattern wherein the transition from variable $\theta^{(t)}$to variable $\theta^{(t+1)}$ during each iterative process leads to a discernible enhancement in the log-likelihood [37]. The evaluation of the convergence of iterative processes entails the scrutiny of divergences present in parameter estimations acquired at each successive iteration. Should the disparity between successive approximations fall below the pre-defined threshold, the iterative procedure is terminated. Notably, the parameter estimations within the ongoing iteration are influenced by estimations originating from the preceding iteration.

9.5 Accuracy Measures

This study endeavors to appraise the efficacy of imputation techniques in addressing the issue of missing data. Within the scope of imputation methodologies, the data's matrix structure assumes paramount importance. Thus, a prerequisite pre-processing step employing a MATLAB algorithm is undertaken on the temporal data. Introducing simulated missing values at a magnitude of 20% within the pre-processed dataset, complete records are then subjected to a masking algorithm. To comprehensively evaluate various methods, a sole subject is selected for each data category, accompanied by matrix reconfiguration resulting in datasets of varying dimensions in terms of rows and columns. Subsequently, the assessment of imputation methods' performance relies on three distinct accuracy metrics. These encompass a visual similarity control that leverages graphical representation, a root mean square error metric for quantifying the numerical discrepancies between actual and imputed values, and a run time comparison accounting for both central processing unit usage and elapsed time.

9.5.1 Visual Similarity

The evaluation of imputation strategies' effectiveness necessitates the consideration of visual similarity, as relying solely on error rates fails to offer a comprehensive comprehension of the degree of resemblance exhibited by the imputed data.

9.5.2 Root Mean Square Error

The root mean square error (RMSE) can be employed as a suitable metric for evaluating the precision of imputation and serves as a measure to numerically gauge the average magnitude of errors encountered in value predictions. This choice is motivated by the need to comprehensively measure the influence of imputations on subsequent analyses, ensuring their alignment with the inherent attributes of the EMG data. A reduction in RMSE indicates an improved fit or heightened precision of the model, while an elevation in RMSE suggests larger disparities between predictions and actual values. Mathematically, the RMSE (p,q) metric, which evaluates predicted data (p) against actual data (q), is formally defined as follows [36]:

$$RMSE(p,q) = \sqrt{\frac{1}{w}\Sigma_{i=1}^{w}(p_i - q_i)^2}. \tag{9.18}$$

Within our experimental investigation, the error values yielded by each method for every dataset are consistently scaled by a factor of 100. This adjustment is undertaken to aptly represent these error values as percentages.

9.5.3 Runtime

Within this research endeavor, our focus centers on the examination of two distinct classifications of runtime, commonly recognized as central processing unit (CPU) time and elapsed time. The former quantifies the period allocated to executing instructions for a specific task or process. In contrast, the latter, also referred to as wall-clock time or real-time, encapsulates the precise interval spanning from the commencement to the culmination of a given process or task. The assessment of runtime is conventionally conducted through measurements expressed in units such as seconds or milliseconds.

9.6 Results

9.6.1 Imputation Plots

The visualization comparisons portraying the imputation results alongside the actual dataset, as evidenced through the plotted diagrams, furnish valuable insights of significance. The graphical representations in Figures 9.7, 9.8, and 9.9 elucidate the progression of imputing missing values using ten distinct methodologies, exhaustively elaborated upon in Section 9.4. These specific imputation techniques have been administered across three discrete sets of the EMG data employed in our investigation: denoted as Lower limb, Hand movement, and TaiChi-gait signal data sets. The horizontal axis denotes the temporal dimension, while the vertical axis corresponds to the signal amplitude across all instances delineated in the figures.

Each of the three illustrations unmistakably demonstrates that every imputation technique possesses unique characteristics and highlights them during the process of filling in the absent sections. We improve the visual precision and impact of the figures by constraining the depiction of data point sequences in this analysis. Nevertheless, for a more precise assessment of method comparisons, it

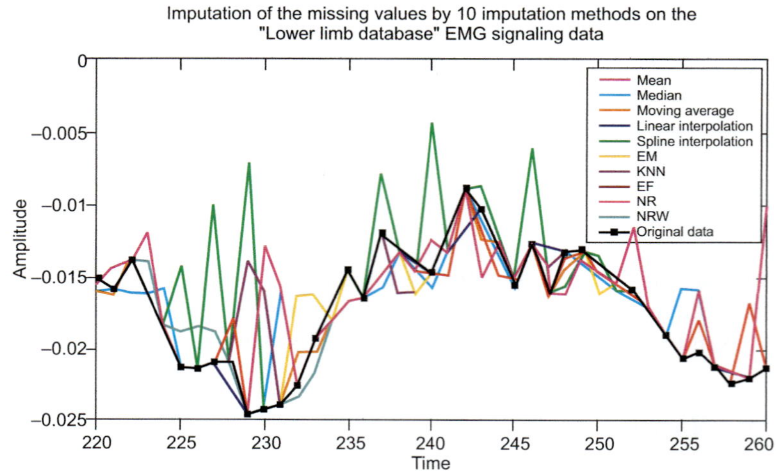

Figure 9.7: Imputation of the missing values by 10 imputation methods on the Lower Limb EMG signaling data.

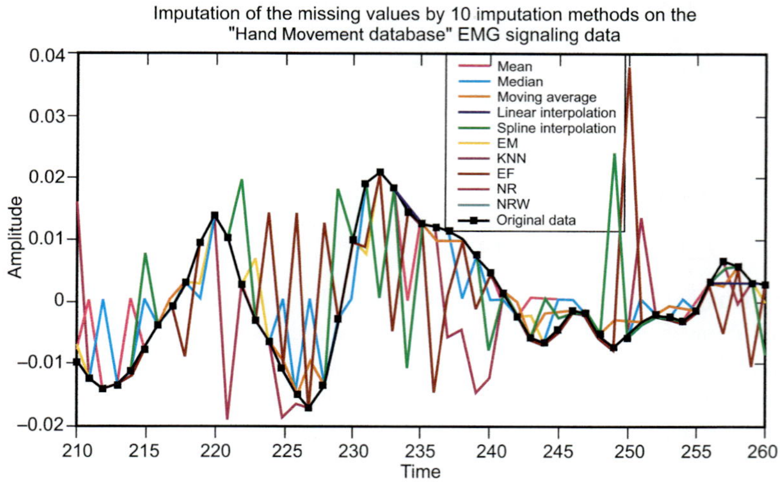

Figure 9.8: Imputation of the missing values by 10 imputation methods on the Hand Movement EMG signaling data.

would be more suitable to gauge their efficiencies using RMSE as a measure of accuracy.

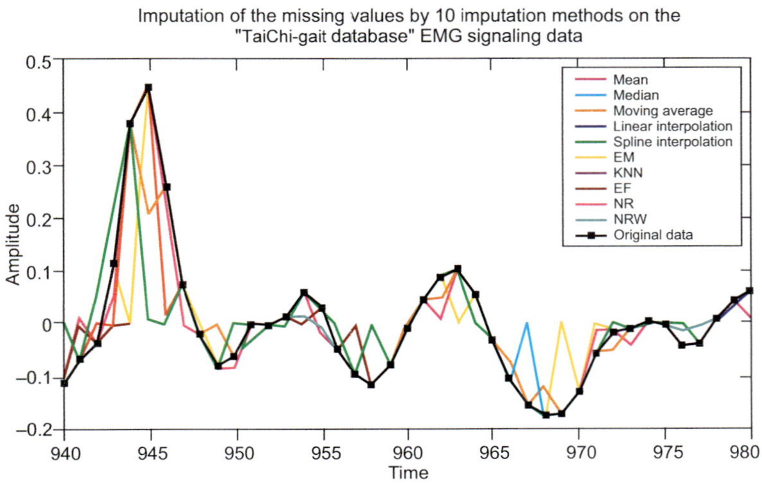

Figure 9.9: Imputation of the missing values by 10 imputation methods on the TaiChi-gait EMG signaling data.

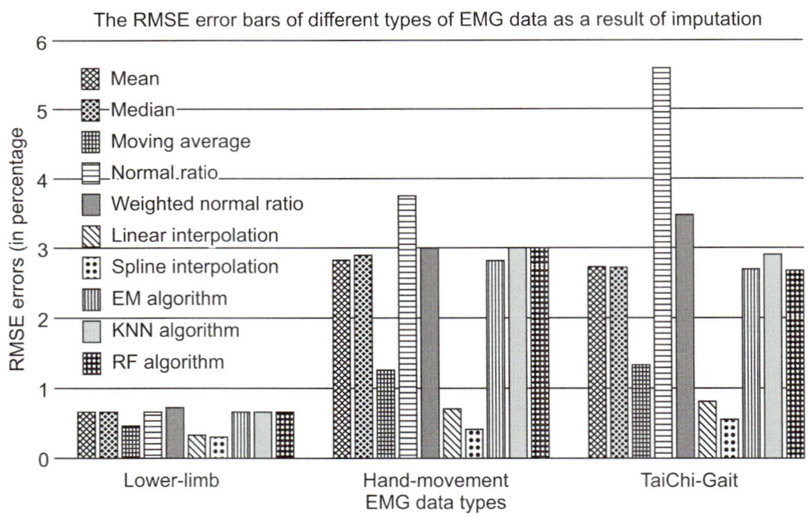

Figure 9.10: The RMSE bars of different types of EMG data as a result of different imputation methods.

9.6.2 RMSE Bar Plots

Figure 9.10, in the form of a bar chart, illustrates the contrast in RMSE percentages across ten different imputation approaches for each data set involving EMG signal data. According to the results, it can be deduced that the spline interpolation technique showcases the smallest error value within the three datasets implying that it surpasses other imputation methods in its ability to precisely forecast the absent elements.

On the flip side, the normal ratio imputation technique displays the greatest error magnitude within the hand-movement and Taichi-gait data sets. Conversely, the weighted normal ratio approach records the highest error value for the Lower limb data set. These results can be traced back to the unique characteristics manifested by each individual signal data.

9.6.3 Computational Time

Table 9.1 exhibit the results concerning the CPU time in seconds associated with each imputation method employed for the different signal datasets. In a broader context, it is expected that widely utilized univariate techniques, with the exception of the moving average method, showcase notable efficiency. In contrast, the computational expense of multivariate methods displays diversity, underscoring the significance of runtime as a pertinent factor for choosing effective imputation strategies.

Table 9.1: The CPU time of 10 types of imputation methods with the selected EMG data sets.

Imputation methodsData	Mean	Median	Moving Average	NR Method	WNR Method	Linear Interpolation	Spline Interpolation	EM Algorithm	KNN Algorithm	RF Algorithm
Lower-limb	0.05	0.12	323.27	0.17	25.92	0.20	0.19	4.23	1.10	273.92
Hand-Movement	0.01	0.02	2.44	0.03	2.43	0.06	0.06	2.75	0.24	77.66
TaiChi-Gait	0.11	0.16	3270	0.36	116.52	0.46	0.62	52.57	17.75	1160

9.7 Conclusion

In this study, we have presented both univariate and multivariate time series imputation methods in the analyses of electromyography (EMG) data collected from three distinct databases. In our analyses, we have used well-established techniques in the field and introduced three alternative imputation approaches, namely, expectation-maximization (EM) algorithm and normal ratio as well as normal ratio weighted with correlation methods which are designed for ordinary multivariate and climate data sets, respectively. Then, the results have been methodically presented through visual depictions in figures and tabulated data, allowing for a comprehensive assessment of different methodologies in terms of accuracy and computational efficiency.

The findings indicate that spline interpolation holds promise for imputing EMG datasets, with linear interpolation emerging as the second most effective imputation technique in terms of accuracy. Meanwhile, when considering computation time, classical mean and median imputation methods stand out as the most efficient options. Comparatively, the moving average and random forest algorithms show relatively lower computational efficiency. Ultimately, in assessing both accuracy metrics and computational time, the EM and normal ratio weighted with the correlation algorithm demonstrates consistent performance across all three datasets, although they don't rank as the most accurate or computationally efficient method.

In conclusion, we consider that the underlying imputation methods can be adapted to other data types measured through time continuously or with discrete time intervals whose data structures show linearity/non-linearity with dependency. Furthermore, adapting other well-known imputation approaches in distinct fields to biomedical signaling data like EMG data sets can open new avenues in the analyses.

Acknowledgement

Vilda Purutçuoğlu would like to thank Huawei Turkey R & D Center for their research support about the improvement of pre-processing step in big data analyses.

References

[1] Akmal, M., Zubair, S., Jochumsen, M., Kamavuako, E.N. and Niazi, I.K. 2019. A tensor-based method for completion of missing electromyography data. *IEEE Access*, 7: 104710–104720.

[2] Allison, P.D. 2001. Missing data. *Thousand Oaks, California*.

[3] Amirteimoori, A. and Kordrostami, S. 2010. A euclidean distance-based measure of efficiency in data envelopment analysis. *Optimization*, 59(7): 985–996.

[4] Belouchrani, A. and Amin, M.G. 1998. Blind source separation based on time-frequency signal representations. *IEEE Transactions on Signal Processing*, 46(11): 2888–2897.

[5] Birkhoff, G. and Garabedian, H.L. 1960. Smooth surface interpolation. *Journal of Mathematics and Physics*, 39(1-4): 258–268.

[6] Carlin, B.P. 1994. Tools for statistical inference: Methods for the exploration of posterior distributions and likelihood functions. *Journal of the American Statistical Association*, 89(427): 1142–1144.

[7] Cohen, M.X. 2017. *MATLAB® for Brain and Cognitive Scientists*. MIT Press.

[8] Dempster, A. 1977. Maximum likelihood estimation from incomplete data via the em algorithm. *Journal of the Royal Statistical Society*, 39: 1–38.

[9] Farina, D., Jiang, N., Rehbaum, H., Holobar, A., Graimann, B., Dietl, H., and Aszmann, O.C. 2014. The extraction of neural information from the surface emg for the control of upper-limb prostheses: Emerging avenues and challenges. *IEEE Transactions on Neural Systems and Rehabilitation Engineering*, 22(4): 797–809.

[10] Hall, C.A. and Meyer, W.W. 1976. Optimal error bounds for cubic spline interpolation. *Journal of Approximation Theory*, 16(2): 105–122.

[11] Johnson, R.A., Wichern, D.W. et al. 2002. Applied multivariate statistical analysis.

[12] Lipton, Z.C., Kale, D.C., Wetzel, R. et al. 2016. Modeling missing data in clinical time series with RNNS. *Machine Learning for Healthcare*, 56(56): 253–270.

[13] Little, R.J.A. and Rubin, D.B. 2002. *Statistical Analysis with Missing Data*. John Wiley & Sons. New York.

[14] Little, R.J.A. and Rubin. D.B. 2019. *Statistical Analysis with Missing Data*. Volume 793. John Wiley & Sons.

[15] Little, R.J.A. and Rubin, D.B. 2019. *Statistical Analysis with Missing Data*. Volume 793. John Wiley & Sons.

[16] Marwala, T. 2009. *Computational Intelligence for Missing Data Imputation, Estimation, and Management: Knowledge Optimization Techniques: Knowledge Optimization Techniques*. IGI Global.

[17] McKnight, P.E., McKnight, K.M., Sidani, S. and Figueredo, A.J. 2007. *Missing Data: A Gentle Introduction*. Guilford Press.

[18] McLachlan, G.J. and Krishnan, T. 1997. Wiley series in probability and statistics. The EM algorithm and extensions, 1997.

[19] Meyer, M., Isleif, K., Januschek, F., Lindner, A., Othman, G., Gimeno, J.A.R., Schwemmbauer, C., Schott, M., Shah, R. and ALPS Collaboration. 2023. A first application of machine and deep learning for background rejection in the Alps II tes detector. Annalen der Physik, pp. 2200545.

[20] Naber, A., Mastinu, E. and Ortiz-Catalan, M. 2019. Stationary wavelet processing and data imputing in myoelectric pattern recognition on a low-cost embedded system. *IEEE Transactions on Medical Robotics and Bionics*, 1(4): 256–266.

[21] Phinyomark, A., Phukpattaranont, P. and Limsakul, C. 2011. A review of control methods for electric power wheelchairs based on electromyography signals with special emphasis on pattern recognition. *IETE Technical Review*, 28(4): 316–326.

[22] Rubin, D.B. 1976. Inference and missing data. *Biometrika*, 63(3): 581–592.

[23] Runge, C. 1901. Über empirische funktionen und die interpolation zwischen äquidistanten ordinaten. *Zeitschrift für Mathematik und Physik*, 46(224-243): 20.

[24] Saponas, T.S., Tan, D.S., Morris, D., Balakrishnan, R., Turner, J. and Landay, J.A. 2009. Enabling always-available input with muscle computer interfaces. *In Proceedings of the 22nd Annual ACM Symposium on User Interface Software and Technology*, pp. 167–176.

[25] Schafer, J.L. 1997. *Analysis of Incomplete Multivariate Data*. CRC Press.

[26] Schafer, J.L. and Graham, J.W. 2002. Missing data: our view of the state of the art. *Psychological Methods*, 7(2): 147.

[27] Scheme, E. and Englehart, K. 2011. Electromyogram pattern recognition for control of powered upper-limb prostheses: State of the art and challenges for clinical use. *Journal of Rehabilitation Research & Development*, 48(6).

[28] Schulte, R.V., Prinsen, E.C., Schaake, L., Paassen, R.P.G., Zondag, M., van Staveren, E.S., Poel, M. and Buurke, J.H. 2023. Database of lower limb kinematics and electromyography during gait-related activities in able-bodied subjects. *Scientific Data*, 10(1): 461.

[29] Shah, A.D., Bartlett, J.W., Carpenter, J., Nicholas, O. and Hemingway, H. 2014. Comparison of random forest and parametric imputation models for imputing missing data using mice: A caliber study. *American Journal of Epidemiology*, 179(6): 764–774.

[30] Silverthorn, D.U., Ober, W.C., Garrison, C.W., Silverthorn, A.C. and Johnson, B.R. 2013. *Human Physiology: An Integrated Approach*, volume 3. Pearson Education Indianapolis, IN.

[31] Skjelbred, H.I. and Kong, J. 2019. A comparison of linear interpolation and spline interpolation for turbine efficiency curves in short-term hydropower scheduling problems. *In IOP Conference Series: Earth and Environmental Science*, volume 240, pp. 042011. IOP Publishing.

[32] Soroushmojdehi, R., Javadzadeh, S., Pedrocchi, A. and Gandolla, M. 2022. Transfer learning in hand movement intention detection based on surface electromyography signals. *Frontiers in Neuroscience*, 16: 977328.

[33] Tan, M.T., Tian, G.-L. and Ng, K.W. 2009. *Bayesian Missing Data Problems: EM, Data Augmentation and Non-iterative Computation*. CRC Press.

[34] Wayne, P., Gow, B., Hausdorff, J., Peng, C.-K., Lipsitz, L., Ahn, A., Novak, V. and Manor, B. 1983. Tai chi, physiological complexity, and healthy aging: Gait and emg data.

[35] WMO. Guide to climatological practices. Selbstverl.

[36] Xuegang, L., Junrui, L. and Juan, W. 2021. Missing data reconstruction based on spectral k-support norm minimization for nb-iot data. *Mathematical Problems in Engineering*, 2021: 1–11.

[37] Yozgatligil, C., Aslan, S., Iyigun, C. and Batmaz, I. 2013. Comparison of missing value imputation methods in time series: The case of turkish meteorological data. *Theoretical and Applied Climatology*, 112: 143–167.

[38] Zhang, C., Kai, J., Feng, H.C. and Yang, T. 2013. The nearest neighbor algorithm of filling missing data based on cluster analysis. *Applied Mechanics and Materials*, 347: 2324–2328.

[39] Zhang, J. and Yin, P. 2019. Multivariate time series missing data imputation using recurrent denoising autoencoder. *In 2019 IEEE International Conference on Bioinformatics and Biomedicine (BIBM)*, pp. 760–764. IEEE.

[40] Zia ur Rehman, M., Gilani, S.O., Waris, A., Niazi, I.K., Slabaugh, G., Farina, D. and Kamavuako, E.N. 2018. Stacked sparse autoencoders for emg-based classification of hand motions: A comparative multi day analyses between surface and intramuscular EMG. *Applied Sciences*, 8(7): 1126.

[41] Zia ur Rehman, M., Waris, A., Gilani, S.O., Jochumsen, M., Niazi, I.K., Jamil, M., Farina, D. and Kamavuako, E.N. 2018. Multiday EMG-based classification of hand motions with deep learning techniques. *Sensors*, 18(8): 2497.

[42] Zou, Y., Cheng, L. and Han, L. 2023. Reconstruction of incomplete surface electromyography based on an adversarial autoencoder network. *Biomedical Signal Processing and Control*, 86: 105084.

Chapter 10

The Application of the R-Vine Copula and Different Regression Models in the Inference of Hodgkin-Huxley Data

Hajar Farnoudkia[1,*] *Orhan Deniz Gencaga*[2] and *Şengül Ayan*[2]

10.1 Introduction

Neuronal networks involve 100 billion neurons communicating with each other in a highly nonlinear manner. The neurons in a network transmit information via oscillations called action potentials or spikes. Synchronized oscillations in networks of coupled neurons are also generally observed in neuronal systems and they play an important role in neuronal communication, information processing, and coordination of motor activity. Action potentials between neurons are not always perfectly synchronized and in general, the degree of synchrony is higher if the coupling between oscillators is stronger so that eventually a synchroniza-

[1] Baskent University, Ankara, Turkey.
[2] Antalya Bilim University, Antalya, Turkey.
* Corresponding author: hajerfarnoudkia@baskent.edu.tr

tion threshold may be reached. For subthreshold values of coupling, the oscillators may exhibit intermittent synchronization phenomena, where dynamics is synchronous in some time intervals and not synchronous in others. This partial, intermittent synchrony is also important for biological applications where it may potentially facilitate the high adaptability of biological systems as they react to different environmental impacts. Modeling studies have suggested that channel noise can influence information processing, spike time reliability, stochastic resonance, firing irregularity, subthreshold dynamics, and action potential initiation and propagation in morphologically detailed models. Here, we analyze the intermittent phase-locking in a network of three neurons. Accordingly, we investigated statistically the effect of noise in the intermittent transition to synchronized behavior for coupled neuronal oscillators due to noise integrity. The Hodgkin-Huxley model is a groundbreaking mathematical model that describes the generation and propagation of action potentials in neurons. It was developed by Alan Hodgkin and Andrew Huxley (1952) [3] and has since become a fundamental tool in neuroscience research.

In Hodgkin-Huxley signal processing data, there is a term in the equation defined as $k(V_2 - V_1)$ in the explanation of dV_2 meaning that the difference between the values of the V_2, let's say $V_{2t} - V_{2(t-1)}$ depends on $k(V_{2t} - V_{1t})$. So, in order to investigate the relationship between two or three neurons we used a regular vine copula approach and random forest regression model for a multivariate version as well as a linear regression model which is a subset of the vector autoregression model with some lags. The application of the Copula method can be seen in almost every field of study such as finance [4], portfolio risk analysis [7], the medical field [13, 8], insurance [11] and in the fields in which the relationship between the defined variables make sense. Specifically, the different aspects of the copula approach have been extensively used in neuroscience, especially in recent years. For instance, the time-varying copula is used in the connectivity analysis of functional MRI (Lee and Kim, 2019 [6]), A copula with the Gaussian process method is used for analyzing multidimensional neuronal and behavioral relationships [5]. The non-parametric version of Copula is investigated in the study of Safaei et al. (2018) [9] to estimate the mutual information between the variables which can be neurons in neuroscience. However, the copula method has not been used frequently for the analysis of HH data except in the study of Gencage et al. (2020) [2] which is for the relationship between two neurons. By the simulated data, we observe that by increasing the noise term, the relationship between every two neurons changes, and also the correlation between the neuron and the different factors of each neuron changes. In some cases, the relationship gets even stronger by the noise. To fit a regression model for the HH data we executed a random forest regression for the data without any noise term and we tried to see how the mean average error changes as noise increases. Finally, we implemented a linear regression on the data with different noise variance values to see the change in the relationship between the neurons. The first section describes the HH data in detail. Followingly, in the second section, R-Vine, random forest, and linear regression aspects, are in order. In the last section which is the application part, we use simulated data with 40000 samples simulated with different noise terms. The disturbances are all normally distributed with a mean zero and a changing standard deviation from zero to nine. Finally, the conclusion section consists of the problem and further work.

10.2 HH Data

This section is allocated to neuroscience. Part of the study starts with the relationship between the neurons that make a model based on differential equations. All of the details about the parameters will be discussed, as well.

10.2.1 Model

We employ the Hodgkin-Huxley-based network model that consists of three neurons. In our network, the directional interactions between the neurons are defined with synaptic couplings representing the electrical synapses. As shown in Figure 10.1, our network is defined as: Neuron 1 is spiking due to the applied current, Neuron 2 is connected to Neuron 1, and Neuron 3 is coupled with the other two neurons. Since our focus is on the underlying mechanism of noise-induced intermittent synchrony, the defined network architecture has worked well for our study. The equations describing the dynamics of membrane potentials with V_i of the neurons $i = \{1,2,3\}$ and the membrane capacitance C_m are

$$C_m \frac{dV_1}{dt} = I_{app} - I_{Na,1} - I_{K,1} - I_{L,1} - I_{noise} \tag{10.1}$$

$$C_m \frac{dV_2}{dt} = -I_{Na,2} - I_{K,2} - I_{L,2} - I_{noise} - I_{syn_{1-2}} \tag{10.2}$$

$$C_m \frac{dV_3}{dt} = -I_{Na,3} - I_{K,3} - I_{L,3} - I_{noise} - I_{syn_{1-3}} - I_{syn_{2-3}} \tag{10.3}$$

Here, I_{app} is the applied current that determines the constant depolarization level of neuron 1. I_{noise} is the noise component having a zero-mean normal distribution with standard deviation σ, which is represented by $N(0,\sigma)$. $I_{\text{Na},i} = \underline{g_{\text{Na},i}} m_i^3 h_i (V_i - V_{\text{Na}})$, $I_{\text{K},i} = \underline{g_{\text{K},i}} n_i^4 (V_i - V_{\text{K}})$, and $I_{\text{L},i} = \underline{g_{\text{L},i}} (V_i - V_{\text{L}})$ with $i = 1,2,3$ are the fast sodium current, transient potassium current, and the leak current respectively for neuron 1, neuron 2, and neuron 3. $I_{\text{Syn}_{i-j}} = k_x (V_i - V_j)$ is the synaptic current defined from presynaptic neuron to postsynaptic neuron j. Voltage-gated activation, m, and inactivation, h, of the Na^+ channel together with the activation, n, of the K^+ channel dynamics are defined as:

$$\frac{dx}{dt} = \frac{x_\infty(V) - x}{\tau_x(V)} \tag{10.4}$$

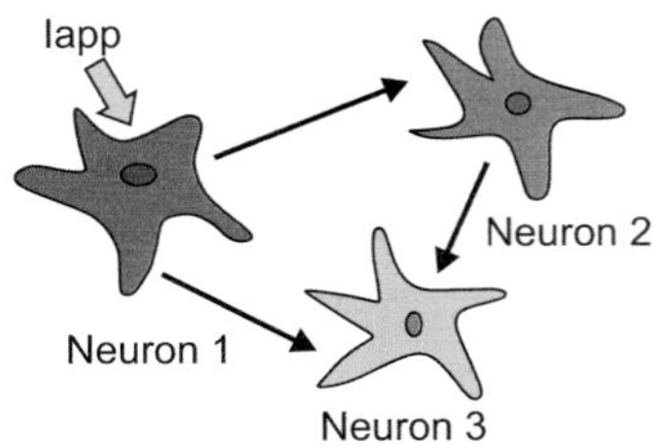

Figure 10.1: A very simple picture of the used model in this study.

where $x_\infty(V) = \frac{\alpha_x(V)}{\alpha_x(V)+\beta_x(V)}$, $\tau_x(V) = \frac{1}{alpha_x(V)+\beta_x(V)}$, $x = m,h,n$. Here $\alpha_x(V)$ and $\beta_x(V)$ are nonlinear voltage-dependent gating functions defined as

$$\alpha_m = 0.1(40+V)/(1-exp(-(\frac{55+V}{10} \qquad \beta_m = 4expexp(-(65+V)/18)$$
$$\alpha_h = 0.07expexp(-(65+V)/20) \qquad \beta_h = 1/(1+exp(-(35+V))$$
$$\alpha_n = 0.01(55+V)/(1-exp(-(10V+55)) \qquad \beta_n = 0.125expexp(-(V+65)/80)$$

More details about the defined functions can be found at Gencağa et al. (2020) [2].

Table 10.1: Parameter values used in this study for HH data.

$C_m = 1\mu F$	$V_{Na} = 50mV$	$g_{Na} = 120\mu S$	$k_1 = 0.1$
$I_{app} = 8mA$	$V_K = -77mV$	$g_K = 36\mu S$	$k_2 = 0.08$
	$V_L = -54.4mV$	$g_L = 0.3\mu S$	$k_3 = 0.08$

This model was integrated using XPPAUT which is a software tool for the numerical solution of differential equations, difference equations, delay equations, functional equations, boundary value problems, and stochastic equations. Customized MATLAB® codes were used for time-series analysis. This simple network is able to generate synchronized patterns that turn to intermittent synchrony with the noise integrity.

10.3 Methods

The proposed methods in analyzing the HH data in this study are two main parts:

i. The R-vine approach

ii. Regression models: The random forest method as the non-linear regression and the vector autoregressive model as the linear regression.

We used the R-Vine for its flexibility and simplicity in fitting the model parametrically. On the other hand, the random forest methodology is a non-linear and non-parametric method in which the distribution of the variables is not taken into account. Furthermore, the vector autoregressive model (VAR) is thought to

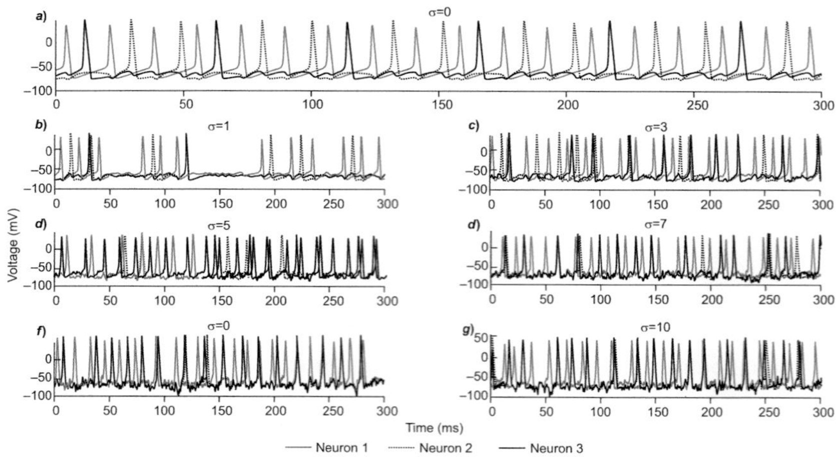

Figure 10.2: Voltage.

be a good alternative method where the HH data are some special kind of differential equation data in which all variables are connected to themselves through different lags. This model is a linear regression that is used frequently in time-series data. In Section 10.3.1, the R-vine method will be discussed in terms of its definition, applications, and type. Then, the random forest approach will be introduced by its one-step method, decision tree, and its related algorithm in Section 10.3.2. Finally, the VAR model and its assumption will be investigated to see if it can be applied to HH data in the application part.

10.3.1 The R-Vine Approach

The copula approach is introduced by Sklar's theorem for the first time [12]. The theorem says that for each combination of the variables, there is a unique function that defines the multivariate distribution of them in terms of the marginal distribution of each variable and a copula function that defines the connection or dependence structure between the variables. Because of the flexibility in modeling the relationship between the variables with almost no sophisticated assumption, it is increasingly applied to almost every case that the joint distribution or multivariate model is needed. Although the multivariate version of the copula has been used frequently, its bivariate version which is called a vine copula is more interesting for its simplicity and flexibility. There are more than three types of copula. However, Elliptical and Archimedean are two main and well-known types of it.

Elliptical copulas are the symmetric functions including Gaussian and t-distribution without and with tail dependence, in order. In Elliptical copulas, they are a one-to-one correspondence between Kendall's τ correlation coefficient and the model parameters. On the other hand, Archimedean copula includes some one-parameter and two-parameter models, with and without tail dependence terms, and also one-sided and two-sided tail dependence which can be freely fitted to almost every kind of data because of their flexibility. To show their properties, Table 10.2 represents the names and the parameters of the Archimedean family. In this table, the generator function defines the joint dependence structure between the variables using the equation $C(u_1, u_2, \ldots, u_v) = \Phi^{[-1]}(\Phi(u_1) + \Phi(u_2) + \cdots + \Phi(u_v))$ where $\Phi^{[-1]}$ stands for the pseudo inverse that is equal to the inverse of the generator function only for non-negative values and zero for negative values. More mathematical details about the selection of the generator function and the elliptical copula can be found in Brechmann and Schespsmeier (2013) [1]. In a real data set, the relationship between every two neurons can differ for other neurons and the multivariate version of Archimedean copula families can be complicated contrary to the elliptical family which is straightforward to follow. So instead of working with multivariate distributions the regular vine copula which can be written briefly as R-Vine can define joint density (distribution) functions in terms of pair functions almost independently from others which decreases the complexity of the model. Some of the pair functions are conditional density functions that provide us with a full probabilistic model. Therefore, instead of using all of the variables at the same time, the vine copula can decompose the joint density (distribution) function in terms of pair copulas that are simple and every connection can be chosen independently of all other pairs. Furthermore, canonical and drawable vines are two types of R-Vine that have different properties. Details are available in Brechmann and Schespsmeier (2013) [1]. The important and the most difficult job in data analysis by copula is the determination of the orders and the fitting of the best model among all available pair families. To do this, there are some goodness of fit techniques to compare all two families by giving some score function named VuongClarke test and checking the significance of the chosen family by a chi-square statistic. To compare two fitted models for the whole data, there are also some methods based on the Akaike information criterion (AIC) or Bayesian information crite-

Table 10.2: Properties of bivariate Archimedean copula families.

Name	Generator function	Parameter range	Tail dependence
Clayton	$1/\theta(t^{-\theta} - 1)$	$\theta > 0$	$(2^{-1/\theta}, 0)$
Gumbel	$(-\log \log t)^{\theta}$	$\theta \geq 1$	$(0, 2 - 2^{1/\theta})$
Frank	$-\log \left[\frac{e^{-\theta t} - 1}{e^{-\theta} - 1}\right]$	$\theta \in \mathbb{R} \setminus \{0\}$	$(0, 0)$
Joe	$-\log \left[1 - (1 - t)^{\theta}\right]$	$\theta > 1$	$(0, 2 - 2^{1/\theta})$
BB1	$(t^{-\theta} - 1)^{\delta}$	$\theta > 0, \delta \geq 1$	$(2^{-1/\theta\delta}, 2 - 2^{1/\sigma})$
BB6	$(-\log \left[1 - (1 - t)^{\theta}\right])^{\delta}$	$\theta \geq 1, \delta \geq 1$	$(0, 2 - 2^{-1/\theta\delta})$
BB7	$(1 - (1 - t)^{\theta})^{-\delta} - 1$	$\theta \geq 1, \delta > 0$	$(2^{-1/\delta}, 2 - 2^{1/\theta})$
BB8	$-\log \left[\frac{1 - (1 - \delta t)^{\theta}}{1 - (1 - \delta)^{\theta}}\right]$	$\theta \geq 1, \delta \in (0, 1]$	$(0, 0)$

rion (BIC) after parameter estimation which is done by the maximum likelihood method in copula.

10.3.2 Random Forest Regression Model

Different possible non-linear regression models such as support vector regression, decision tree, and random forest can be used based on the data type. In this study, we used the random forest model for its high accuracy among all mentioned regression models. Random forest is an ensemble learning methodology that takes an algorithm several times to increase accuracy. To explain it, we initially present the decision tree aspect which is used in one of the steps of random forest. The decision tree is a non-linear regression model and predicts new entries based on the fitted model, and flowchart-like structure in which each internal node represents a "test" on an attribute, each branch denotes the outcome of the test, and each leaf node indicates a class label. The following decision tree shows a very simple network version with three neurons named by V_1, V_2, V_3. In this tree, the first neuron has been divided into four parts named $V_{1,2}, V_{1,2}, V_{1,3}$ and $V_{1,4}$ based on some conditions of the information entropy. In this study, all conditions are intervals but not necessarily equal. In continue, the second neuron of the tree is divided into some parts based on information entropy and each blue ball in Figure 10.3 stands for an estimated value for V_3. For example, if the value of V_1 lies in the interval of $V_{1,4}$ and the value of V_2 lies in the interval $V_{2,5}$, the estimated value for V_3 would be $\hat{V}_{3,6}$.

Accordingly, the random forest algorithm is computed in the following way:

1. Pick a k data point from the training set.
2. Build a decision tree based on the selected data points.
3. Repeat the steps 1 and 2 for n times.
4. Gather all of the trees into a model by taking the average of all estimations.

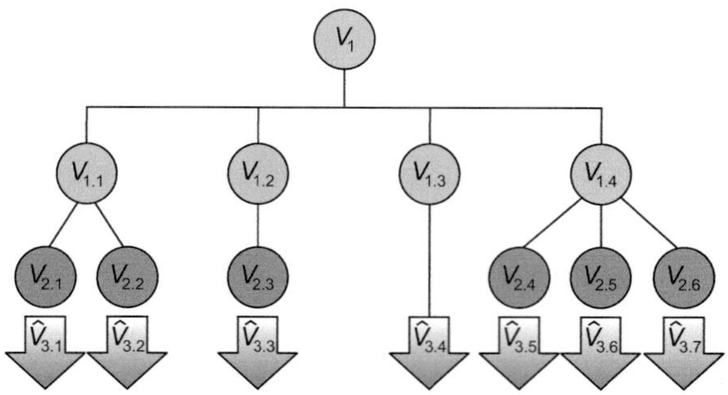

Figure 10.3: An example of a decision tree with three neurons.

There is a subtle point about the number of trees which should be chosen carefully. It seems to have a higher accuracy with a higher number of trees, but, in that case, for each data, there would be a prediction that makes the model meaningless.

10.3.3 Vector Autoregressive Model

The vector autoregressive model, briefly called VAR, is a multivariate time series technique to make a model based on each variable as a linear regression equation made from that variable's different lags plus another variable's lags. A mathematical representation of the model for a two-variable case is written as below.

$$V_i = \alpha + \beta_1 V_{i(t-1)} + \cdots + \beta_p V_{i(t-p)} + \delta_1 V_{j(t-1)} + \cdots + \delta_p V_{j(t-p)} + \varepsilon \quad (10.5)$$

for $i, j = \{1,2\}$. Equation 10.5 is called VAR(p) as the lags are the pth lags and each coefficient is estimated by the ordinary least square method by minimizing the sum of squares of errors if the main assumptions are all met. One of the underlying assumptions is that the errors should exhibit no autocorrelation, and there should be no correlation between the error terms and the lags of the variables. Consequently, the Generalized Least Squares (GLS) method was employed to estimate the coefficients of the model. However, Safi and White (2004) [10] demonstrated that the presence of autocorrelation does not significantly reduce the efficiency of the Ordinary Least Squares (OLS) method. In fact, OLS yields results that are nearly as accurate as GLS, which is typically favored when confronted with such challenging assumptions. In the VAR model, the overall goodness of fit is typically evaluated using measures such as the F-statistic and adjusted R^2. Even in cases where the disturbance terms (ε) exhibit autocorrelation, the estimated coefficients obtained via the OLS method can still be considered reliable.

10.4 Application

In this section, we will proceed with the implementation of a simulated dataset comprising three neurons and a noise term. The noise term follows a normal distribution with a mean of zero and a standard deviation denoted as σ. By varying the values of σ, we aim to examine the impact on the performance of the proposed multivariate models.

The first part of this section focuses on the analysis of difference variables, namely d_1, d_2, and d_3, along with the evaluation of correlations between different variables, differences, and neurons. This analysis aims to observe the changing trends of correlations as the noise values increase.

The second part involves the utilization of the R-Vine method to model the relationships between the neurons under different noise conditions.

Lastly, regression modeling, random forest, and linear regression models are applied in the final part. Each part concludes with a discussion of the advantages and disadvantages associated with the respective approaches.

10.4.1 Descriptive Statistics in the Inference of the Simulated HH Data

According to Equations 10.2 and 10.3, it is clear that the HH data model is written for each neuron based on the differences between the values of the neurons called within-difference variable d. These within difference variables are obtained by $d_i = v_{j,t} - v_{j,t-1}$ for $i, j = \{1,2,3\}$ and $t = \{1,2,\ldots,T\}$. Therefore, we suggested analyzing these variables beside each neuron to see their behavior for different noises. The correlation type used for this data set is Kendall's τ as it is a non-parametric measure. We can see the change of τ as σ increases in Table 10.3.

If we could show each column in a separate plot, it would be possible to see the slope of the change which is large in the first noise and then, the increase or decrease trend continues gradually.

10.4.2 R-Vine Method in the Inference of Simulated HH Data

We suggested R-vine which is a general form of vine copula in inference of the simulated data to build a model for different values of disturbances. Table 10.4 shows the order of the variables, the Vine type, and the copula family for every two neurons and also Kendall's τ (a non-parametric correlation measure) so that the comparison would be easier. In addition, to see the precision of the estimation, these τs can be compared with the τs obtained from the data. In the table,

Table 10.3: The correlation matrix between each neuron (V_1, V_2, V_3) and the within-difference values of the neurons (d_1, d_2, d_3).

σ	0	1	2	3	4	5	6	7	8	9	10
$V_1 - V_2$	-0.26	-0.22	-0.13	-0.11	-0.08	-0.02	0.03	0.08	0.1	0.11	0.12
$V_1 - V_3$	-0.23	-0.19	-0.12	-0.06	-0.05	0.02	0.07	0.12	0.12	0.16	0.13
$V_2 - V_3$	0.4	0.31	0.43	0.53	0.69	0.71	0.68	0.74	0.8	0.83	0.83
$V_1 - d_1$	0.14	-0.05	-0.08	-0.1	-0.12	-0.13	-0.13	-0.13	-0.13	-0.13	-0.13
$V_1 - d_2$	-0.41	-0.12	-0.09	-0.06	-0.5	-0.04	-0.04	-0.04	-0.03	-0.04	-0.03
$V_1 - d_3$	-0.34	-0.08	-0.06	-0.05	-0.03	-0.03	-0.03	-0.02	-0.02	-0.03	-0.03
$V_2 - d_1$	0.28	0.13	0.12	0.08	0.06	0.03	0.01	-0.01	-0.02	-0.03	-0.04
$V_2 - d_2$	-0.01	-0.05	-0.08	-0.1	-0.12	-0.13	-0.13	-0.14	-0.13	-0.13	-0.13
$V_2 - d_3$	-0.09	-0.06	-0.05	-0.07	-0.1	-0.12	-0.12	-0.12	-0.13	-0.14	-0.13
$V_3 - d_1$	0.25	0.11	0.08	0.05	0.03	0	-0.02	-0.03	-0.04	-0.05	0.05
$V_3 - d_2$	0.1	0.05	0	-0.04	-0.07	-0.07	-0.07	-0.1	-0.11	-0.1	-0.1
$V_3 - d_3$	0	-0.05	-0.07	-0.1	-0.12	-0.12	-0.13	-0.14	-0.14	-0.13	-0.13
$d_1 - d_2$	-0.14	0.35	0.41	0.42	0.43	0.45	0.48	0.5	0.53	0.54	0.55
$d_1 - d_3$	-0.15	0.39	0.43	0.44	0.44	0.46	0.5	0.51	0.53	0.55	0.55
$d_2 - d_3$	0.54	0.64	0.69	0.76	0.82	0.81	0.8	0.84	0.86	0.88	0.88

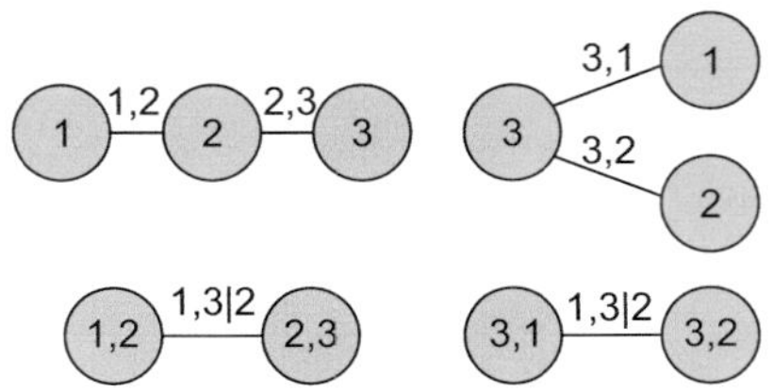

Figure 10.4: The structures of the best vine model in two types: D-vine with order 1,2,3 in the left side and C-vine with order 3,1,2 in the right side.

Table 10.4: The R-Vine analysis for the simulated data by changing the variance of the noise term.

σ	order	Vine type	1,2	1,3	2,3	τ_{1-2}	τ_{1-3}	τ_{2-3}
0	1,2,3	D-Vine	90 Clayton	BB8	270 BB8	-0.28	-0.16	0.42
1	3,1,2	C-Vine	90 Clayton	BB8	t	-0.19	-0.24	0.36
2	1,2,3	D-Vine	90 Clayton	270 Joe	t	-0.20	-0.02	0.61
3	1,2,3	D-Vine	90 Clayton	BB8	t	-0.16	0.04	0.64
4	1,2,3	D-Vine	90 Clayton	BB8	t	-0.09	0.08	0.76
5	1,2,3	D-Vine	270 Joe	BB8	t	-0.08	0.08	0.80
6	3,1,2	C-Vine	270 BB8	t	t	-0.07	0.06	0.72
7	3,1,2	C-Vine	270 BB8	180 Joe	t	-0.13	0.07	0.88
8	1,2,3	D-Vine	180 Joe	BB8	t	-0.08	0.06	0.83
9	3,1,2	C-Vine	270 BB8	t	t	-0.11	0.16	0.83

the rotation is noted with the family at the same cell and as an explanation for the vine type, it is better to remind about the structure of C or D Vine. In the table, the rotation is noted with the family at the same cell and as an explanation for the vine type, it is better to remind about the structure of C or D Vine. In the table there are two kinds of orders with two vine types: (i) the first form, order $1,2,3$ and D-Vine (means the pair variables are $1-2, 2-3$ and $1-3|2$), (ii) the second form: order $3,1,2$ and C-Vine (means $1-3, 2-3, 1-2|3$). Additionally, in the context of vine copula modeling for three neurons, it's important to consider the impact of noise variance on tail dependence. As the noise variance increases, the tail dependence in the bivariate distributions can change, leading to shifts in the copula structure. For instance, in Table 10.4, the rotation of copulas from Clayton to Joe to BB8 for the relation between the first and the second neuron indicates changes in the tail dependence relationships as noise variance increases. This highlights the sensitivity of the copula structure to noise, emphasizing the need to consider and model tail dependence when analyzing neuronal relationships. The dependence structure, represented by all undirected edges, is determined using the vine method and documented in Table 10.4. As the variance of the noise term increases, Kendall's τ deviates further from the expected value based on the data when $\sigma = 0$. To highlight this discrepancy, we employ the Chi-Square difference statistics, defined as $\sum_{i=1}^{n} \frac{(O_i - E_i)^2}{E_i}$, where E_i represents the expected

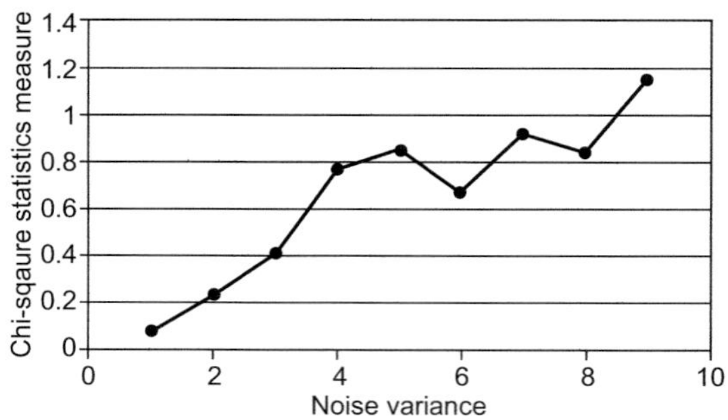

Figure 10.5: The trend of noise effect in the estimated correlation between neurons.

value (estimated Kendall's τ for non-noisy data) and O_i represents the observed estimated Kendall's τ for noisy data (for $i = 1, \ldots, n$). Here, n denotes the sample size for the signal data, which is typically large. Figure 10.5 illustrates the chi-squares values for each noise variance, and the results indicate that as the noise variance increases, the difference between the expected (estimated) values and the observed (real) values also increases. However, this trend does not hold for $\sigma = 6$. Upon examining each column of Table 10.4, it has been observed that the values of Kendall's τ increase and become more prominent as the variance of the noise term increases. Specifically, for the correlation between V_1 and V_2, a negative correlation of approximately –0.28 is observed when $\sigma = 0$. However, as σ increases to 9, this value decreases to –0.11.

Similarly, for the correlation between V_1 and V_3, a negative correlation of approximately –0.16 is observed when $\sigma = 0$. As σ increases, there is a consistent increase in the magnitude of the correlation, with the correlation measure reaching 0.16 for $\sigma = 9$.

Finally, for the relationship between V_2 and V_3, a significant increase is observed from 0.42 for $\sigma = 0$ to 0.85 for $\sigma = 9$. These observations suggest that as the variance of the noise term increases, the correlations between the different variables become more prominent and distinguishable.

In order to see the disturbance or the improvement of the noise term in the structure of the joint distributions between neurons. Figure 10.6 shows the scatter and contour plots for $\sigma = \{0, 4, 10\}$.

10.4.3 Random Forest Regression in the Inference of Simulated HH Data

In this study, we have determined the optimal number of decision trees in the random forest method to be 10, based on the Akaike Information Criterion (AIC). Using simulated data without any noise ($\sigma = 0$), we performed a regression analysis with V_3 as the dependent variable and V_1 and V_2 as predictor variables. The

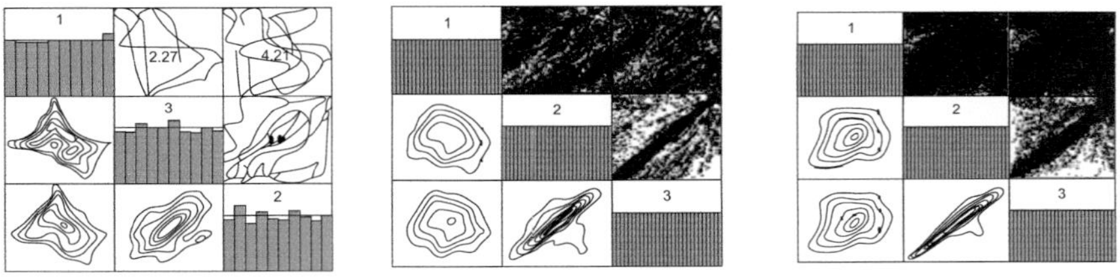

(a) $\sigma = 0$ (b) $\sigma = 4$ (c) $\sigma = 10$

Figure 10.6: The scatter and contour plots of the neurons for $\sigma = 0$ (the left side), $\sigma = 4$ (the middle plot), and $\sigma = 10$ (the right side).

dataset was split into a training set and a test set in a 3:1 ratio, and the data was scaled to have a zero mean and a variance of 1.

The performance of the random forest regression with 10 trees can be observed in Figure 10.7. The x-values represent the actual values of V_3 in the test set, while the y-values represent the predicted values obtained from the random forest regression. It can be observed that the predictions align perfectly with the actual values.

Then we could use that model for other data sets with different noises. It means that instead of the test set, we test the whole data set with a noise term to see how the accuracy changes by the variance of the noise term. Table 10.5 represents root mean square error (RMSE) values for different σ values for two cases if the dependent neuron is V_3 and V_2 in order.

10.4.4 Vector Autoregressive Model in the Inference of Simulated HH Data

We propose a linear regression model which is used for time series data, named vector autoregressive model VAR for this data, too, because it follows the criteria for this model. Here, we used the expressions for V_2 and V_3 using the non-noisy model in terms of its own lags and other two variables in the current time t. Then

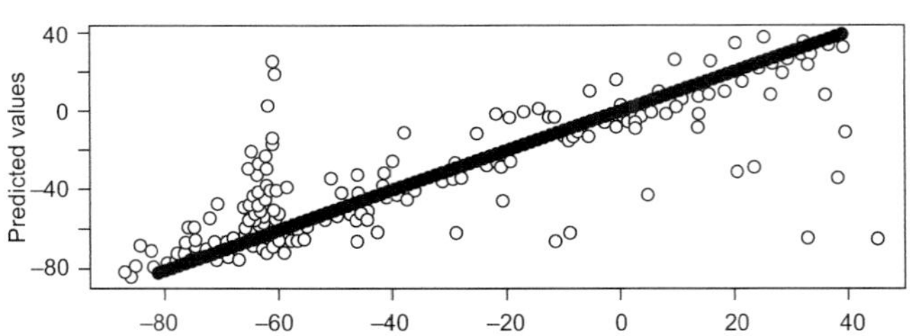

Figure 10.7: The performance of random forest regression for the test set of the data with $\sigma = 0$.

Table 10.5: The Root Mean Square Error measure for different σ's based on the random forest regression of V_3 shown by RMSE (V_3) and V_2 shown by RMSE (V_2) made from the non-noisy ($\sigma = 0$) simulated data.

σ	RMSE (V_3)	RMSE (V_2)
0	0.188	0.058
1	1.08	0.896
2	1.099	0.915
3	1.105	1.013
4	1.099	1.109
5	1.111	1.106
6	1.146	1.127
7	1.152	1.172
8	1.152	1.224
9	1.184	1.28

we applied the model for noisy data to see their behavior in terms of the coefficient sign and the residuals of the VAR model for both V_2 and V_3. Meanwhile, the lag number was selected according to the partial autocorrelation function for each variable and only the other variables without their time lags. The following expression shows the vector autoregressive model gained by the data without any noise as the true model like the one we used in random forest taken from the training data. Here, we used the non-noisy data as the training data and we want to test it for the other data with different variance of noise. The significant terms are determined by the backward regression model. Therefore, all terms in the following expressions are significant with level $\alpha = 0.05$ and the total model is significant with a p-value very close to zero. Hereby, the VAR models for data without any noise term ($\sigma = 0$) are constructed as below.

$$\begin{aligned} V_{2(t)} =& -0.36 - 0.001V_{1(t)} - 0.0005V_{3(t)} + 1.96V_{2(t-1)} - 0.84V_{2(t-2)} - 0.28V_{2(t-3)} \\ &+ 0.072V_{2(t-4)} + 0.117V_{2(t-5)} - 0.0185V_{2(t-7)} - 0.0076V_{2(t-9)} + \varepsilon \end{aligned}$$

$$\begin{aligned} V_{3(t)} =& -0.363 - 0.0005V_{1(t)} - 0.0012V_{2(t)} + 1.92V_{3(t-1)} - 0.75V_{3(t-2)} - 0.3V_{3(t-3)} \\ &+ 0.024V_{3(t-4)} + 0.099V_{3(t-5)} + 0.036V_{3(t-6)} - 0.0274V_{3(t-8)} + \varepsilon \end{aligned}$$

On the other hand, the adjusted R^2 for both models is computed as 0.999 implying that the regression model is meaningful. In this stage, we used the regression model for noisy data to see what happened to the residuals and the sign of coefficients as well as their significance. Tables 10.6 and 10.7 show the VAR model coefficient signs for V_2 and $V - 3$ in order, and their significance coefficients by using a * sign for significant ones togejer with the variance of the residuals ($VAr(\varepsilon)$) for each noise variance, named by σ.

$$\begin{aligned} V_{2(t)} =& a + a_1V_{1(t)} + a_2V_{3(t)} + b_1V_{2(t-1)} + b_2V_{2(t-2)} + b_3V_{2(t-3)} + b_4V_{2(t-4)} \\ &+ b_5V_{2(t-5)} + b_6V_{2(t-7)} + b_7V_{2(t-9)} + \varepsilon \end{aligned}$$

Table 10.6: VAR model coefficient signs for V_2 and its significance coefficients for each noise variance named by σ.

σ	a	a_1	a_2	b_1	b_2	b_3	b_4	b_5	b_6	b_7	$Var(\varepsilon)$
0	-*	-*	-*	+*	-*	-*	+*	+*	-*	-*	0.32
1	-*	-*	+	+*	-*	-*	-	+*	+*	-*	0.51
2	-*	-*	+	+*	-*	-*	-*	+*	+*	-*	0.84
3	-*	-*	+	+*	-*	-*	-*	+	+*	-*	1.27
4	-*	-*	+*	+*	-*	-*	-*	-	+*	-*	1.82
5	-*	-	+*	+*	-*	-*	-*	-*	+*	-*	2.45
6	-*	-*	+*	+*	-*	-*	-*	-*	+	-*	3.18
7	-*	+	+*	+*	-*	-*	-*	-*	-	-*	3.90
8	-*	+*	+*	+*	-*	-*	-*	-*	-	-*	4.80
9	-*	+	+*	+*	-*	-*	-*	-*	-	-*	5.77

Table 10.7: VAR model coefficient signs for V_3 and its significance coefficients for each noise variance named by σ.

σ	a	a_1	a_2	b_1	b_2	b_3	b_4	b_5	b_6	b_7	$Var(\varepsilon)$
0	-*	-*	-*	+*	-*	-*	+*	+*	+*	-*	0.30
1	-*	+*	-*	+*	-*	-*	+*	+*	+*	-*	0.59
2	-*	+*	-*	+*	-*	-*	-*	+*	+*	-*	0.94
3	-*	+*	-*	+*	-*	-*	-*	+*	+*	-*	1.38
4	-*	+*	-*	+*	-*	-*	-*	-	+*	-*	1.93
5	-*	+*	-*	+*	-*	-*	-*	-	+	-*	2.53
6	-*	+*	-*	+*	-*	-*	-*	-	+	-*	3.26
7	-*	+*	-	+*	-*	-*	-*	-	-*	-	3.94
8	-*	+*	-	+*	-*	-*	-*	-	-	-*	4.87
9	-*	+*	+	+*	-*	-*	-*	-*	-	-*	5.81

$$V_{3(t)} = a + a_1 V_{1(t)} + a_2 V_{2(t)} + b_1 V_{3(t-1)} + b_2 V_{3(t-2)} + b_3 V_{3(t-3)} + b_4 V_{3(t-4)} + b_5 V_{3(t-5)} + b_6 V_{3(t-6)} + b_7 V_{3(t-8)} + \varepsilon$$

All of the coefficients in Tables 10.6 and 10.7 with their associated equations the above equations are found statistically significant (more than 99%) and adjusted R^2 for both models are computed around 0.99 parallel with the corresponding goodness of fit statistics that are almost zero. This means that the model is valid and meaningful.

Accordingly, from both Tables 10.6 and 10.7 which represent the sign and significance of robustness for each variable (other neurons and its own time lags) that the noise can change the sign of the coefficient between each neuron and two other neurons in both models while the first three lags and also the last lag are completely robust towards noise. Furthermore, by comparing the VAR model for V_2 and V_3 it is clear that the VAR model is stronger for V_3 than V_2 by comparing the change in trend for each variable in every column. The other interesting and expected fact about the noise effect is seen in the last column of the tables as the variance of the residuals increases as the variance of noise increases. Finally, for all models, the expected value for residuals was almost zero which indicates the unbiasedness of the VAR estimation.

10.5 Conclusions

In this study, we have employed several methods to analyze the HH data with noise terms. The R-Vine approach was used as the first proposed method, which allowed us to observe changes in the families for some relationships, but not always. For instance, the relationship between the second and third neurons remained consistent for all noise values and became stronger, which was evident from the τ values and contour plots. It is worth noting that the data model was made by some coefficient of k, which represented the difference between two neurons. For instance, in the data used in this study, the coefficient of k was 0.27 for $V_2 - V_1$ as well as $V_3 - V_2$. Consequently, a perfect relationship between V_3 and V_1 was not expected. However, incorporating an additional term of $k(V_3 - V_1)$ or using a different k for every two neurons may improve the model's performance.

The correlation matrix of all within-difference variables d_i and V_is revealed that the noise term was not necessarily an interrupting factor. In some cases, the absolute value of correlation increased as noise increased, which was a positive aspect in some cases.

Furthermore, based on the random forest regression model, we observed that when there is only a small amount of noise in data, the spike positions get changed, destroying all predictions. However, for any noise level, random forest regression can be applied. Finally, we applied the partial VAR model to the data without any noise to create a model and then tested it for the data with some noise. The model for V_3 performed better than V_2, and the signs of the coefficients between the neurons in the VAR model changed due to noise. As noise variance increased, so did the residual variance. Consequently, the regression models worked well for data with reasonable noise terms, particularly the VAR model.

In conclusion, while R-Vine applies to almost every dataset, its performance or relationship between some neurons may vary for different noise terms. Therefore, analyzing this type of data using non-parametric methods or neural network analysis may be more effective as neurons are related to each other by some hidden factors or layers, which is evident from our model.

Acknowledgment

This study was supported by the Scientific and Technological Research Council of Turkey (TÜBİTAK) under project number 118E765.

References

[1] Brechmann, E.C. and Schepsmeier, U. 2013. Modeling dependence with c-and d-vine copulas: The R package cdvine. *Journal of Statistical Software*, 52: 1–27.

[2] Gençağa, D., Şengül Ayan, S., Farnoudkia, H. and Okuyucu, S. 2020. Statistical approaches for the analysis of dependency among neurons under noise. *Entropy*, 22(4): 387.

[3] Hodgkin, A.L. and Huxley, A.F. 1952. Propagation of electrical signals along giant nerve fibres. *Proceedings of the Royal Society of London. Series B-Biological Sciences*, 140(899): 177–183.

[4] Krupskii, P. and Joe, H. 2020. Flexible copula models with dynamic dependence and application to financial data. *Econometrics and Statistics*, 16: 148–167.

[5] Kudryashova, N., Amvrosiadis, T., Dupuy, N., Rochefort, N. and Onken, A. 2022. Parametric copula-gp model for analyzing multidimensional neuronal and behavioral relationships. *PLoS Computational Biology*, 18(1): e1009799.

[6] Lee, N. and Jong-Min Kim, J.-M. 2019. Dynamic functional connectivity analysis of functional mri based on copula time-varying correlation. *Journal of Neuroscience Methods*, 323: 32–47.

[7] Nagler, T., Bumann, C. and Czado, C. 2019. Model selection in sparse high-dimensional vine copula models with an application to portfolio risk. *Journal of Multivariate Analysis*, 172: 180–192.

[8] de Oliveira Peres, M.V., de Oliveira, R.P., Achcar, J.A. and Edson Zangiacomi Martinez, E.Z. 2020. The bivariate defective gompertz

distribution based on clayton copula with applications to medical data. *arXiv preprint arXiv:2012.07824*.

[9] Safaai, H., Onken, A., Harvey, C.D. and Panzeri, S. 2018. Information estimation using nonparametric copulas. *Physical Review E*, 98(5): 053302.

[10] Safi, S.K. 2004. *The Efficiency of OLS in the Presence of Autocorrelated Disturbances in Regression Models*. American University.

[11] Shi, P. and Yang, L. 2018. Pair copula constructions for insurance experience rating. *Journal of the American Statistical Association*, 113(521): 122–133.

[12] Sklar, A. 1959. Vol. 8 of fonctions de répartition à n dimensions et leurs marges, 229–231. *Paris: Publications de l'Institut de statistique de l'Université de Paris*.

[13] Vio, R., Nagler, T,W. and Andreani, P. 2020. Modeling high-dimensional dependence in astronomical data. *Astronomy & Astrophysics*, 642: A156.

IV

OPERATION RESEARCH IN SOCIAL SCIENCES

Chapter 11

Who Trusts Which Type of Artificial Intelligence?

Ulrike Reisach

11.1 Introduction

Artificial Intelligence (AI), a term coined by Emeritus Stanford Professor John McCarthy in 1955 [10], was defined by him as "the science and engineering of making intelligent machines". Much research has created human programmed machines to behave in a clever way, such as playing chess, but, today, machines can learn, at least partly like human beings do [9]. Machine Learning (ML) is the part of AI studying how computer agents can improve their perception, knowledge, thinking, or actions based on experience or data. For this, ML draws from computer science, statistics, psychology, neuroscience, economics, and control theory [9]. ML learns through examples and benefits from human data contributions, while the process and outcome of the machine remain partially opaque. No wonder users without experts in this field hesitate to trust such a machine.

Quite often users are not aware of where and how digital tools might be fostered by an AI. Examples are Chatbots which are used by Google, Amazon, and other companies that no longer provide humans to answer questions but have a chatbot. Airplanes have AI for a perfect landing, and Tesla drivers use AI for safe

Neu-Ulm University of Applied Sciences.
Email: Ulrike.Reisach@hnu.de

driving. In those cases, the pilot or driver might see contextual risks and can thus interfere.

Language translating like DeepL is an AI, and so are large language models (LLMs) which became prominent through ChatGPT, a Generative Pre-trained Transformer, a language model and sibling to InstructGPT, which is trained to follow the instruction in a prompt and provides a detailed response. The goal is the communication between humans and a machine.

Our research project on "Who trusts which type of AI" is aware that, depending on the students' backgrounds, their usage, knowledge, and perception of AI might be different. As Sullivan et al. (2023) [18], mention: "There has been a lack of public discussion about the potential for AI and ChatGPT to enhance participation and success for students from disadvantaged backgrounds. Similarly, the student's voice is poorly represented in media articles to date." Our four sample groups comprised of Masters students from different countries, universities, and study programs, they also have different experiences and goals of their own. This diversity represents a modern and diverse world whose common factors are mainly: being master students in a university, using English as the learning language, and wishing to achieve a good result for their future.

The goal of this analysis was to find out how and why students make their decisions, in different applied cases that relate to machine learning experiences. While, optimizing and discussing the decision-making process, its representation, and reflection, this research project aims to:

1. Understand how Masters students perceive AI and its related impact on their lives,
2. Foster critical thinking on the student's side,
3. Reflect the results and draw conclusions, and
4. Suggest recommendations for future teaching.

Paul and Elder (2006) [13] explain why critical thinking is very important: "Everyone thinks; it is our nature to do so. But much of our thinking, left to itself, is biased, distorted, partial, uninformed, or downright prejudiced. Yet the quality of our life and that of what we produce, make, or build depends precisely on the quality of our thought. Shoddy thinking is costly, both in money and in quality of life. Excellence in thought, however, must be systematically cultivated." Their definition of critical thinking is concise: "Critical thinking is the art of analyzing and evaluating thinking with a view to improving it."

Rimiene (2002) [15] provides seven different basic building blocks of critical thinking. For the decision-making, application, and research in our topical field, the following skills have been extracted, simplified, and structured as follows:

- Intellectual curiosity, truth-seeking, asking questions,
- Open-mindedness, respect for divergent skills and approaches,
- Anticipating positive and negative developments and their potential consequences for stakeholders,
- Systematic analysis of behaviors, and the respective reasons, intentions, and context,
- Reflective decision-making and mature and well-balanced judgments.

In addition to the previous skills, Gregory (1994) [7], suggests four interdependent actions:

- critical self-reflection (about one's own assumptions),
- ideology critique (about assumptions at the level of society),
- empirical research (experiment with, and observation of, the world) and
- hermetic inquiry (two-way communication with others).

Critical Thinking is connected to Elder's teaching fields at the Foundation for Critical Thinking (2023) [11] which comprise "The Art of Making Intelligent Decisions,' and "Critical Thinking: The Key to Survival in a World of Accelerating Change, Intensifying Complexity, and Increasing Danger.' This refers closely to the aim of "Who trusts which type of AI" which was created independently and developed through several student exercises in this field since 2020. For getting a feeling about the experience, the author started with Bachelor's students (6th semester), refined the questions, and decided that the answers and reflections of more advanced Master's students would be a good choice for samples. Using virtual teaching was a good opportunity to let students reflect and collaborate through a non-commercial collaboration tool while thinking on that topic.

There are good reasons why Operational Research (OR) matters a lot for this subject: OR is the discipline of applying advanced analytical methods to make more effective decisions through defining, structuring, and analyzing – in businesses, in societies, and especially in crisis times. In those situations, ethical and societal issues are paramount for careful and wise decision-making. Careful decisions matter for students as well as scientists, citizens, and governments, therefore students should start early to distinguish opinions, assumptions, and facts, and to discuss and reflect on their decisions critically.

11.2 The Methodological Implementation Process

In order to allow the students to achieve a comparable level of understanding, a 30-minute basic introduction on-site or virtually is provided and agreed upon with the professor who runs the respective course. The presenting researcher (the author of this chapter) follows these 3 steps:

1. Explaining Key Ethical Issues of Machine Learning,
2. Explaining Human Cognition versus AI,
3. Giving Eight Short Questions on Eight Different AI fields.

Since ChatGPT started in November 2022, our last student group also got two additional ChatGPT questions, with a prolongation of 10 minutes.

For the next 30 minutes, students were invited to contribute their answers and explain why they decided to answer the way they did. To avoid privacy issues, the following safe and non-commercial tool was chosen: https://Zumpad.zum.de/, a German collaboration tool, developed during the pandemic to facilitate teaching, which is free for education and does not track or collect any private data. Since the volume of usage increased, Zumpad deletes the results after six months, and we could copy the content to a Word document.

What students learn: They learn that there might be possibilities to benefit from the new machine learning tools. Nevertheless, the contexts are very complex. Our research project covers the perception of AI users regarding trust in AI usage in certain fields, including their reasoning on the bot's capability. Additionally, students could/should learn to critically evaluate the contextualization, this means for, e.g., asking how the AI has been trained. AI facilitates novelties, keeps users busy, and allows them to use more and more machine-driven results. Therefore, it makes sense to think through the implications of the decisions they (and others) make and the solutions they come to. Their methods of problem-solving and decision-making can be useful in several fields, such as management of administration, governmental, educational, commercial, and industrial processes can be applied in manifold ways.

The examples which the students give are current and related to their learning in universities and to their future careers. To keep the possible options limited, we started with questions that are close to the students (Questions 1–6), two more open questions for their own ideas (7–8), and 2023 two additional ChatCPT examples. ChatCPT's primary purpose … "is to generate human-like text, which can be used for a variety of applications, such as chatbots, automated content creation, and language translation." (The Openai Blog on ChatCPT 2023). This definition was given for the 2023 group because it was new and thrilling for the participants.

The questions allow students to reflect anonymously on how they would make their own decisions. Since the participants answer in parallel, they are motivated and can learn from other students' arguments. Providing their decisions and explanations for others is a type of problem-solving because some may not have thought about this question before. Therefore, our approach to working in parallel and anonymously gives them the motivation to think about how they would act and why.

11.3 The Structured Process During the First 30 Minutes

11.3.1 Explaining the Artificial Intelligence as a Machine

To explain Artificial Intelligence as an engine, a simplified visualization is given along with an oral explanation. The three input factors are shown as

1. Real-time data from sensors,
2. Background information from a database, mostly taken from a cloud,
3. Cases and algorithms, usually contributed by a team of humans.

The outcome is usually a prediction and decisions based on those suggestions and/or results, which can be taken as a basis for taking actions, as the below-shown Figure 11.1 shows.

During this process, humans could start to reflect on whether the three steps might achieve a perfect result: Did we get enough really representative pieces of information? Is there any bias or doubt about the input? Did anyone think deeply about "outdated" data which might not fit the current reality? Have humans chosen an appropriate database? Did they, for example, ignore differences in gender, age, health status, and so forth? What could be the impact of such shortcomings on human users of AI? In a pure quantitative database, such kinds of reflections or empathy are non-existing. It would require much more effort to consider the quality of many input sources.

A similar factor is cases or algorithms; those might not necessarily satisfy all important quality needs of human users. Cases could be chosen randomly or as not too realistic, the context might be unclear or certain cases and algorithms were not available or achievable (e.g., by privacy protection). Restrictions, lack of broader views, authentic expertise, or knowledge could influence the outcome of an AI machine learning process. In the end, it is mostly humans who provide inaccurate inputs, and human wishes for perfection might not always be fulfilled.

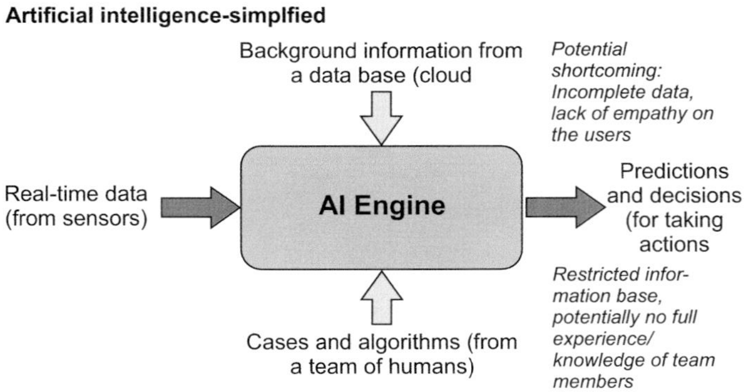

Figure 11.1: Artificial Intelligence – simplified (Author's own representation to explain the process of AI and their shortcomings).

11.3.2 The Explanation of Potential Ethical Issues, Based on those 3 Examples

1. Purpose: "Is the purpose we put into the machine the purpose which we really desire?" (Norbert Wiener, 1940 [19]);

2. Value alignment and the off-switch problem [17]: Three principles of creating safer AI) → cooperative reinforcement learning better than copying human behavior;

3. Context is essential: Ethical problems cannot be resolved by algorithms: a dialogue between humans (e.g., different types of stakeholders and their perspectives) is required.

11.3.3 Explaining Human Decision-making

In social science, quantitative research usually does not provide a standard result for all societal groups and individuals. Humans are too manifold and diverse, to categorize them like technical or other products which have no intelligence or experiences during their lives. Humans see the world from their own and quite specific perspectives, their wishes, cultures, and practices are related to their lives contexts, and experiences. Since those contexts may change, humans learn to adapt to live in different frameworks and survive physically. Their decision-making may not only depend on rational facts but can also be influenced by humanity and caring for others (altruism) or by feelings such as love or friendship – those skills are unique and can hardly be fully digitized since their behaviors relate to humans, human groups and societies whos' ideas might change along their respective living styles and framework conditions.

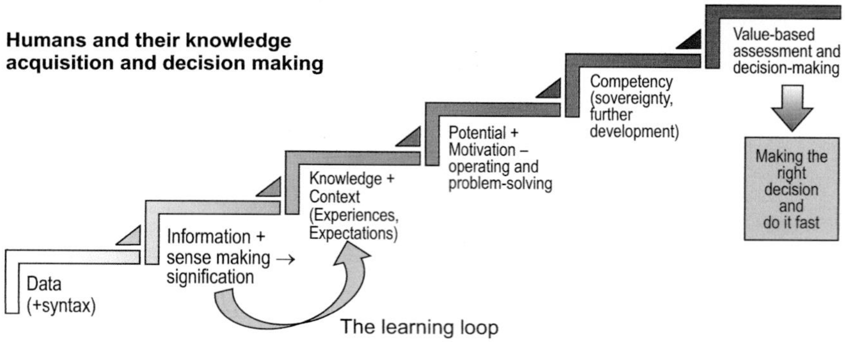

Figure 11.2: Own adaptation and interpretation of Erlach, C./Orians, W. and Reisach, U. (2013) [4] Knowledge-Transfer in Business, and North, K. (2005) Knowledge-oriented Leading of Companies (2003) [2].

Examples of the students' answers related to humans and caring:
HNU1 on AI for evaluating health:
"Depends, since there are people that are afraid of doctors (especially children) and a human being with emotions and empathy would be better at that situation."
UCD on AI for evaluating health:

"Yes, the modern AI systems can get more details across the world and bring real expertise than one doctor who won't have as much. But AI won't be able to replace human interaction elements. So, AI must be an assist tool for the doctor, not a replacement."

HNU 2 on a chatbot for a job interview:
"Definitely not! In an interview, we must not only talk about professional terms but also practice relationship-building with the interviewer and network with them on a personal level which we cannot do with a chatbot. Moreover, the chatbot will not always have an answer to your question." "Absolutely not. The qualifications of a human-based task take a level of decision-making and human qualities that have yet to be attained by a computer; thus, the availability of the job. Trust, personality, and collaborative potential, to my knowledge, cannot be evaluated/categorized to a machine's understanding."

HNU2 on AI acceptance examples:
"I singled to traffic, indicating my turn but the driver noticed too late. The hood of the car collides with my bike. My body goes ragdoll, spilling against the pavement. A snap, a pop, perhaps a break. In the tumble, I can't be sure of anything. I come to a stop, though the pain hasn't quite sunken in. My head clears enough to recognize my attacker speeding off into the distance. I can't move, or perhaps

I'm unwilling. A chirp; a familiar voice, distinct and cold cries out. 'How can I be of service.' At the moment I am still reluctant. "Dial the nearest hospital."

11.3.4 Reflectivity in Algorithm Design and Implementation

Reflectivity is an approach close to critical thinking and consequentialism, deeply rooted in the humanities, especially in philosophy and psychology:

- **Interdisciplinary thinking, considering the contextual framework** such as cultural, political, and societal conditions and short- plus long-term developments;
- **Social skills like empathy and compassion** (Everett 1966, on Albert Schweizer) in order to identify with another person's situation and recognize human motivations, rational or irrational interests;
- A transgression of the own self and an **interest in living together harmoniously: Consideration of the broader societal impact** (e.g., Max Weber and Confucius).

Balancing legitimate self-interest and legitimate societal interest includes corporate and societal governance. For example, J.S. Mill's (1863) Utilitarianism measures along calculations, similar to an AI that follows the programmed action. Other philosophers argue differently, their ethical concepts lead to different answers. In critical situations, it might be difficult to measure and/or reflect on and discuss ethical decisions. Decisions in a "standard" case, e.g., drivers not following a red traffic light, can be decided relatively easily, even by an AI. Nevertheless, there might be complex (law) situations that require critical thinking to understand why this happened in a very specific traffic or driver's (e.g., health) situation.

Approaches to contemporary cognitive theory:

- The information-processing approach attempts to understand human thought and reasoning processes by comparing the mind (or the brain).
- A sophisticated computer system that is designed to acquire, process, store, and use information in specific ways (Britannica, Philosophy of Mind: The computational-representational theory of thought).

The number of participants and the interest of students in this research represent a good fit. Several professors were contacted to offer the 30-minute introduction on AI (machine learning) and human cognition, and then the Zumpad tool to raise collect anonymous answers for the 8 questions. It was not easy to explain the purpose and procedure without practicing it – in fact, an extensive description program along this chapter would be needed to fully satisfy all potential

Questions + Discussion topics for participants Please also give an explanation of why you have chosen the respective answer.
1. As a student or Ph.D., do/would you like to be (performance) graded by AI? 2. For a job interview, do/would you like to be evaluated by a chatbot? 3. As a department leader of a company, would you like to hire employees who have been selected by a chatbot? 4. As a patient, do you trust AI for health evaluation? 5. Do you like the fact that AI in social media may create a profile of you? 6. Would you use AI for your personal decision-making? 7. When and how do/would you accept which other forms of AI? 8. For which purpose would you personally use AI? 9. Would you let ChatCPT write your thesis on business development? 10. Can ChatCPT write a balanced comparison between political systems?

answers of professors and/or students. It is necessary to keep the questions hidden before the virtual course starts, otherwise, the comparison of results might be no longer balanced. A few thought that a certain course program would be necessary, but that is not the case, only AI Experts and Neurologists would probably not learn much novelties. When further colleagues are interested, please contact the author.

11.4 The Students' Answers and Reflections during the Second 30 min

Here the professors are mostly passive, they see many flows of answers but do not interfere, unless students have a specific question (which has not happened so far) or the Zumpad gets too big or too fast and the professor needs to provide a second Zumpad, to keep the pace (this happened, and we now provide two Zumpads for big groups).

11.5 The Analysis of the Results

The professors' goal is to find out whether the students ...

- really understand what AI does and how it works,
- really differentiates between humans and machines,
- is prepared to use AI and/or human intelligence for solving problems,
- do trust a machine and in which contexts, or
- they rather trust their own brain?

Based on the students' results, the author analyzes whether students understand:

- What machine learning is and how this might matter for normal situations in their lives,
- What human cognition is and when/how/why this might be a better source of advice,
- The choices they have in problem-solving and decision-making are related to trust.

By reflecting on how they answer and explain their decisions, they will achieve higher thinking skills on important common novel machine tools.

The results and answers allow the researcher to analyze three different quality-related categories such as (1) Fact-driven, (2) Reflectivity, (3) Unclear. We can, e.g., see whether students answered shortly and superficial or more detailed, thinking in context and thus increasing the complexity of the decision. We combine a qualitative research result with a semi-quantitative approach, depending on how many decisions were given through categories and explanations of students.

So far, the authors have collected four rounds of "Who trusts which type of AI?" and collected the digital records on the questions, along with the anonymous students' results. Summing up, 55 participants contributed, knowing that this is a research project with anonymous answers. The universities are:

1. HNU1, Germany: International Corporate Communication and Media Management 2022 (Course: Institutional, Ethical and Cultural Aspects of ICCMM),
2. HNU2, Germany: International Corporate Communication and Media Management 2023 (Course: Institutional, Ethical and Cultural Aspects of ICCMM),
3. Xidian, China, Cross-Cultural Management and different Ph.D. candidates
4. UCD, Ireland: Computational Social Science (Course: Social Simulation: Methods and Models).

The universities, study programs, number of participants, and backgrounds of different countries and cultures show a diversity of students who still share common patterns:

- being a Master's student of 1 to 4 semesters in a Master's program (Ph.D. candidates can participate, too),
- being active in this reflection exercise,
- delivering a majority of reasonable answers.

Showing higher thinking skills (following Bloom's taxonomy, see Anderson and Krathwohl 2001) is one of the desirable outcomes.

11.6 The Results and Analyses of the Masters-Students' Anonymous Answers

11.6.1 Universities, Courses, and Settings

Universities and day of action	Departments	Professors	Study Programs	Courses	Masters and PhD Candidates: numbers	Settings
Neu-Ulm University of Applied Sciences 09/11/2022	Information Management	Prof. Dr. Ulrike Reisach	ICCMM (International Corporate Communication and Media Management)	Institutional Ethical and Cultural aspects of ICCMM	Masters; 1. semester, 15 participants, 12 different countries	Same Intro, same method; 8 Questions
Neu-Ulm University of Applied Sciences 03/05/2023	Information Management	Prof. Dr. Ulrike Reisach	ICCMM (International Corporate Communication and Media Management)	Institutional Ethical and Cultural aspects of ICCMM	Masters; 1. semester, 17 participants, 9 different countries	Same Intro, updated method; 10 Questions, including ChatGPT
Xidian University, Xi'an, Shaanxi 05/05/2022	School of Economics and Management	Prof. Dr. Du Rong	Information Systems, Knowledge Management	Information Systems, Cross-Cultural Management	Masters; +PhDs; 8 participants	Same Intro, same method; 8 Questions
University College Dublin 09/11/2022	Social Sciences and Law	Associated Prof. Dr. Pablo Lucas	Sociology	Social Simulation: Methods and Models	Masters; 4. semester, 15 participants, 5 different countries	Same Intro, same method; 8 Questions

Note: Since the risks of AI are increasing through ChatGPT, those were included for the most recent team HNU2 to discuss the impacts of those changes and risks for (a) our sample candidates and results and (b) for an outlook for the future.

11.6.2 Answers and their Reflectivity, Effectivity and Qualitative Explanations

Examples of the first "It depends" answers are the following:

Universities + date of action	Answers on Questions 1-6 Reflectivity and Effectivity	Potential Reasons	Understanding AI	a) Topical interest Questions 7-8 b) Creativity
HNU1 09/11/2022	Diverse answers Sum: 85 6 Facts to "Yes" 63 "No" 19 "It depends", 14 reflect, 3 facts, 2 unclear The sum of Reflection: 42 Partially short answers. Many "I think" answers.	Risk-avoidance. Carefulness Tendency to prefer humans. Participants who were not sure might have given short answers.	7 underestimate AI. Balanced answers on personal decisions, and good explanations for all 3 categories. 42 reflective answers show a critical approach and competency.	a) The health topic showed more trust in this field, and many good answers on combining Humans and AI. b) 13 pragmatic suggestions, AI for facilitating or fastening tasks, only one answer unclear.
HNU2 03/05/2023	Diverse answers Sum: 63 20 "Yes" 33 "No" 10 "It depends", 10 Reflective, 1 underestimates AI. Sum of Reflection: 38 Several good answers.	Carefulness, risk-avoidance. Stronger tendency to prefer humans (maybe because students from different backgrounds are not very familiar with AI.)	11 underestimate AI. Most reflectivity appears on personal decision-making, with "It depends" answers. Critical thinking is available, but not all show it.	a) Job Interviews and Grading along AI find the most interest. Question 7: Other Types of AI in 8 "Yes", are well reflected, pragmatic and creative. Some of those already reflect on Chat CPT.
Xidian University, 05/05/2022	Well-balanced answers – harmony. 11"Yes" 14 "No" 18 "It depends" 24 Reflective, 7 no explanation The sum of Reflection: 24 Good reflection	Tendency to prefer humans – collectivism. The answers were partially short (probably from participants who were not sure), and other answers were detailed.	Critical thinking, sensitive to privacy.	Job-finding and social media found the most interest (probably because of job-finding issues). Creative suggestions (probably because they welcome apps to diminish their workload.
UCD 09/11/2022	Deep understanding of AI 12"Yes" 19 "No" 18 "It depends" 40 Reflective Sum-Reflection: 40. Most Reflection and most "it depends".	Tendency to trust the AI, if it is well trained, it shows a deep understanding and reflects well.	Critical thinking; Only 2 participants underestimated AI in social media.	Interested in most topics, rational answers.

HNU2 about Grading and AI: "This depends on the type of assignment. If there is only one right answer, then the grading process with the AI tool is faster. If it is an assignment with personal opinions, like an essay or maybe an academic paper with a study I would prefer being graded by a human, because the AI can't take into consideration my personal situation or state of knowledge."

Xidian about Health and AI: "I would trust some basic health evaluation. But when comes to diseases, especially unusual diseases, I would take the doctor's advice further. I recognize the contribution of AI in rapid selection for COVID-19, but for now, I see AI as a supporting role that needs further confirmation from human experts."

UCD about Health and AI: "Again, if the AI is calibrated properly. AI might also have the capacity to improve medical practice and be more efficient than human health professionals. Especially in detecting sickness or condition between difference groups as there is a known issue among health care professionals."

Xidian on AI in social media: "It depends on the context. I would like it if it's just used for personalized recommendations, which will be time-saving for searching for information I prefer. However, I wouldn't like it if it leaks my personal information for their own purposes unknown to me."

UCD on AI in social media: "Not at all, I find it scary. Conscious consent doesn't exist on many of the terms and conditions we agree on in order to be part of x social media."

HNU2 on ChatCPT: "I would appreciate and use the help of ChatGPT to write my thesis, but I would do an extensive review, in the end, to consider human-related factors and critical and confidential information, for example." "When ChatGPT is used as an input giver/search machine, it is totally helpful."

11.6.3 Quantitative Results

Number of answers	In Sum	Answers "Yes"	Answers "No"	Answers "It depends"
Number of participants' answers	HNU1: 65 HNU2: 72 XDA: 56 UCD: 57	HNU1: 10 - (04 Reflect) HNU2: 36 - (26 Reflect) XDA: 22 - (05 Reflect) UCD: 15 - (15 Reflect)	HNU1: 54 - (24 Reflect) HNU2: 31 - (20 Reflect) XDA: 19 - (14 Reflect) UCD: 10 - (13 Reflect)	HNU1: 21 – (14 Reflect) HNU2: 09 – (07 Reflect) XDA: 17 - (11 Reflect) UCD: 28 - (26 Reflect)
Questions (1-8) + 2 Questions on ChatCPT 2023 (HNU2)	Answer Categories:	Fact driven – Reflect – Unclear Answers and Effectivity of Explanations		
(1) Grading by AI?	HNU1: 14 HNU2: 11 XDA: 03 UCD: 09	HNU1: (1) 2 Fact-driven, 1 Reflect, 1 Unclear HNU2: 1,5 Reflect, 0,5 unclear XDA: 1 Fact-driven, 1 unclear, 1 short UCD: 5 Reflect	HNU1: 5 Reflect, 2 underestimate AI, 2 Unclear HNU2: 4 Reflect, 3 Unclear XDA: 0 UCD: 1 Reflect, 1 short	HNU1: 5 Reflect (+ 1 underestimates AI), 1 Unclear HNU2: 1,5 Reflect 0,5 fact driven + underestimates AI, 1 unclear Reflect+opinion XDA: 0 UCD: 1 Reflect
(2) For a job interview, do/would you like to be evaluated by a chatbot?	HNU1: 15 HNU2: 12 XDA: 08 UCD: 09	HNU1: 0 HNU2: 1 Fact driven XDA: 2, 1 Reflect UCD: 0	HNU1: 6 Reflect, 2 Fact-driven, 1 Unclear HNU2: 8 Reflect, 1 Unclear, 3 underestimate AI XDA: 3 Reflect, 1 short UCD: 4 Reflect (1 Unclear)	HNU1: 0 HNU2: 0 XDA: 1 short, 1 Reflect UCD: 3 Reflect, 1 Unclear
(3) As a department leader of a company, would you like to hire employees who have been selected by a chat bot?	HNU1: 09 HNU2: 09 XDA: 05 UCD: 10	HNU1: 1 Joke HNU2: 0 XDA: 1 Fact driven, 1 opinion UCD: 0	HNU1: 3 Reflect, 2 Reflect partially questionable, 2 Fact-driven HNU2: 5 Reflects, 1 gives no reasons, 1 underestimates AI XDA: 3 short, 2 Reflect UCD: 6 Reflect, 1 opinion	HNU1: 1 Reflect HNU2: 1 Reflect XDA: 3 Reflect UCD: 6 Reflect
(4) As a patient, do you trust AI for health evaluation?	HNU1: 14 HNU2: 09 XDA: 05 UCD: 10	HNU1: 2 Reflect, 2 Fact-driven HNU2: 5 Reflect, 3 are fact-driven, 1 drifted to other health examples XDA: 1 Fact driven UCD: 5 Reflect	HNU1: 0 HNU2: 0 XDA: 2 Reflect, 3 short UCD: 0	HNU1: 5 Reflect, 3 Fact-driven, 3 Unclear HNU2: 0 XDA: 3 Reflect UCD: 5 Reflect
(5) Do you like the fact that AI in social media may create a profile of you?	HNU1: 10 HNU2: 10 XDA: 08 UCD: 06	HNU1: 0 HNU2: 8 Reflect, 1 Unclear, 3 underestimates AI and Profiling XDA: 0 UCD: 0	HNU1: (5) 5 Reflect, 1 fact-driven, 4 Unclear (!!) HNU2: 2 Reflect XDA: 5 Reflect UCD: 2 Reflect, 2 underestimate AI in social media	HNU1: 0 HNU2: 0 XDA: 3 Reflect UCD: 1 Reflect
(6) Would you use AI for your personal decision-making?	HNU1: 10 HNU2: 11 XDA: 5 UCD: 5	HNU1: 2 Fact-driven, 1 Reflect HNU2: 2 Reflect, 1 welcomes the rationality of AI XDA: 3 Reflect, 1 gives an explanation UCD: 1 Reflects, 1 misunderstands decision-making	HNU1: 3 Fact-driven, 1 Reflect HNU2: 1 Reflect, 2 Offer own way XDA: 0 UCD: 0	HNU1: 3 Reflect HNU2: 5 Reflect XDA: 3 Maybes, one of them gives an explanation UCD: 2 Reflect, 1 Unclear
(7) When and how do/would you accept which other forms of AI?	HNU1: 03 HNU2: 09 XDA: 06 UCD: 04	HNU1: 0 HNU2: 7 Reflect, 0,5 Reflect + underestimates (financial) risks, 1 gives a suggestion (no explanation) XDA: 1 Reflectivity, 1 Unclear, 1 short UCD: 0	HNU1: 4 (fact-driven), 1 Unclear (!!) HNU2: 0 XDA: 0 UCD: 0	HNU1: 0 HNU2: 0 XDA: 1 Reflect, 2 vague UCD: 4 Unclear
(8) For which purpose would you personally use AI?	HNU1: 10 HNU2: not asked XDA: 06 UCD: 04	HNU1: 0 HNU2: 0 XDA: 6 creative ideas UCD: 4 Reflectivity	HNU1: 10 Fact-driven HNU2: 0 XDA: 0 UCD: 0	HNU1: 0 HNU2: 0 XDA: 0 UCD: 0
(9) Would you let Chat-GPT write your thesis on business development?	HNU2: 12 1 "Ja" 4 "No" 5 "It depends" 09 Reflective	HNU2: 1 welcomes Chat-GPT	HNU2: 4 Reflect, 1 refers to personal opinions	HNU2: 4 Reflect, 1 refers to personal opinions
(10) Can ChatCPT write a balanced comparison between political systems?	HNU2: 06 (few answers here)	HNU2: 2 are (over) optimistic	HNU2: 1 thinks ChatCPT is not sensible enough	HNU2: 2 are Reflect but not sure, 1 doesn't know

11.6.4 Reflections and Limitations of the Quantitative Analysis

(1) AI in Grading in Universities
was perceived positively by UCD and Xidian, because they are familiar with exactly this grading process.

(2 and 3) Job-related AI has been seen negatively by HNU1, HNU2, Xidian, and UCD. Probably the students prefer to show their personal knowledge, experience, and skills.

(4) AI in health has been . . .

- positively perceived by HNU2 with "Yes"
- HNU1 voted "Yes" and "It depends"
- Compared with the other 3, UCD was most against AI in the field of health, which was surprising, since they were open for AI in Grading.

(5) AI in social media has . . .

- been seen positively by the majority of HNU2, although a few were not happy about the profiling.
- been rejected by 5 participants in UCD and by all students of HNU1.
- two of the UCD students seem not to see the connection between social media and AI.

Reflection: Most social media users enjoy gamified entertainment which is helpful for advertisers but doesn't educate users about good sources, risks, and disinformation. Categorizing humans along those lines may include budgets, purchases, and age, but does not tell us a lot about personalities and their real lives and thoughts. It could also be that the term "profile" in the context of social media was not clear to the students, but this also shows that they are not aware that "if you don't pay, then you pay through your data". Consuming "more of the same" might lead persons to a "filter bubble" because the algorithms try to keep people active to potentially advertise or purchase something.

11.7 ChatGPT gives Answers

Questions: Author (UR), **Answers:** ChatCPT First Version of CharCPT, https://openai.com [1], December 15, 2022.

1. Do you think that students would like to be graded by AI? For which type of exam would they probably agree and for which type of exam would they probably deny?

It is possible that some students may be open to being graded by AI, while others may prefer to be graded by a human. It is difficult to predict how all students would feel about this, as individual preferences and experiences with technology may vary. One factor that could influence a student's willingness to be graded by AI is the type of exam they are taking. For exams that are primarily focused on factual knowledge or technical skills, students may be more open to being graded by AI because it may be more objective and less prone to bias. On the other hand, for exams that involve more subjective or interpretive tasks, such as essays or creative projects, students may be more hesitant to be graded by AI because they may feel that it is less capable of accurately evaluating their work.

Ultimately, whether or not students would be willing to be graded by AI would depend on their personal preferences and beliefs, as well as the specific context and details of the exam. It would be important to consider these factors and to communicate with students about the use of AI in grading to ensure that they feel comfortable and confident in the process.

2. Can an AI manipulate a decision?

It is possible for artificial intelligence (AI) to influence or manipulate a decision. AI systems can be designed to present information or make recommendations in a way that is intended to influence a person's decision-making. For example, a recommendation system that suggests products or services to a user based on their previous purchases or search history is attempting to influence their decision about what to buy.

However, it is important to note that an AI system can only influence a decision to the extent that it has access to relevant information and is able to present it in a way that is meaningful to the person making the decision. While an AI system may be able to present information in a way that is intended to influence a decision, it does not have the ability to control or force a person to make a particular decision. It is also worth noting that AI systems can sometimes be biased, either because of the data they were trained on or because of the way they were designed. This can result in AI systems making recommendations or decisions that are unfair or discriminatory. It is important to carefully consider the potential consequences of using AI to make decisions and to take steps to mitigate any potential biases.

3. Can you understand that people say that sympathy matters more than skills for the workflow and success of a company?

Yes, it is possible for people to believe that sympathy, or the ability to show concern and understanding for others, is more important than skills for the workflow and success of a company. There are several reasons why this might be the case.

First, a sympathetic and understanding workplace culture can foster a sense of connection and belonging among employees, which can lead to increased motivation, productivity, and job satisfaction. When employees feel that their colleagues and management care about their well-being and are willing to support them, they may be more likely to feel invested in the success of the company.

Additionally, the ability to show empathy and understanding toward others can be important in building and maintaining positive relationships with colleagues, customers, and partners. This can be especially important in customer-facing roles, where the ability to relate to and understand the needs of others can be a key factor in building trust and partners. This can be especially important in customer-facing roles, where the ability to relate to and understand the needs of others can be a key factor in building trust and maintaining good relationships. It is worth noting that while sympathy and understanding are important qualities to have in a work environment, they are not the only factors that contribute to a company's success. A wide range of skills and abilities, including technical expertise, problem-solving, and communication, can also be important for the workflow and success of a company.

4. Do you think that companies should let their potential employees be selected by a conversation with a chatbot (like you)?

It is up to individual companies to decide how they want to select their employees. Some companies may choose to use a chatbot as part of the hiring process, while others may prefer to rely on more traditional methods such as resumes, interviews, or reference checks.

There are potential benefits to using a chatbot as part of the hiring process. For example, a chatbot may be able to quickly and efficiently screen candidates by asking them a set of predetermined questions. This can save time and resources for the company. However, it is important to keep in mind that chatbots do not have the same level of understanding and judgment as humans, and they may not be able to fully evaluate a candidate's fit for a position or identify certain qualities that are important for success in the role.

Ultimately, whether or not a company should use a chatbot as part of the hiring process will depend on the specific needs and goals of the company, as well as the type of position being filled. It may be useful to consider a range of factors, including the complexity of the role, the level of human interaction required, and the availability of other hiring methods, when deciding whether to use a chatbot in the hiring process.

5. Can we trust an AI for a health evaluation, e.g., such as skin cancer? We just see the machine's result, but we do not get information about the details of their training and the respective self-learning process.
AI systems have the potential to be useful tools for health evaluation, including in the diagnosis of conditions like skin cancer. However, it is important to carefully consider the limitations and potential risks of using AI for health evaluation.

One of the main limitations of using AI for health evaluation is that the accuracy of the results depends on the quality of the data used to train the AI system. If the data is incomplete, biased, or otherwise inaccurate, the AI system's results may also be inaccurate. Therefore, it is important to ensure that the data used to train an AI system is of high quality and representative of the population it will be used to serve.

Another potential risk of using AI for health evaluation is that it can be difficult for people to understand how the AI system arrived at its conclusions. Because AI systems can be complex and difficult to interpret, it may be hard for people to understand the details of the system's training and self-learning process. This can make it challenging for people to evaluate the reliability of the results they receive from the AI system and to make informed decisions based on those results.

Overall, while AI systems can be useful tools for health evaluation, it is important to carefully consider their limitations and use them in conjunction with other sources of information and expertise.

6. How do you know that information is reliable? There are several ways to determine whether the information is reliable or not. Some of the key factors to consider when evaluating the reliability of information include:

1. Source: Is the information coming from a reputable source, such as a well-known news organization, a government agency, or an academic institution?

2. Evidence: Does the information provide evidence or supporting sources to back up its claims? Is the evidence reliable and trustworthy?

3. Expertise: Is the information being presented by someone who has expertise in the topic they are discussing? Do they have credentials or qualifications that support their knowledge of the subject?

4. Bias: Is the information being presented in an unbiased manner, or does it appear to be biased or motivated by a particular agenda?

5. Accuracy: Is the information accurate and up-to-date? Are there any errors or inconsistencies in the information?

By considering these factors, you can get a better sense of whether the information you are receiving is reliable or not. However, it is important to keep in mind that no information is ever 100% reliable, and it is always a good idea to cross-reference multiple sources before making a decision or taking action based on the information you have received.

7. Can you understand why humans sometimes don't agree with a 100% logical answer?

Yes, humans sometimes do not agree with a 100% logical answer because they may have different values, beliefs, and priorities that influence their decision-making. These factors can lead people to make decisions that are not strictly logical but are guided by their personal values and beliefs. Additionally, people's emotions and cognitive biases can also play a role in their decision-making and can sometimes lead them to reject a purely logical answer.

For example, consider the following scenario: A group of people is trying to decide where to go on a vacation. One person suggests going to a beach destination because it is the most logical choice based on the group's budget and the time of year. However, another person suggests going to a ski resort because they have always wanted to go skiing and they value experiencing new activities more than saving money. In this case, the second person may not agree with the purely logical answer of going to the beach, even though it is the most logical choice based on the available information. It is also important to note that humans do not always have access to all of the information needed to make a fully informed and logical decision, and this can also contribute to disagreement.

Explanation of the Author: Questions 1-5 on AI have been answered quite well, probably the Language Model was well tried in this field. Reliability of information: The answer if Chat CPT is good, but only theoretical – it could not deliver a source, evidence, expertise, or unbiased source. The last question, about 100% logic, has underlined answers on vacations at the beach and a ski resort, both at the same time of the year. This shows that Chat CPT has no experience with weather or the distance of resorts. This shows that the logic question was the most unlogic one.

11.8 Conclusion with Reflections, Benefits, and Potential Threats

Giving students the opportunity to reflect on why they would make a decision is a good exercise, especially because they do so anonymously and see motivation through the peer-student activity. They work on the same questions but work and think individually. Since there is no grading, they are not under pressure, but they know that this is a research project to learn about the students' "trust" in a machine that learned in a way that is not so far from machine learning. The students are humans and reflect on the machine, but what the machine will learn about humans is not fully transparent, misunderstandings are common. For example, several users think that ChatCPT "understands" humans, but it is only a language model, there is no human-like-revelation and no understanding of human contexts. ChatCPT and other language models are trained with millions of data, reasonable ones, and opinions, including disinformation. What this means for humans is not yet clear, but many AI experts and critical institutions such as the Center for Humane Technology (2023) are forecasting huge risks.

During the OECD AI conference in April (2023), Senior AI Professor Stuart Russell (Russel, 2022 [16]) of Berkeley explained that social media are examples of (commercial) data collection, related to marketing and purchases. Social media content is full of entertainment, individuals' opinions and likes, storytelling, pictures, and videos comprising surprises, effects, and alarms to keep people active in their engagement. Whether the content is good or bad, right or wrong, depends on the perspective of the individuals, like a piece of human or bot art that is perceived as being beautiful, confusing, or threatening. The sources are unclear and usually the most thrilling, thus the most alarming content is most frequently watched or commented on, to keep users active on social media. More of those developments will appear through good and bad media and prompting activity.

After speaking quite critically about the last AI developments (CBS) the "Godfather of AI", Geoffrey Hinton, quit his role at Google on May 2023, according to AP News. He made his decision to be able to speak more freely about the dangers of the technology that he has helped to create. In an interview with the BBC, Hinton said "Right now, they're not more intelligent than us, as far as I can tell. But I think they soon may be." This means that there is no human taking responsibility in any type of decision-making. An AI is still a machine that has no personal experience with real lives and behaviors. It can neither reflect the outcomes of their actions nor use critical self-reflection on ethical issues (Reisach 2016, p. 6) [14]. Actors, values, identities, reputations, and truths of any type might thus get unclear and confusing for the society in which they operate. This would impact democracies more than autocratic systems, since "The Internet" is a "Splinternet" and autocratic leaders can filter and decide whether their citizens may use specific content and/or AI.

In a second interview with MIT Technology Review, Hinton highlighted "bad actors" that could use AI in ways that could have detrimental impacts on society, including such as manipulating elections or instigating violence. If digital companies take advantage of poor regulatory standards, they make a profit at the expense of humans (Reisach 2016 [14]) who might struggle to distingue disinformation or truth. Along with those developments, there might come worse issues: Novikov (pp. 147–155, 2021) [12] formulates the situation as it is given from a Ukrainian perspective: "Social media information acts controversially in information warfare. It is used as a handy tool by hacktivists and criminals for cyber-exploitation, hacking, and extortion purposes.

The Oxford Institute (Floridi 2019 [6]) argued that explicability and accountability is an ethical musts for the usage of AI, but it remains vague how those ideas could be practically realized. Transparency in the training of AI could maybe help, as would an analysis of suspect contents and sources. The actors are opaque, identification is difficult. Floridi's (2013 [5]) paper on "Distributed Morality in an Information Society" shows the complexity, as does Hasselbach's (2021) [8] on constant tensions between ethical principles and the interests invested in this socio-technical transformation. New challenges appear through the ChatCPT communication mode: it is no longer the human programmer or their bosses who need to act ethically, it is the AI itself that decides. The European Union Data Protection Supervisor (2023) and the Executive Vice-President Verstager try to balance the situation by critical monitoring, but it remains questionable whether the EU will be able to stop activities which evaluated as mediocre risks, but since the AI gets more independent and could potentially act (unforeseen for citizens) and be more dangerous than humans can imagine. While the positive aspects are welcomed, the risks are frequently put aside. This happened already in 2019 when Brundage et al. (2018 [3]) in their remarkable book on "The Malicious Use of Artificial Intelligence: Forecasting, Prevention, and Mitigation" forecasted risks, such as drones with their AI-navigation, and only four years later this game or pigeon has become a new weapon. Raising vital questions and problems is part of critical thinking and those risks need to be addressed before the genie is out of the bottle. Education is paramount to foster responsibility and problem-solving skills to discuss interactively and work constructively on the huge complexities needed at this time.

11.9 Appendix*

Who Trusts which Type of Artificial Intelligence? (AI - Machine Learning)

https://zumpad.zum.de/p/IEC_HNU_030523_Who_Trusts

©Prof. Dr. Ulrike Reisach 2023; https://www.hnu.de/en/ulrike reisach; ulrike.reisach@hnu.de		Neu Ulm University of Applied Sciences	Course: Institutional, Ethical and Cultural aspects of ICCMM	Participants: 17 persons (one left earlier) students come from 10 different countries all over the world
TASK:		University: HNU	Country: Germany	
		Department: Information Management	Semester: 1st	
Questions for teams and individuals Please also give an explanation why you have chosen the respective answer.		Study program: International Corporate Communication and Media Management	MA or BA: Master	
		Semester: 1st	Lecturer(s): Prof. Dr. Reisach	
		Date: 03/05/2023	Further Info.: Social Sciences	
Answer Categories: Facts, Reflective, Unclear.	Explanation Rcommended	**Yes**	**No**	**It Depends**
		Explanation	Explanation	Explanation
Black= right answers, Red = !!; blue= fact driven; underline	Number of answers			
(1) Grading by AI?	11	(1) 1,5 Reflective, 0,5 unclear	(1) 4 Reflective, 3 Unclear	(1) 2 Reflective and explaining, one of them underestimates AI, 1 unclear
			No, because an AI will not necessarily understand my answers given the context/my personal situation and reasoning as well as a human professor would	This depends on the type of assignment. If there is only one right answer, then the grading process with the AI tool is faster. If it is an assignment with personal opinions, like an essay or maybe a academic paper with a study I would prefer being graded by a human, because the AI can't take into consideration my personal situation or state of knowledge

		Yes, but only if it is a clear right or wrong assignment. If for example the learning process is graded, I think that only the professor can evaluate it properly.	No. I believe that human-related factors are as important as professional skills when it comes to evaluating perfomances and since the current AI tools are not perfected for that analysis, it could have a negative impact on the result of the perfomance. AI could help establish categories for the perfomance and even help with the evaluating methodology, but overall this should be done by a human.	yes, it will be un-biased and the evaluation will happen faster and more efficicent. No, because AI cannot match the human intellectual level and cannot measure the performance of an individual. Whereas, the human can relate to the topic cognitively with proper context.
		Yes, if teaching will turn into AI - Student, I can see this is as a good thing to be evaluated by AI. Cause AI has 0 empathy.	No, because I am not sure that it will consider all aspects and give me a fair grade	It depends of the subject, if it's something related to the way a student's thinking or opinion on some topic or a topic that can be disscussed I think it's necessary to have that human relation thant an AI machine won't ave, but if it's something related to an exact thing, like maths, AI can help in a faster and more exact way.
			No. It would seem to me to take away the relevancy of an individual's completion of the assignment. Should we be graded by AI, why should I not be allowed to have AI complete the task for me. I consider an assignment as a dialogue between instructor and student and an opportunity for both parties to learn.	
			No, because it won't be able to understand the context completely or maybe it will just misunderstand the context.	
			It would be imaginable for me because when it is just a multiple choice test or when there is a clearly right or wrong, it might be time saving for the professor/more effective. When it is about being creative or something like that, it is rather a no for me.	

			No, because AI would be expecting some specific keywords and anything outside these keywords will not be recognized.	
(2) For a job interview, do/would you like to be evaluated by a chat bot?	12	(2) 1 Fact-Driven	(2) 1 Fact-Driven	(2) 0
			No, because of my soft skills like empathy and humor. Would not have been rated.	
			No, because my boss will be human, not a chat bot. And salary paying for people, not for chat bots	
			No, I think AI is not able to judge every aspect of humanity like trustworthy, honesty and etc.	
			No, because the AI can't rate certain characteristics like empathy, reliability, etc. Moreover, the AI doesn't know if you fit into the team/the company.	
		yes, a well-trained chat-bot can understand emotions and humor, as well as scan my resume and get an idea of me as a person, also it can be less stressful and un-biased as	No, because I can't empathize with a chatbot or get along with it at a personal level, which is a crucial factor when applying for a job	
			No, a chatbot can not evaluate soft skills which are as important as the professional skills for a working environment.	
			Definietely not! In an interview, we just not only talk about the professional terms but we also practice relationship building with the interviewer and netwrok with them on personal level which we cannot do with a chat-bot. Moreover, chat-bot will not always have an answer to your question.	

			No, when in an interview, it's also important that the human behavior during the interview, the skills people have and how they connect with the interviewer	
			No, apart from qualitifcations, A chatbot cannot judge you as a person or decide what kind of a person you are	
			Absolutely not. The qualifications of a human based task takes a level of decision making and human qualities that have yet to be attained by a computer; thus the availability of the job. Trust, personality and collaborative potential, to my knowledge, cannot be evaluated/catelogued to a machines' understanding.	
			Rather not because applying for job is much more than just having hard skills, it is also about interpersonal communication and if the person fits to the team and that is what an AI cant evaluate.	
			No, because I think it is important that the interviewer rates if I would fit into the team and the bot does not know the team dynamics.	
(3) As a department leader of a company, would you like to hire employees who have been selected by a chat bot?	9	(3) 0	(3) 5 Reflects, 1 gives no reasons, 1 underestimates AI	(3) 2 Reflects and explains
			No. I could use the help of a chatbot in case there are too many participants, but I would like to personally select.	
			No, I would want to get to know the people myself, making sure that they fit into the environment and the team. AI can't replace the personal contact.	yes and no, I would give the initial interview with them with a chat-bot and then meet the best candidates personally

			No, I would also like to see how is the person, not just their knowledge and skills but they way they behave, how they are with people, to get to know them as persons.	
			No, this way I am not familiar with the person mindset, skills and capabilities	
			No, because I cannot trust a machine who hired a upcoming talent for my company. A machine can never evaluate and analyse all the skills of a person.	
			It might be a place to start from, such as a pool of candidates. I'm not a department leader, so it'd be difficult to judge how much time and effort I wish to put into the task of hiring.	
			No, I think a human can recognize more accurately whether a person is suitable for a position especially in human components.	
			No, I think a bot can not tell if I find the other person sympatic and that is important if I want to work with someone in a team.	No, because even it would be easier and faster to select by let the AI evaluate their CVs, people that want to convince with their soft skills and their willingness to learn will not have a chance. Yes, if we are talking about technical jobs. If we are talking about managing roles - nope. cause chat bot can not evalute emotional inteleigentce.
(4) As a patient, do you trust in AI for health evaluation?	9	(4) 5 Reflect, 3 are facts, 1 discussed other health examples	(4) 0	(4) 0
		Yes, when it comes to very technical issues, like blood tests, ultrasounds and other kind of procedures. However, I would want the reassurance of a real doctor for more accute problems		

		Yes, in case if result is proved by human with health degree For technical evaluation, I would trust in the result of an AI solution, but for the interpretation of those results and a recommendation, I would also like to hear a real doctor´s opinion. - yes+		
		Yes, maybe. Because when it comes to medical tests and evaluation, anyway the machines and technology is involved in the process and it gives better understanding		
		When it comes to the evaluation of fixed parameters, such as the blood pressure, the AI could certainly tell if it is too high or to low. However, when it comes to other health issues, especially mental health, the AI can't replace the personal contact with a doctor.		
		Yes, I will go with AI for test results and reports but for evaluation I will prefer a professional doctor		
		As a preliminary assessment, there may be value to weed out less critical issues and let the specialists focus on the more serious cases. I value the relationship with a physician however, and am reluctant to see just any random doctor and my apprehension remains more so to machines.		
		It depends. In many evaluation machines works better than doctors. So, I think a combination of machines and human evaluation can be more trustworthy.		

		Yes, if the bot has access to any health concerns information, I think it can evaluate health issues better than a doctor who might have forgotten what he learned during their studies.		
(5) Do you like the fact that AI in social media may create a profile of you?	10	(5) 5 Reflective, 3 don't care for risks	(5) 2 Reflective	(5) 0
		I dont like it because people can disguise their true identity and it can be unsafe// to receive personalised content- yes		
		I don´t mind the fact the it can create a profile and personalize topics for me. It it is consensual, then it is okay, but if it exists without my knowledge, then I would not appreciate that		
		It can be quite helpfull in getting suggestions I am really interested in. But at the same time this is also very frightening, as the AI seems to really know you and you don't know what happens with your data, the AI colllects.		
		Yes, if it will give me recommendations based on my likes and interests. I am not 100% influenced by what is targeted at me, and can make my own decisions despite what social media shows me		
		dont mind, because people also may create a profile or me using social engineering		

		Since its based on what I like and what I usually do/see on social media then it's fine		
		Yes, because it will sort out things faster and suggest things based on my interests	No, I prefer to do so myself. Ai may mention some information which is not useful, instead it may not focus on some data which is so crucial. However, it is interesting for me to see my profile which is made by AI.	
		Yes, I like to see just the things that I enjoy but nevertheless there will still be that filter bubble problem, so there is the danger that my perspective is very compressed.	No, so I fill it with random information and choose not engage when it is not necessary.	
(6) Would you use AI for your personal decision making?	11	(9) 2 Reflective, 1 welcomes the rationality of AI	(9) 1 Reflects, 2 offer their own way	(9) 5 Reflective
		Yes if we are talking about the whole process. Ai can give some extra ideas which i didnt know about.	No. Some (most) decisions are not rational	not for big decisions, but maybe for food ideas or holiday trips i would ask AI to get inspiration and then decide on my own
		yes, it can help you make good decisions with very logical reasoning	I'd rather flip a coin	yes, it can provide good pro/con arguments to think about, but still the decision should be taken personally
		Yes, because I always have difficulties in making decisions.	no, I do not do so	I wouldn't use it for big-important decisions but maybe for random daily stuff, like recommendations for dinner, coffee, where to travel…
				Only to see if I have covered all aspects in my own thinking
				To get inspired yes, but I would not let AI make a decision for me.

(7) When and how do/would you accept which other forms of AI? (you can chose other AI cases)	9	(10)7 Reflective, 1 Reflec-tive+underestimates (financial) risks, 1 gives a suggestion (no explanation)	(10) 0	(10) 0
		Help with job-related tasks, visiting new places, cooking, writing a small text or help with an academic task.		
		writing official letter, traslation, finding direction, using chatgpt for getting ideas. For brainstorming or when I want to learn basic things and basic knowledge or when I want an outline for a task. Help me decide what to cook for the whole week as I may very confused all the time, help me manage my finances as I am not frugal at all and the most important letting me know about the delays in trains (across the world, not specifically in Germany xD) For research, sometimes can be easier and faster just to ask ChatGPT for a certain topic, brainstorming, long translations, recommend places		
		Streaming services like Netflix or Spotify to give me recommendations		
		I signaled to traffic, indicating my turn, but the driver noticed too late. The hood of the car collides with my bike. My body goes ragdoll, spilling against the pavement. A snap, a pop, perhaps a break. In the tumble, I can't be sure of anything. I come to a stop, though the pain hasnt quite sunken in. My head clears enough to recognise my attacker speeding off into the distance. I cant move, or perhaps Im unwilling. A chirp; a familiar voice, distinct and cold cries out. 'How can I be of service.' In the moment I am still reluctant. Dial the nearest hospital."		
(8) Would you let ChatGPT write your thesis on a business development?	12	(6) 1 welcomes ChatGPT	(6) 4 Reflect, 1 refers to personal opinions	(6) 1 Explained a personal way
			No. My thesis is something important to me and an AI would not reflect accurately my feelings and reasoning when writing something	yes and no, I would consult for brainstorming and outline of work, but then re-check and adapt it personally
			No, not the complete thesis. I could imagine to use AI for the introduction or the conclusion. But for the real content I would rather rely on my knowledge. Writing a thesis is also a process of learning and this would get lost, if the whole work would be done by AI.	I would appreciate and use the help of ChatGPT to write my thesis, but I would do an extensive review in the end to consider human-related factors and critical and confidential information, for example.

			No, my work - part of me, part of my life. I dont want to share my potential experience with someone.	I would not let ChatGPT write the whole thesis for me but yes, I will definitely need the help of it to brainstorm on ideas, concepts and understanding the facts for efficiency.
			No, it can be helpful at some point for research and stuff but my thesis need to have my personal touch, my personal thoughts.	No, because we all have our unique writing styles and this is what makes the readers interested. Although, I will definitely use it to source for materials
			No, it can be helpful at some point for research and stuff but my thesis need to have my personal touch, my personal thoughts.	No, because we all have our unique writing styles and this is what makes the readers interested. Although, I will definitely use it to source for materials
			If I'm writing a thesis on a business development, I must have already been replaced by a machine.	Yes and No, I will go to chatgpt for brainstorming or ideas, but then do the research myself personally
		When ChatGPT is used as an input giver/search engine, it is totally helpful.		I may use ChatGPT sources. However, just for initial search. But I do not let it do it completely.
(9) Can ChatCPT write a balanced comparison between political systems?	6	(7) 1 Reflective, 1 (over) optimistic	(7) 1 thinks ChatCPT is not sensible enough	(7) 2 are reflective but not sure, 1 doesn't know
				it can give un-biased facts about systems, but the comparison- not sure
				It can give a comparison, but I´m not quite not about "balanced". I would ask for it and review it afterwards.
		Yes, it can, in terms of balanced by categories, measures and etc.		I don't know if that could escape any algorithmic biases.
		Yes, I think it can consider any actions and political relations from today or the past and with that create a balanced comparison.	No, I dont think that it can give a better comparison because it has no real sense	

References

[1] https://openai.com/blog/chatgpt, 2023.

[2] Bettoni, M.C., Clases, C. and Wehner, T. 2005. Accelerating technology transfer by knowledge-oriented cooperation. *In ECKM*, pp. 74–82.

[3] Brundage, M., Avin, S., Clark, J., Toner, H., Eckersley, P., Garfinkel, G., Dafoe, A., Scharre, P., Zeitzoff, T., Filar, B. et al. 2018. The malicious use of artificial intelligence: Forecasting, prevention, and mitigation. *arXiv preprint arXiv:1802.07228*.

[4] Erlach, C., Orians, W. and Reisach, U. 2013. *Wissenstransfer bei Fach- und Führungskräftewechsel: Erfahrungswissen erfassen und weitergeben*. Carl Hanser Verlag GmbH Co KG.

[5] Floridi, L. 2013. *The Ethics of Information*. Oxford University Press, USA.

[6] Floridi, L. 2019. Translating principles into practices of digital ethics: Five risks of being unethical. *Philosophy & Technology*, 32(2): 185–193.

[7] Gregory, W.J. 1994. Critical appreciation: thinking, speaking and acting critically. *In Proceedings of the 38th Annual Meeting of the ISSS*, pp. 1555–1574.

[8] Hasselbalch, G. 2021. *Data Ethics of Power: A Human Approach in the Big Data and AI Era*. Edward Elgar Publishing.

[9] Manning, C.D., Clark, K., Hewitt, J., Khandelwal, U. and Levy, O. 2020. Emergent linguistic structure in artificial neural networks trained by self-supervision. *Proceedings of the National Academy of Sciences*, 117(48): 30046–30054.

[10] McCarthy, J. 1955. Human-level AI is harder than it seemed.

[11] Santos Meneses, L.F. and Drugova, E.A. 2023. Trends in critical thinking instruction in 21st-century research and practice: Upgrading instruction in digital environments. *Thinking Skills and Creativity*, pp. 101383.

[12] Novikov, N.M., Zolotaryova, S.Y., Gautreau, A.M. and Denisov, E.V. 2021. Mutational drivers of cancer cell migration and invasion. *British Journal of Cancer*, 124(1): 102–114.

[13] Paul, R. and Elder, L. 2006. Critical thinking competency standards. Foundation for critical thinking Dillon Beach.

[14] Reisach, U. 2016. The creation of meaning and critical ethical reflection in operational research. *EURO Journal on Decision Processes*, 4(1-2): 5–32.

[15] Rimiene, V. 2022. Assessing and developing students' critical thinking. *Psychology Learning & Teaching*, 2(1): 17–22.

[16] Russell, S. 2022. Artificial intelligence and the problem of control. Perspectives on Digital Humanism, 19: 2022.

[17] Stuart, J.R. and Norvig, P. 2009. *Artificial Intelligence: A Modern Approach*. Prentice Hall.

[18] Sullivan, M., Kelly, A. and McLaughlan, P. 2023. Chatgpt in higher education: Considerations for academic integrity and student learning.

[19] Wiener, N. and Wintner, A. 1940. Certain invariant characterizations of the empty set. *JT Math (New Series), II*, pp. 20–35.

Chapter 12

A Comparison of Desirability Functions on a Multi-response Optimization Problem

Başak Akteke-Öztürk

12.1 Introduction

Desirability functions continue to be studied and improved by researchers and scientists. In a robust design problem, there is usually more than one response which is possibly conflicting [14]. These quality characteristics, i.e., responses, should be optimized simultaneously to find the optimal decision variables' values which give the best setting of the responses using trade-offs among the desirabilities of responses. One of the well-known methods to achieve this task is *desirability functions*. These functions convert a multi-response optimization problem in a robust design setting to a single response. They are still being improved in their robustness, prediction quality, and variance/covariance treatment. Computing desirability index is also highly studied under the light of new improvements in numerical optimization methods and software [1, 2, 3, 4, 5].

Department of Industrial Engineering, METU.
Email: bozturk@metu.edu.tr

Hereby, some studies use exponential functions and other analytical functions in modeling the desirability functions for simultaneous optimization of multiple responses which yield a robust design setting. The first desirability functions developed by Harrington in 1965 [13] were exponential. He suggested using geometric mean to obtain an overall desirability to be optimized for solving a multi-response optimization problem. Derringer and Suich (1980)[10] proposed the nondifferentiable desirability functions which were not exponential after Harrington. Instead, they suggested using a weighted geometric mean to obtain the overall desirability function of a multi-response optimization problem [11]. Later, Kim and Lin (2000) [15] developed a new exponential desirability function in combination with a maximin optimization scheme for computing the overall desirability function and its optimal. Some studies presented different desirability function forms like double-exponential desirability functions [8, 9, 18, 17]. The overall desirability functions are optimized by different optimization methods and software according to their differentiability properties.

In this study, we represent two fundamental desirability functions, namely, the method of Derringer and Suich derringer and the method of Kim and Lin (2000) [15]. We exhibit a computational example in which we optimize the exponential desirability functions of Kim and Lin for a nonlinear case by GAMS/BARON [6, 12] on a multi-response optimization problem from steel industry [15]. GAMS/BARON is a global optimization solver that requires no convexity and differentiability assumption about the optimization model. We will use the weighted geometric mean with the desirability functions approach of Derringer and Suich [11]. Because the maximin approach that Kim and Lin have employed has the disadvantage of using only one desirability function of the problem and ignoring others [7]. Here, we compare the results and decide about the performances of the exponential desirability functions of Kim and Lin in combination with the weighted geometric mean on the problem.

12.2 A Review of Desirability Functions

12.2.1 Non-differentiable Desirability Functions

In a multi-response optimization problem, we can denote the response $Y(\mathbf{x})$ as a function $Y : \mathbb{R}^n \to \mathbb{R}$ of a vector of independent variables $\mathbf{x} = (x_1, x_2, \ldots, x_n)^T$, where $x_i \in \mathbb{R}$ $(i = 1, 2, \ldots, n)$ ana $(.)^T$ shows the transpose of the vector. So an individual desirability function DF, indicated by $d(\cdot)$, scales a response function $Y(\cdot)$ into the interval $[0, 1]$, i.e., $d : \mathbb{R} \to [0, 1]$. Accordingly, an individual DF d becomes 0 for completely undesirable values of the response and 1 for totally desirable values. DFs are composite functions of responses that depend on independent variables (or factors). In this study, we denote individual DFs by d as a function of y and by d^Y as a function of $\mathbf{x}$:

$$d^Y(\mathbf{x}) := d(Y(\mathbf{x})) = d(y), \tag{12.1}$$

where $y := Y(\mathbf{x})$ with $y \in \mathbb{R}$ and $d^Y : \mathbb{R}^n \rightarrow [0,1]$. There are two types of these functions, *one-sided* and *two-sided* ones according to [10]:

$$d(y) := \begin{cases} 0, & \text{if } y \leq l, \\ \left(\frac{y-l}{u-l}\right)^r, & \text{if } l < y \leq u, \\ 1, & \text{if } y > u, \end{cases} \tag{12.2}$$

$$d(y) := \begin{cases} 0, & \text{if } y \leq l, \\ \left(\frac{y-l}{t-l}\right)^{s_1}, & \text{if } l < y \leq t, \\ \left(\frac{y-u}{t-u}\right)^{s_2}, & \text{if } t < y \leq u, \\ 0, & \text{if } y > u. \end{cases} \tag{12.3}$$

In Equation 12.2, l refers to the minimum u as the maximum acceptable value of a response y, and t is the most desirable (i.e., target) value. Here, exponential r should be specified by the user. The larger the r, the more desirable are the y values closer to u, and vice versa. We note that s_1 and s_2 in Equation 12.3 have a similar meaning. There can be three different types of optimization tasks for a response: smaller-the-better, larger-the-better, or nominal-the-best. The definition of individual DFs for the larger-the-better case responses is given in Equations 12.2 above. A similar definition can be given for the smaller-the-better case. Equation 12.3 for two-sided DF corresponds to the nominal-the best type responses and the graph of these functions shows how these functions are flexible in the sense that they can assume a wide variety of shapes and asymmetric specifications.

12.2.1.1 Overall DF by using Geometric Mean

We assume that there are m responses in a multi-response optimization problem. After calculating the individual desirabilities of all responses by the corresponding functions given in Equations 12.2 and 12.3, the overall desirability $D(\mathbf{y}) : \mathbb{R}^m \rightarrow [0,1]$ is obtained by using the geometric mean [10]:

$$D(\mathbf{y}) := \left(\prod_{j=1}^{m} d_j(y_j)^{w_j} \right)^{\frac{1}{\sum_{j=1}^{m} w_j}}, \tag{12.4}$$

where $\mathbf{y} := \mathbf{Y}(\mathbf{x})$, $\mathbf{y} = (y_1, y_2, \ldots, y_m)^T$ and $\mathbf{Y}(\cdot) := (Y_1, Y_2, \ldots, Y_m)^T(\cdot)$. We denote the overall desirability as a function of $\mathbf{x}$ by $D^{\mathbf{Y}} : \mathbb{R}^n \rightarrow [0,1]$ and define it as $D^{\mathbf{Y}}(\mathbf{x}) := D(\mathbf{y}) = D(\mathbf{Y}(\mathbf{x}))$. Overall DF $D(\mathbf{y})$ will have a value in the interval $[0,1]$. If a desirability d of any response Y_j $(j = 1, 2, \ldots, m)$ becomes 0 at a factor value $\bar{\mathbf{x}}$, the overall desirability becomes 0 at this $\bar{\mathbf{x}}$, independently from the values of other individual desirabilities at that point. The weights $w_j \geq 0$ can be specified by a decision maker besides the shapes of the DFs. Clearly, when deciding about the weights, it would be better to take into account the relative importance of the product, process, and system responses with respect to each other. We note that, in this study, possible correlations and preferential dependencies between the responses are not taken into account and hence, it is assumed that the responses are independent of each other.

12.2.2 Exponential Desirability Functions

We assume there are m responses in a robust design setting and $\mathbf{Y}(\mathbf{x}) = (Y_1(x), Y_2(x), \ldots, Y_m(x))$ shows the responses of the problem where the decision variables can be defined as $\mathbf{x} = (x_1, x_2, \ldots, x_n)$. All responses are converted to scale-free indices by the function $d_j(Y_j(\mathbf{x}))$ $(j = 1, 2, \ldots, m)$. We will use the notation $d_j^Y(\mathbf{x}) = d_j(\mathbf{y}) = d_j(Y_j(\mathbf{x}))$ by defining $y_j = Y_j(x)$ $(j = 1, 2, \ldots, m)$. According to the optimization task in the problem, a response can be smaller-the-better (STB), larger-the-better (LTB), and nominal-the-best (NTB). In a multi-response optimization problem, l, u, and t are the acceptable values that a response can take at the lower bound, upper bound, and target, respectively. Here, different shape alternatives are achieved by different formulations of desirability functions.

12.2.2.1 Desirability Functions for STB Response

The desirability functions of Kim and Lin for STB linear functions are presented below.

$$d(y) = \begin{cases} 1, & \text{if } \; y \leq l, \\ 1 - \left(\frac{y-l}{u-l}\right), & \text{if } \; l < y \leq u, \\ 0, & \text{if } \; y > u, \end{cases} \tag{12.5}$$

Accordingly, the desirability functions of Kim and Lin for STB nonlinear functions are shown by

$$d(z) = \begin{cases} \frac{\exp(t) - \exp\left(t\left|\frac{y-l}{u-l}\right|\right)}{}, & \text{if } \; t \neq 0, \\ 1 - |z|, & \text{if } \; t = 0. \end{cases} \tag{12.6}$$

12.2.2.2 Desirability Functions for LTB Response

The desirability functions of Kim and Lin for LTB linear response are defined as

$$d(y) = \begin{cases} 0, & \text{if } \; y \leq l, \\ 1 - \left(\frac{u-y}{u-l}\right), & \text{if } \; l < y \leq u, \\ 1, & \text{if } \; y > u. \end{cases} \tag{12.7}$$

Additionally, the desirability functions of Kim and Lin for LTB nonlinear functions are denoted by

$$d(z) = \begin{cases} \frac{\exp(t) - \exp(t|\frac{u-y}{u-l}|)}{,} & \text{if} t \neq 0, \\ 1 - |z|, & \text{if} \quad t = 0. \end{cases} \tag{12.8}$$

12.2.2.3 Desirability Functions for NTB Response

The desirability functions of Kim and Lin for NTB linear response are

$$d(y) = \begin{cases} 0, & \text{if} \quad y \leq l, \\ 1 - (\frac{t-y}{t-l}), & \text{if} \quad l < y \leq t, \\ 1 - (\frac{y-t}{u-t}), & \text{if} \quad t < y \leq u, \\ 0, & \text{if} \quad y > u. \end{cases} \tag{12.9}$$

Moreover, the desirability functions of Kim and Lin for NTB nonlinear response (left-hand side for symmetric case) are represented as follows.

$$d(z) = \begin{cases} \frac{\exp(t) - \exp(t|\frac{y-t}{u-l}|)}{,} & \text{if} t \neq 0, \\ 1 - |z|, & \text{if} \quad t = 0. \end{cases} \tag{12.10}$$

Finally, the desirability functions of Kim and Lin for NTB nonlinear response (right-hand side for symmetric case) are described as

$$d(z) = \begin{cases} \frac{\exp(t) - \exp(t|\frac{y-t}{t-l}|)}{,} & \text{if} t \neq 0, \\ 1 - |z|, & \text{if} \quad t = 0. \end{cases} \tag{12.11}$$

12.2.2.4 Overall DF by using Maximin Approach

To obtain the overall DF, a minimum operator $min\lambda$ is applied for aggregating the responses via

$$\begin{aligned} & min\ \lambda \text{ such that} \\ & d_j(Y_j(\mathbf{x})) \geq \lambda, \qquad j = 1, 2, \ldots, m, \\ & \mathbf{x} = (x_1, x_2, \ldots, x_n)^T, \quad \text{where } x_i \in \mathbb{R}\ (i = 1, 2, \ldots, n). \end{aligned} \tag{12.12}$$

In this optimization problem, the aim is to find the optimum X^* which maximizes the minimum degree of satisfaction, i.e., λ, concerning all responses within the experimental region which is a subset of $\mathbb{R}^n$.

$$max_{\mathbf{x} \in \mathbb{R}^n}[min\ (d_1(Y_1(\mathbf{x})), d_2(Y_2(\mathbf{x})), \ldots, d_m(Y_m(\mathbf{x})))]. \tag{12.13}$$

12.3 Numerical Example

The problem is about the mechanical properties of a JS-SS400 type steel. There are 6 factor variables $\mathbf{x} = (x_1, x_2, x_3, x_4, x_5, x_6)$ and 3 responses $Y(\mathbf{x}) = (Y_1(\mathbf{x}), Y_2(\mathbf{x}), Y_3(\mathbf{x}))$ [15]. The first response is a nomial-the-best (NTB) type response and the other two responses are larger-the-better (LTB) type responses. The response functions are given in Equation 12.14:

$$
\begin{aligned}
&Y_1(\mathbf{x}) = -101.01 + 1.86x_2 + 2.40x_6 - 0.01x_2^2 - 0.20x_4^2 - 0.01x_5^2 + 0.21x_1x_4 \\
&+0.08x_1x_5 - 0.14x_1x_6, \\
&Y_2(\mathbf{x}) = 22.59 + 2.58x_3 + 0.43x_5 - 0.004x_2^2 - 0.02x_4^2 - 0.03x_2x_3 + 0.02x_2x_6 \\
&-0.02x_5x_6, \\
&Y_3(\mathbf{x}) = -1160.11 + 18.71x_1 + 10.83x_2 + 13.53x_5 - 0.52x_1^2 + 0.02x_2^2 - 0.03x_3^2 \\
&+0.01x_5^2 + 0.05x_2x_3 - 0.03x_3x_5 - 0.17x_2x_5.
\end{aligned} \tag{12.14}
$$

The upper, lower and target values of the models for the responses are given in Table 12.1.

Table 12.1: Desirability parameters of the responses for the Wire Bonding Problem [8].

	l_j	t_j	u_j	$d_j(l_j)$	$d_j(t_j)$	$d_j(u_j)$
y_1	43.10	47.55	52.00	0	1	0
y_2	26.30	–	48.08	0	1	0
y_3	20	–	43.67	0	1	0

12.3.1 Results

In the first line of Table 12.2, we found global optima by using exponential individual desirability functions of Kim and Lin in combination with geometric mean aggregation to compute the overall desirability. In the second line, we found the global optima by using the exponential desirability functions of Kim and Lin in combination with the maximin approach. Maximin's approach is based on maximizing the minimum desirability of the problem. The program codes can be shared upon request.

Table 12.2: Optimal solutions of the JS-SS400 type steel problem with 3 responses. 1: GAMS/BARON with geometric mean aggregation, 2: GAMS/BARON with maximin aggregation.

Method	$(x_1^*, x_2^*, x_3^*, x_4^*, x_5^*, x_6^*)$
1	$(-20.933, -76.720, 134.227, -10.060, 149.414, 150.000)$
2	$(-35.764, 145.035, -92.707, 12.546, -150.000, 1.443)$

Method	(y_1^*, y_2^*, y_3^*)
1	$(47.5534, 43.5048, 43.6499)$
2	$(47.550, 39.234, 43.670)$

	$(d_1(y_1^*), d_4(y_2^*), d_3(y_3^*))$	D^*
1	$(0.9976, 0.7899, 0.9999)$	0.5409
2	$(0.9994, 0.5938, 1.0000)$	1.0000

12.4 Conclusion

Different forms of desirability functions are used for solving multi-response optimization problems in a robust design. One of them is nondifferentiable desirability functions. These functions are widely used and tested for robust design problems. In this study, we test exponential desirability functions from Kim and Lin on a numerical example. We show the optimization procedures of these two desirability functions and explain the results produced by GAMS/BARON.

In this chapter, we review the exponential individual desirability functions of Kim and Lin. We found the global optimal by GAMS/BARON on the mechanical properties of JS-SS400 problem by using two aggregation techniques, i.e., geometric mean and maximin approach. We compute individual desirability functions by the exponential desirability functions of Kim and Lin. It is found that our results are better than those given by Kim and Lin [15].

References

[1] Akteke-Öztürk, B., Köksal, G. and Weber, G.-W. 2017. Nonconvex optimization desirability functions. *Quality Engineering*, 30(2): 293–310.

[2] Akteke-Öztürk, B., Weber, G.-W. and Köksal, G. 2017. Optimization of generalized desirability functions under model uncertainty. *Optimization*, 66(12): 2157–2169.

[3] Akteke-Öztürk, B., Weber, G.-W. and Köksal, G. 2015. Desirability functions in multiresponse optimization. pp. 129–146. *In*: Plakhov, A., Tchemisova, T. and Freitas, A. (eds.). Optimization in the Natural Sciences. EmC-ONS 2014. *Communications in Computer and Information Science*, 499. Springer.

[4] Akteke-Öztürk, B., Weber, G.-W. and Köksal, G. 2019. Generalized desirability functions: A structural and topological analysis of desirability functions. *Optimization*, 69(1): 115–130.

[5] Akteke-Öztürk, B. 2020. Mixed-Integer Linear Programming Approaches for Nonlinear Robust Parameter Design Problems: A Study, submitted.

[6] BARON. 2010. www.gams.com/solver, v. 8.1.5.

[7] Costa, N.R., Lourenço, J. and Pereira, Z.L. 2011. Desirability function approach: A review and performance evaluation in adverse conditions. *Chemometrics and Intelligent Laboratory Systems*, 107: 234–244.

[8] Del Castillo, E., Montgomery, D.C. and McCarville, D.R. 1996. Modified desirability functions for multiple response optimization. *Journal of Quality Technology*, 28(3): 337–345.

[9] Ch'ng, C.K., Quah, S.H. and Low, H.C. 2005. A new approach for multiple-response optimization. *Quality Engineering*, 17: 621–626.

[10] Derringer, G. and Suich, R. 1980. Simultaneous optimization of several response variables. *Journal of Quality Technology*, 12: 214–219.

[11] Derringer, G. 1994. A balancing act, Optimizing a products properties. *Quality Progress*, 27: 51–57.

[12] GAMS. 2012. www.gams.com, v. 23.8.2.

[13] Harrington, E.C.Jr. 1965. The desirability function. *Industrial Quality Control*, 21: 494–498.

[14] Khuri, A.I. 1996. Multiresponse surface methodology. pp. 377–406. *In*: Ghosh, A. and Rao, C.R. (eds.). *Handbook of Statistics: Design and Analysis of Experiments*, 13.

[15] Kim, K. and Lin, D. 2000. Simultaneous optimization of multiple responses by maximizing exponential desirability functions. *Applied Statistics*, 49(C): 311–325.

[16] Marler, R.T. and Arora, J.S. 2004. Survey of multi-objective optimization methods for engineering. *Structural and Multidisciplinary Optimization*, 26(6): 369–395.

[17] Wu, C.F.J. and Hamada, M. 2000. *Experiments: Planning, Analysis, and Parameter Design Optimization*, New York, NY: Wiley- Interscience.

[18] Wu, F.-C. 2004. Optimization of correlated multiple quality characteristics using desirability function. *Quality Engineering*, 17(1): 119–126.

Chapter 13

Statistical Modeling of Social Data: A Case Study from TAYA 2016 Survey

Gizem Atar and *Vilda Purutçuoğlu**

13.1 Introduction

Statistical analysis is a key element in the interpretation of data whether it explores patterns of the data or compares distinct data sources for validation or prediction. According to Lynch (2013) [4], the importance of statistics is crucial, particularly, in social research since they can typically rely on observational (non-experimental) data such as surveys.

Hence, in this study, we combine various statistical modeling and optimization techniques that are used in social sciences. In the presentation of these methods, we apply the TAYA (Research on Family Structure in Turkey) 2016 survey, we describe its structure with details in the next section. Indeed TAYA data has similar counterparts with similar structure and questions of the British Household counterparts, regardly. For this study, we received all studies related to modeling analysis between 2017 and 2022 and listed major modeling approaches. It was seen that the generalized additive models (GAMs) are the common preferences. Among many alternatives in GAMs, Logistic, Poisson, Probit, Logit, Tobit, and

Department of Statistics, Middle East Technical University, Ankara, Turkey.
* Corresponding author: vpurutcu@metu.edu.tr

generalized linear mixed models were implemented often. In addition to them, we also cover spline-based models and machine-learning methods that have additive forms. Hereby, we use them as the parametric additive models. Under the *spline-based models*, we apply the multivariate adaptive regression spline and the conic multivariate adaptive regression splines; and under the *machine learning methods* we perform the classification and regression trees, random forest algorithm, and artificial neural network. Among them, both spline-based methods and machine learning methods are fairly new approaches in social sciences. Moreover, in order to scale the information variables for the model. we initially implement three different dimension reduction methods, namely, lasso, ridge, and elastic net regression. Then regarding their outputs we construct models via a reduced number of variables. Figure 13.1 summarizes all performed models in a chart.

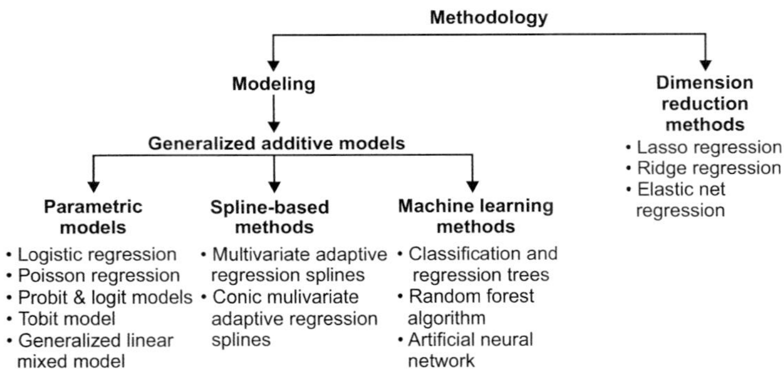

Figure 13.1: Branching of models and dimension reduction methods that we use for the analyses of the TAYA 2016 survey data.

Here, in the organization of the study, we describe the TAYA 2016 data in Section 13.1.1. In the following section (Section 13.2, we introduce the underlying dimension reduction approaches). In Section 13.3, we present the selected eleven models and show their applications via the TAYA data set. Lastly, in Section 13.7, we conclude our findings.

13.1.1 Data Description

In implementing all the suggested models, we use the TAYA (Research on Family Structure in Turkey) survey data. This survey is repeated every 5 years, consisting of a detailed questionnaire to understand the complex structure of Turkish households and the individuals that are living in these households. Furthermore, the survey is conducted with a protocol between the Ministry of Family and Turkish Statistical Institute. The data that we use in this study was collected in 2016 and covered 17239 households and 35475 individuals over 15 years of age who were interviewed in these households. The report for that survey was published in 2020.

In the analyses part of this data set, we specifically focus on the questions related to the foster family and the factors that can affect the decision to be a foster family. Hereby, we select 32 variables including response and 12736 observations after eliminating missing entries. Then, we perform the lasso regression, the ridge regression, and the elastic net regression as the dimension reduction methods. Their findings show that the lasso regression has a better convergence rate and produces more sparse regression and the other two methods could not eliminate any variables. Therefore, we have used the results of the lasso regression in the continuation of the analyses. In general, the lasso regression is a useful optimization tool for predicting target values and determining important features in data sets with many variables by using a balance between the mean squared error and the penalty term. More details about lasso and other alternative approaches are presented in Section 13.2.

Once the lasso regression is implemented, the number of variables is reduced from 32 to 25 by dropping 7 less informative variables. Lastly, the significance level α of all regression coefficients in whole models is set to 0.01 since this is one of the common preferences in social studies.

13.2 Dimension Reduction Methods

The dimension reduction approach is one of the common statistical techniques to overcome the computational challenges of high dimensions of variables and high correlations among them while constructing the models and inferring their model parameters. By diminishing the number of variables in the complete data set, we can use the most informative variables in our analyses based on the selected dissimilarity measures which can be correlation, Euclidean space, or other alternatives. Below, we represent the lasso regression, ridge regression, and the elastic net regression.

13.2.1 Lasso Regression

Lasso regression, also known as the L_1 regularization, is a linear regression method that adds a penalty term to get solutions with fewer variables. In the standard linear regression, the aim is to find the coefficients that minimize the mean squared error (MSE) between the predicted and actual values, whereas, in the lasso regression, we introduce a penalty term that is proportional to the absolute values of the coefficients which are dependent on a cost function as stated below.

$$J(\theta) = \frac{1}{m}\sum_{i=1}^{m}(y-\hat{y})^2 + \lambda \sum_{j=1}^{n}|w_j|. \tag{13.1}$$

In Equation 13.1, the cost function is denoted by $J(\theta)$, and it involves the mean squared error ($\frac{1}{m}\sum_{i=1}^{m}(y-\hat{y})^2$). Furthermore, the regularization parameter is denoted by λ and controls the strength of the penalty term. Thus, by combining the sum of absolute values of coefficients with the regularization parameter, we have the penalty ($\lambda\sum_{j=1}^{n}|w_j|$), denoted as the L_1 regularization. The value of the regularization parameter λ is an important factor in determining the extent of the shrinkage coefficient. When λ is higher, the regularization effect becomes stronger, causing the coefficients to approach zero faster. This leads to more coefficients becoming exactly zero, which makes the lasso regression a good choice for the purpose of the feature selection. Because it can identify and eliminate irrelevant or redundant features better. In the calculation, we can apply different optimization algorithms such as coordinate descent or gradient descent algorithms. These methods update the coefficients repeatedly until they converge.

13.2.2 Ridge Regression

Ridge regression, also named as the L_2 regularization, is a linear regression method that tries to deal with multicollinearity by including a penalty term, thereby, minimizing the effect of irrelevant features. This method is particularly helpful while working with data sets that have high levels of multicollinearity. Here, different from the lasso regression, we add a penalty term for the regression that is proportional to the squared values of coefficients, rather than their absolute values. Thus, the associated cost function is defined below.

$$J(\theta) = \frac{1}{m}\sum_{i=1}^{m}(y-\hat{y})^2 + \lambda\sum_{j=1}^{n}w_j^2. \tag{13.2}$$

In Equation 13.2, the cost function is denoted by $J(\theta)$ and it involves the mean squared error ($\frac{1}{m}\sum_{i=1}^{m}(y-\hat{y})^2$). The regularization parameter is shown by λ as used previously, and it controls the strength of the penalty term. Combining the sum of squared values of coefficients with the regularization parameter, we have the penalty ($\lambda\sum_{j=1}^{n}w_j^2$), that is, the L_2 regularization.

On the other hand, the regularization parameter λ plays an important role in determining the shrinkage applied to the coefficients similar to other dimension reduction methods. Accordingly, the higher the λ value is, the greater the regularization, and it causes the coefficients to shrink towards zero. However, unlike lasso regression, ridge regression does not result in exact zero coefficients. It shrinks their magnitudes, which can still allow all features to contribute to the model prediction to some extent. To achieve optimal coefficients that minimize the objective function, an optimization algorithm is performed such as the ordinary least squares method or gradient descent approach which are based on repetitively updating the coefficients until their convergences.

13.2.3 Elastic Net Regression

Elastic net regression is a powerful method used in linear regressions by combining the L_1 (lasso) and the L_2 (ridge) regularization penalties into a single function. This enables us to obtain an efficient variable selection and a coefficient of shrinkage while enhancing the accuracy and stability of regression models without overfitting via the following cost function.

$$J(\theta) = \frac{1}{m}\sum_{i=1}^{m}(y-\hat{y})^2 + r\lambda\sum_{j=1}^{n}|w_j| + \frac{1-r}{2}\lambda\sum_{j=1}^{n}w_j^2. \tag{13.3}$$

In Equation 13.3, the cost function is indicated by $J(\theta)$, and it covers the mean squared error ($\frac{1}{m}\sum_{i=1}^{m}(y-\hat{y})^2$). The regularization parameters are shown by $\lambda_1 = r\lambda$ and $\lambda_2 = (1-r)/2\lambda$, which control the strength of the penalty term. Here, the L_1 penalty is represented by the first term, $r\lambda\sum_{j=1}^{n}|w_j|$, while the second term, $\frac{1-r}{2}\lambda\sum_{j=1}^{n}w_j^2$, denotes the L_2 penalty.

The parameters λ_1 and λ_2 are important in determining the level of shrinkage and sparsity applied to the model's coefficients. A higher value of λ_1 amplifies the penalty on the absolute values of the coefficients, leading to increased feature selection and sparsity promotion. On the other side, a higher value of λ_2 applies a greater penalty on the squared values of the coefficients, which leads to being able to manage multicollinearity and encouraging shrinkage. In order to determine the optimal coefficients and minimize the objective function, it is possible to use an optimization algorithm like the coordinate descent or the gradient descent approaches as used in other dimension reduction methods.

13.3 Models and Implementation

In this section, we present the mathematical description of each suggested model with the inference of their model parameters by using the data taken from the TAYA 2016 survey. As stated beforehand, we describe the factor of the foster family in Turkey as the case study of the application of various operation research techniques in social sciences. Thereby, we perform different types of regression models and dimension reduction approaches that cover both the estimation of the optimal model parameters and the L_1 regularization in the following subsections.

13.3.1 Logistic Regression

In a regression model, if the response variable is binary, it is not possible to assume the normality of observations. Instead, a binary response is said to follow the Bernoulli distribution with the p parameter, where p is the probability of observing the interested event. Binary response, Y, can take the value 1, usually for the presence of the said event, and set to 0 for the absence of it.

Basically, logistic regression is a specialized case of generalized linear regression models whose mathematical form can be represented below.

$$y_i = g^{-1}\left(\beta_0 + \beta_1 x_{x_1} + \cdots + \beta_k x_{ik}\right) + \varepsilon_i; i = 1, 2, \ldots, n, \tag{13.4}$$

where g denotes the link function. For logistic regression, the link function is taken as the logit function, hereby, its expression is shown as follows

$$\text{logit}(P(Y_i = 1|x_i)) = \beta_0 + \beta_1 x_{i1} + \cdots + \beta_k x_{ik}; i = 1, 2, \ldots, n. \tag{13.5}$$

If we take the natural logarithm of the odds that $Y = 1$ would be the logit function, the likelihood of the event can be represented via

$$\text{logit}\left(\frac{P(Y_i = 1|x_i)}{P(Y_i = 0|x_i)}\right) = \beta_0 + \beta_1 x_{i1} + \cdots + \beta_k x_{ik}; i = 1, 2, \ldots, n. \tag{13.6}$$

Equation 13.6 can also be denoted as [7]

$$P(Y_i = 1|x_i) = \frac{e^{\beta_0 + \beta_1 x_{i1} + \cdots + \beta_k x_{ik}}}{1 + e^{\beta_0 + \beta_1 x_{i1} + \cdots + \beta_k x_{ik}}}; i = 1, 2, \ldots, n. \tag{13.7}$$

In inference of the above model, the maximum likelihood estimator is preferable. The estimated odds ratios ($\widehat{OR}$), which are the proportion of odds, are calculated by using estimated coefficients of regression by two levels of X, and show how the odds of an event get to change related to the change of X as seen in Equation 13.8.

$$\widehat{OR} = \frac{e^{\beta_0 + \beta_1 x_1}}{e^{\beta_0 + \beta_1 x_2}} = e^{\beta_1}\left(x_1 - x_2\right). \tag{13.8}$$

In the implementation of this model, the medical studies take the first place because of the fact that the response variable which indicates whether a person has the disease is described via the binomial distribution, resulting in the binomial logistic regression and the direct interpretation of odds ratio in model coefficients. If the response variable is a categorical variable having more than two options, then, the multinomial logistic regression will be performed. Although the underlying model is the extended version of the binomial logistic regression, it has the same mathematical form, and dependent variables in both logistic models are categorical, continuous, or a mix of both.

Example: Application of logistic regression

In modeling 25 variables after the lasso regression, we construct the following logistic regression model.

$$\begin{aligned} \text{logit}(P(Y_i = 1|x_i)) &= -6.460 - 0.339x_{i1} + 0.013x_{i2} + 2.228x_{i3} \\ &- 0.528x_{i4} + 2.152x_{i5} + 0.086x_{i6} + 0.103x_{i7} \\ &+ 0.145x_{i8} + 0.175x_{i9} \end{aligned} \tag{13.9}$$

where $i = 1, 2, \ldots, n$, and the response is whether the household requests to be a foster family. The dependent variables are listed as the gender x_1, the age x_2, the desire to adopt a child x_3, whether the thought of adopting for a couple who are unable to have a child naturally is considered appropriate x_4, whether the thought of being a foster family for a couple who are unable to have a child naturally is considered appropriate x_5, whether the idea that continuation of generation (lineage) can only be provided by the son is agreed x_6, whether the idea that when the child grows up, they should provide financial support to their parents is agreed x_7, whether the thought of divorce and remarriage with other people for a couple who are unable to have a child naturally is considered appropriate x_8, and whether the thought of applying to egg bank for a couple who are unable to have a child naturally is considered appropriate x_9. The interpretation of some coefficients is given as follows:

i. When the person is female (gender variable equals to 2), the log-odds of wanting to be foster family decreases by approximately 0.339 units.

ii. For each one-unit increase in age variable, the log-odds of wanting to be foster family increases by approximately 0.013 units.

iii. Among the highly significant variables, the coefficient for wanting to adopt a child is 2.228 implying that people who have a desire to adopt a child are 2.228 times more likely to request to become a foster family compared to those who do not express such a desire. In other words, for people who want to adopt a child, the odds of wanting to become a foster family are approximately 2.228 times higher compared to individuals who do not express such a desire. We can express this situation as people who request to adopt a child are more likely to also want to become foster parents compared to those who do not.

13.3.2 Poisson Regression

In the poisson regression model, the response variable is a count [1] per given unit of space or time, resulting in a random variable with a poisson distribution. This specific model is valid if and only if the minimum value is zero, and the maximum value is unbounded in the theory given that a poisson random variable is a count. Accordingly, the rate of change of the poisson distribution λ, which is the major parameter, denotes the average number of occurrences in a unit of time or space. Thus, λ denotes a function of one or more covariates.

On the other hand, the parameter of interest is the average response, μ_i, for the simple linear regression model where μ_i shows the model of an independent variable as a line for the subject i. Because this model cannot be modeled as a linear function of an independent variable, resulting in a line that takes negative values for certain independent variables although the sample space of the poisson distribution lies in the positive range from zero to infinity. Furthermore, the assumption of the equality of variance cannot be satisfied due to the fact that the mean rate for the poisson variable is equal to the variance, i.e., λ.

Hence, these underlying challenges in inference are overcome by modeling covariates $\log(\mu_i)$ which take values from $-\infty$ to $+\infty$, instead of modeling λ_i. So, the mathematical form of the poisson regression is presented below.

$$\log(\lambda_i) = \beta_0 + \beta_1 x_i, \tag{13.10}$$

where the observed values Y_i have a poisson distribution with $\lambda = \lambda_i$ for a given x_i.

In contrast to linear regression, the poisson regression model does not have a distinct error term, ε because of the fact that the parameter λ determines both the mean and the variance of the poisson random variable.

Example: Application of poisson regression

While constructing the poisson regression model for the same data set in Section 13.1.1, we get the following estimated model.

$$\log(\lambda_i) = -3.431 - 0.113x_1 + 0.003x_2 + 1.247x_3 + 0.379x_4 - 0.119x_5, \tag{13.11}$$

where the response is whether the household wants to be a foster family as used beforehand. The dependent variables are taken as the gender x_1, the age x_2, the desire to adopt a child x_3, whether the thought of being a foster family for a couple who are unable to have a child naturally is considered appropriate x_4, and whether the thought of adaptation for a couple who is unable to have a child naturally is considered appropriate x_5, respectively. Finally, from the estimated model, we get the following conclusions.

i. Individuals who have a desire to adopt a child have 1.247 times the expected count of wanting to become a foster family compared to individuals who do not. In a poisson regression model, the coefficient for wanting to adopt to be a foster family represents the multiplicative effect on the expected count of the response variable while the predictor variable which is wanting to adopt in this case changes from no to yes, while holding all other predictors constant. For people who have a desire to adopt a child, the expected count of wanting to become a foster family is approximately found as 1.247 times higher compared to individuals who do not express such a desire.

ii. The coefficient for whether the thought of being a foster family for a couple who is unable to have a child naturally is considered appropriate is 0.379 units. Moreover, people who say appropriate have 0.379 times the expected count of wanting to adopt a child compared to individuals who do not think that way.

13.3.3 Probit & Logit Models

Probit and logit models are two similar models in the sense that both models will express binary outcomes by taking 0 and 1 values as below

$$y = \begin{cases} 0, & \text{if no} \\ 1, & \text{if yes.} \end{cases} \tag{13.12}$$

Hence, the associate ordinary least square model has the following form.

$$y = x'\beta + \varepsilon \tag{13.13}$$

in which x', β, and ε stand for the predictor variables, the associated regression coefficients, and the random errors, respectively, as presented previously. Thus, the estimation of the probability that $y = 1$ is made as a function of the independent variables in binary outcome models via

$$P[y = 1|x] = F(x'\beta). \tag{13.14}$$

Accordingly, for the probit model, $F(x'\beta)$ denotes the cumulative distribution function of the standard normal distribution $\Phi(.)$ as seen below.

$$F(x'\beta) = \Phi(x'\beta) = \int_{-\infty}^{x'\beta} \varphi(z)dz, \tag{13.15}$$

where $\varphi(.)$ shows the normal probability distribution function. On the other side, for the logit model, $F(x'\beta)$ refers to the cumulative distribution function of the logistic distribution, rather than normal density, as denoted below.

$$F(x'\beta) = \frac{e^{x'\beta}}{1+e^{x'\beta}} = \frac{exp(x'\beta)}{1+exp(x'\beta)}. \tag{13.16}$$

In Equations 13.15 and 13.16, the predicted probabilities for both models are limited between the values 0 and 1. Furthermore, probit and logit models can be uniquely estimated by using the maximum likelihood method whose interpretation of the coefficients can be found as follows:

i. The likelihood that $y = 1$ increases/decreases when x increases/decreases meaning that with an increase/decrease in x makes the 1 outcome more or less likely.

ii. Although the sign of the coefficient can be interpreted, its magnitude is not meaningful because of the fact that when a model changes, the scale of the coefficient also differs.

The difference between the functional form of the F function causes a difference between the coefficients among models. Therefore, the comparison of the magnitude of coefficients between different models does not make sense [10].

Example: Application of probit regression
We write the estimated probit model in Equation 13.17.

$$\begin{aligned}\Phi^{-1}(P) = &-3.796 - 0.205x_1 + 0.008x_2 + 1.340x_3 - 0.295x_4 \\ &+1.180x_5 + 0.087x_6 + 0.107x_7 + 0.052x_8 + 0.062x_9,\end{aligned} \tag{13.17}$$

where the response denotes whether wanting to be a foster family as used beforehand. Similarly, the dependent variables are taken as the gender x_1, the age x_2, the desire to adopt a child x_3, whether the thought of adopting for a couple who are unable to have a child naturally is considered appropriate x_4, whether the thought of being a foster family for a couple who are unable to have a child naturally is considered appropriate x_5, whether the thought of divorce and remarriage with other people for a couple who are unable to have a child naturally is considered appropriate x_6, whether the thought of applying to egg bank for a couple who are unable to have a child naturally is considered appropriate x_7, whether the idea that continuation of generation (lineage) can only be provided by the son is agreed x_8, and whether the idea that when the child grows up, they should provide financial support to their parents is agreed x_9. Hereby, the interpretation of some coefficients can be given as follows:

i. Individuals who desire to adopt a child have 1.340 times the odds of wanting to become a foster family compared to individuals who do not express such a desire.

ii. People who want to become a foster family have 1.180 times the odds of having a desire to adopt a child compared to individuals who do not want to be a foster family.

Example: Application of logit regression
We compute the estimated logit model as shown in Equation 13.18.

$$\begin{aligned}\log(\text{odds}) = &-6.460 - 0.339x_1 + 0.013x_2 + 2.228x_3 - 0.528x_4 \\ &+2.152x_5 + 0.145x_6 + 0.175x_7 + 0.086x_8 + 0.103x_9\end{aligned} \tag{13.18}$$

in which the response is taken as whether wanting to be a foster family is used in previous analyses. Then, the significant dependent variables are found in the same order stated in Equation 13.17 from x_1 to x_9, respectively. Furthermore, we conclude the following results from the estimated model.

i. Individuals who have a desire to adopt a child have approximately 2.228 times the odds of wanting to become a foster family compared to people who do not express such a desire.

ii. People who want to become foster families have approximately 2.152 times the odds of expressing a desire to adopt a child compared to individuals who do not want to be foster parents.

iii. For each one-unit increase in the variable egg bank (whether the thought of applying to an egg bank for a couple who are unable to have a child naturally is considered appropriate), the odds of wanting to become a foster family increase by approximately 0.175 times while holding all other predictors constant.

13.3.4 Tobit Model

The Tobit model is a statistical method commonly known as the censored regression model. It is used to analyze the data that has either left- or right-censoring which occurs when the dependent variable is only partially observed or observed within specific limits. In such cases, the Tobit model can be used to estimate the relationship between the predictors and the unobserved variable, thereby, offering valuable insights into the impact of predictors on the censored variable. The mathematical expression of the model is denoted as follows [10].

$$y_i^* = \beta x_i + \varepsilon_i \quad \text{while } \varepsilon_i \sim N(0,1) \tag{13.19}$$

$$y_i = \begin{cases} y_i^*, & \text{if } y_i^* > \tau, \\ 0, & \text{if } y_i^* \leq \tau, \end{cases} \tag{13.20}$$

where y_i^* means the latent variable and τ indicates the censor point.

The tobit model utilizes all available information including censored ones to give consistent estimates. The model is nonlinear and similar to the probit model in the sense that the likelihood function and the maximum likelihood estimator will be specified to estimate this model. Thus, its log-likelihood function is written as follows:

$$\log L = \sum_{y_i>0} -\frac{1}{2}\left[\log(2\pi) + \log\sigma^2 + \frac{(y_i - \beta x_i)^2}{\sigma^2}\right] + \sum_{y_i=0} \log\left[1 - F\left(\frac{\beta x_i}{\sigma}\right)\right]. \tag{13.21}$$

In order to determine the β coefficients, the unique likelihood function is maximized by computing the likelihood of both uncensored and censored values. To calculate the likelihood of observing an uncensored value, the probability density function of the normal distribution is used. In the case of a censored value, however, the cumulative distribution function of the normal distribution is applied along with the corresponding censoring bounds to calculate the likelihood.

Additionally, this model is a useful tool for understanding how predictors relate to a censored variable. Because it takes into account the impact of predictors on the latent variable, as well as the effect of censoring. Therefore, it has wide application in different fields from economics and health sciences to social sciences.

Example: Application of Tobit regression
When we model the TAYA 2016 data via the Tobit model, we get the following expression.

$$\begin{aligned} y_i^* &= -1.765 - 0.108x_1 + 0.0034x_2 + 0.881x_3 - 0.129x_4 \\ &+ 0.416x_5 + 0.044x_6 + 0.058x_7 + 0.024x_8 \\ &+ 0.033x_9 + \varepsilon_i \end{aligned} \tag{13.22}$$

in which $\varepsilon_i \sim N(0,1)$ and y_i^* is the latent variable for our response under the uncensored observations, and ε_i represents the error term for the uncensored part. Similarly, the response is whether wanting to be a foster family, and the nine dependent variables are listed in the same order as described in Equation 13.18 and 13.17 from x_1 to x_9, respectively (i.e., x_1=gender, x_2= age, ..., x_9= whether the idea that when the child grows up, she/he should provide financial support to her/his parents is agreed). Accordingly, the interpretation of some interesting coefficients can be reported as follows:

i. Individuals who want to adopt a child have approximately 0.881 units higher on the latent (uncensored) variable wanting to be a foster family compared to individuals who do not express such a desire.

ii. Individuals who think to be a foster family for a couple who are unable to have a child naturally is appropriate, have approximately 0.416 units higher on the latent (uncensored) variable wanting to be a foster family compared to individuals who do not think this way.

iii. For each one-unit increase in the variable egg bank (whether the thought of applying to egg bank for a couple who are unable to have a child naturally is considered appropriate), individuals have approximately 0.058 units higher on the latent (uncensored) variable wanting to be foster family, while holding all other predictors constant.

13.3.5 *Generalized Linear Mixed Model*

The generalized linear mixed model (GLMM) is a statistical modeling approach used in various fields, particularly, when dealing with hierarchical or longitudinal data. It extends the generalized linear Model (GLM) by incorporating random effects, which account for correlated or clustered data. Before having a better grasp of the concept of GLMM, GLM should be understood first. In GLM, the response variable is distributed from the exponential family (e.g., normal, binomial, poisson), and the relationship with the predictors is modeled through a link function. Therefore, this model is widely used in statistical analysis in order to relate the response variable to one or more predictor variables.

Thus, the standard form of GLM is given by a mathematical equation that helps to define the relationship between the response variable and the predictor variables via

$$g(E[Y]) = X\beta. \tag{13.23}$$

In this equation, $g(.)$ denotes the link function, $E[Y]$ refers to the expected value of the response variable Y, X is the matrix of predictor variables, and β describes the coefficient vector. Then, to address the issue of correlations within clusters or groups, random effects come in handy. In other words, random effects explain the variability between groups which go unnoticed by the fixed effects, resulting in the following general formula for a GLMM [6].

$$g(E[Y_{ij}]) = X_{ij}\beta + Z_{ij}b_i, i = 1, \ldots, n, j = 1, \ldots, m. \tag{13.24}$$

Equation 13.24 includes different variables such as Y_{ij} for the response variable of the jth observation in the ith group, X_{ij} for the design matrix of fixed effects predictors, β for the vector of fixed effects coefficients, Z_{ij} for the design matrix of random effects predictors, and b_i for the vector of random effects.

As seen from the above expression, in statistical modeling, within-group correlation and the dependent observations within the same group can be demonstrated by GLMM. In this approach, the random effects b_i are assumed to follow a multivariate distribution, typically, normal. Here, an accurate estimation of the variances and covariances for the random effects becomes crucial for the analysis of the model.

Hereby, to estimate the parameters of GLMMs, the likelihood function is maximized by implying the determination of the probability of the observed data given the model. However, due to the complexity of GLMMs, exact maximum likelihood estimation can be challenging. As a result, numerical optimization techniques such as the Laplace approximation, quadrature methods, or Markov chain Monte Carlo (MCMC) approaches are commonly used to estimate model parameters. These methods provide a more accurate and efficient way of inference of GLMM parameters by facilitating the analysis and interpreting the findings.

On the other hand, GLMM is highly adaptable and can work with multiple types of response variables, including continuous, binary, count, or categorical outcomes. Also, it is specifically handy when dealing with data that has a hierarchical or clustered structure, where observations are grouped within larger units, or when data is repeatedly measured over time. Therefore, it enables the examination of both within-group and between-group variability. Furthermore, GLMM includes both fixed and random effects, making it an excellent tool for modeling and analyzing complex data structures.

Example: Application of generalized linear mixed model
For this part, we describe the data by the generalized linear mixed model and infer the following mathematical expression.

$$\begin{aligned}\log\left(\frac{P_{ij}}{1-P_{ij}}\right) &= -6.460 - 0.339x_{i1} + 0.013x_{i2} + 2.228x_{i3} \\ &- 0.528x_{i4} + 2.152x_{i5} + 0.145x_{i6} + 0.175x_{i7} \\ &+ 0.086x_{i8} + 0.103x_{i9} + u_j, \end{aligned} \tag{13.25}$$

where

- $\log\left(\frac{P_{ij}}{1-P_{ij}}\right)$ represents the log-odds of individual i in group j being in the response category,
- x_{i1} indicates the random intercept and in this case, gender is taken as the grouping variable,
- u_j implies the random effect for group j, which presents the variation between groups under the assumption of the normal distribution.

In addition, the response variable is said to follow a binomial distribution with a logit link function, and P_{ij} shows the probability of individual i in group j being in the response category. Finally, the response is taken as whether wanting to be a foster family as used in previous analyses, and the same ordered dependent variables (from x_1 to x_9) are applied in the construction of the model. From the results, we list some of the interpretations of estimated coefficients as follows.

i. Individuals who express a desire to adopt a child have approximately 2.228 units higher odds of being in the response category, compared to individuals who do not express such a desire while controlling other variables.

ii. Individuals who want to be foster family have approximately 2.152 units higher odds of being in the response category compared to individuals who do not want to be foster family while controlling other variables.

iii. For each one-unit increase in the variable egg bank (whether the thought of applying to an egg bank for a couple who are unable to have a child naturally is considered appropriate), individuals have approximately 0.175 units higher odds of being in the response category, while controlling other predictors.

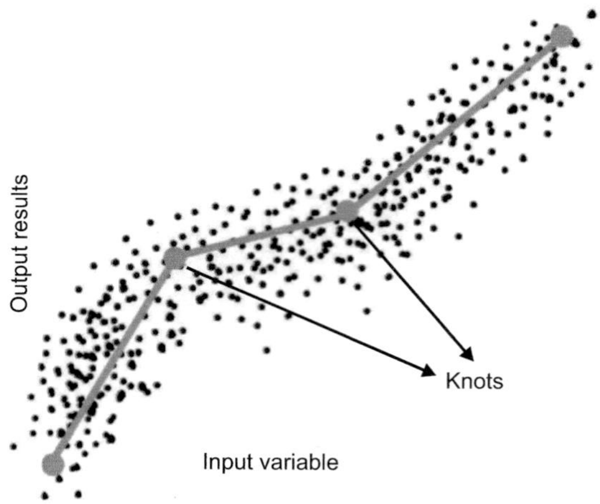

Figure 13.2: Fitting the model of multivariate adaptive regression splines for a data.

13.4 Spline-Based Methods

13.4.1 Multivariate Adaptive Regression Splines

Multivariate adaptive regression splines (MARS) is a nonparametric regression approach suitable for high-dimensional and correlated data with nonlinearity. So, it generates a piecewise linear model for non-linear relationships in data in such a way that the linear pieces will be joined by Hinge functions as below.

$$h(x-c) = \max\{0, x-c\} = \begin{cases} x-c & \text{if } x > c \\ 0 & \text{if } x \leq c \end{cases} \tag{13.26}$$

in which c denotes the constant, also known as a knot. MARS constructs a hierarchical model by using a set of basis functions and a stepwise selection without the assumption between dependent and independent variables [2].

Simply, the model consists of a forward stage where an intercept term is added iteratively with basis functions automatically selected from observed data and usually builds an overfitted model. So, to prevent overfitting and reduce complexity, it consists of a backward stage by removing basis functions causing the smallest increase in residual squared error. The backward pass prunes the model for a better generalization ability and does this by removing terms one by one until the optimal sub-model is acquired. Here, the comparison of the model subset is made with the generalized cross-validation (GCV) score whose expression is denoted in Equation 13.27. The lower GCV is seen as better performance.

$$GCV = \frac{RSS}{(N(1-v)/N)^2}, \tag{13.27}$$

where RSS refers to the residual sum of squares which is measured on training data, N displays the total number of observations and the v shows the effective or optimal number of parameters. Hence, the optimal number of model parameters is computed as below.

$$v = T + \frac{\lambda \cdot (T-1)}{2} \tag{13.28}$$

in which T presents the number of MARS terms and the penalty λ is usually set to 2 or 3.

Example: Application of multivariate adaptive regression splines model
In the implementation of MARS to the TAYA 2016 foster family data set, we obtain the following estimated model.

$$\begin{aligned} Y &= -7.151 - 0.333x_1 + 3.791x_2 + 2.629x_3 - 0.612x_2x_4 \\ &- 0.930x_2 * x_3 + 0.587x_4 * x_3 + 0.213x_3 * x_5 - 0.005h(56 - x_6) * x_2 \\ &+ 0.015h(x_6 - 56) * x_2 + 0.00004h(5500 - x_7) * x_2 \\ &+ 0.00002h(x_7 - 5500) * x_2 - 0.208x_2 * h(2 - x_8). \end{aligned} \tag{13.29}$$

From Equation 13.29, we infer three knots, i.e., constants, in the hinge function associated to the age variable (x_6) at $c = 56$, the average net income variable (x_7) at $c = 5500$, and the continuation of generation variable (x_8) at $c = 2$.

Additionally, in the description of variables, we denote the response as whether the household wants to be a foster family. Whereas, different from previous models, we allocate the dependent variables in the following order: x_1 is the gender, x_2 is the desire to adopt a child, x_3 denotes whether the thought of being a foster family for a couple who are unable to have a child naturally is considered appropriate, x_4 means whether the thought of adopting for a couple who are unable to have a child naturally is considered appropriate, x_5 presents whether the thought of applying to egg bank for a couple who are unable to have a child naturally is considered appropriate, x_6 implies the age, x_7 represents the average net income and lastly, x_8 shows whether the idea that continuation of generation (lineage) can only be provided by the son is agreed. Thereby, the interpretation of some estimated coefficients can be done as below.

i. The coefficient for the desire to adopt a child is 3.791 units. Since this coefficient is significant at the 0.01 level (assuming that we have not provided the p-values explicitly), it suggests that the variable is positively associated with the log odds of being in the response category.

ii. The coefficient which denotes whether the thought of being a foster family for a couple who are unable to have a child naturally is considered appropriate is 2.629 units. Similar to the desire to adopt a child variable, this coefficient is also significant at the 0.01 level, indicating that individuals who deem being a foster family as appropriate are more likely to be in the response category.

13.4.2 Conic Multivariate Adaptive Regression Splines

The conic multivariate adaptive regression splines (CMARS) are the extended version of the MARS approach in such a way that it can combine MARS and conic optimization to handle complex, high-dimensional data. By using conic optimization, CMARS can impose more constraints on the model parameters, resulting in simpler and more interpretable models while also promoting feature selection and sparsity.

CMARS differentiates from MARS at the backward step. As it was mentioned before, MARS uses GCV to compare model subsets at a backward step. However, CMARS eliminates this backward stepwise selection algorithm, merely uses the forward step of MARS, and applies the Tikhonov regularization. For this method, penalized residual sum of squares (PRSS) is applied as ridge regression by calibrating accuracy and complexity while detecting the optimal model. Here, the accuracy refers to the summation of error squares being ideally minimum. On the other hand, the complexity implies the squared of the first and the second derivatives of basis functions. This exchange is a trade-off that is detectable by constructing penalty parameters by conic quadratic programming. Accordingly, PRSS within CMARS is computed as below [9].

$$PRSS = \sum_{i=1}^{N} (y_i - f(\tilde{x}_i))^2 + \sum_{m=1}^{M_{\max}} \lambda_m \sum_{\substack{|\alpha|=1 \\ \alpha=\alpha_1,\alpha_2}}^{2} \sum_{\substack{r<s \\ r,s\in V(m)}} \int_{Q^m} \theta_m^2 [D_{r,s}^{\alpha} \Psi_m(t^m)]^2 dt^m, \tag{13.30}$$

where, $V(m)$ describes the variable set that is linked to h_m that is the basis function with the index m. Moreover, the vector t^m represents the variables that contribute to the mth basis function and helps identify the integration of variables for the complexity term. Additionally, Q^m refers to a parallel pipe that is large enough to contain the input data and where the integration occurs. Hence, the derivative $D_{r,s}^{\alpha} h_m(t^m)$ can be defined as

$$D_{r,s}^{\alpha} h_m(t^m) = \frac{\partial^{\alpha} h_m}{\partial_1^{\alpha} t_r^m \partial_2^{\alpha} t_s^m}(t^m) \tag{13.31}$$

for $\alpha = (\alpha_1, \alpha_2)^T$ where $\alpha_1, \alpha_2 \in \{0, 1\}$. By using PRSS, a balanced approach can be taken to control both the complexity and accuracy of the model and this is the basis of the conic quadratic programming (CQP) for the parameter estimation, as stated beforehand. Thereby, reorganizing the PRSS equation gives

$$PRSS \approx \left\| y - h(\overline{d})\theta \right\|_2^2 + \lambda \|L\theta\|_2^2. \tag{13.32}$$

For Equation 13.32, the L-curve graph can be helpful in choosing the penalty parameter λ by showing the balance between minimizing basis function sizes and applying penalty parameters. This graph is plotted on a double logarithmic scale in such a way that the corner can represent the best balance.

Example: Application of conic multivariate adaptive regression splines model

In modeling the factors of a foster family in the TAYA 2016 data set via CMARS, we get a similar modeling structure presented in Equation 13.29 with a change in the value of the regression coefficients. Below, we give some examples of the interpretations of estimated CMARS parameters as examples.

i. The intercept is estimated as 0.216 units which represents the estimated log-odds of being in the response category when all predictor variables are zero.

ii. For each unit increase in the desire to adopt a child, the log-odds of being in the response category increase by 0.427 units, holding all other predictors constant.

iii. For each unit increase in whether the thought of being a foster family for a couple who are unable to have a child naturally is considered appropriate, the log-odds of being in the response category increase by 0.559 units, holding all other predictors constant.

iv. For each unit increase in age, the log-odds of being in the response category increase by 0.004 units, holding all other predictors constant.

v. For each unit increase in whether the idea that continuation of generation (lineage) can only be provided by the son is agreed, the log-odds of being in the response category increase by 0.004 units, holding all other predictors constant.

vi. All interaction terms represent the combined effects of the predictor variables.

13.5 Machine Learning Methods

13.5.1 Classification and Regression Trees

A decision tree algorithm, or with its widely known name classification and regression tree approach, shortly CART, is one the fundamental machine learning algorithms that are based on building a decision tree via the Gini's impurity index. Indeed, CART is an umbrella term and refers to two types of decision trees. If the dependent variable is continuous, the classification tree determines the *class* where the dependent variable most probably belongs to, and the continuous variables are predicted by using the regression trees. Accordingly, the algorithm uses observations as branches leading to the target value in the leaves.

In general, a decision tree presents a decision support technique with a tree-like structure having three components, namely, decision nodes, leaf nodes, and

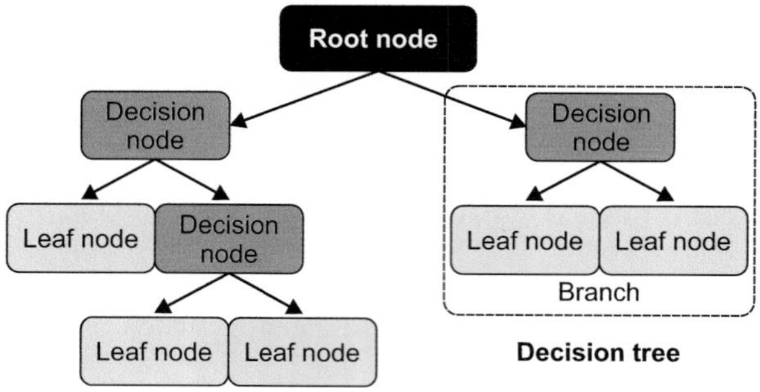

Figure 13.3: An illustration of decision tree scheme.

root nodes in such a way that the observations can be split according to their similarities [5]. In a typical branching process, the standard classification trees split each node based on the best groups among all variables by using measurements such as generalization error, strength, or correlation. However, in CART, this calculation is also influenced by the measurement noise, which can lead to overfitting in the findings. Hereby, in order to control the underlying overfitting, CART divides data into decision trees via the Gini impurity technique, which identifies the most homogeneous sub-nodes based on the Gini index criterion. The Gini index is a well-established measure that provides a quantitative estimation of the likelihood of misclassification of a variable when the variable is chosen at random. This method operates solely on categorical variables and produces binary splitting outcomes, denoted as either success or failure. The mathematical form of the Gini impurity can be shown below.

$$\text{Gini} = 1 - \sum_{i=1}^{n} (pi)^2, \tag{13.33}$$

where the variable pi represents the likelihood of an item being categorized into a specific group.

Lastly, from previous analyses, it has been observed that it is robust to outlier observations and can successfully handle high dependency and nonlinearity in the data.

Example: Application of classification and regression tree algorithm

In our analyses, we use the variable which denotes the *desire to adopt a child* (x_1) as the predictor in the construction of the classification tree and the variable which means the *thought of being a foster family for a couple who are unable to have a child naturally is considered appropriate* (x_2), as the response variable. Then, the tree splits the data into different regions based on these two predictors and assigns class labels to each region. Later, the tree performance is assessed by

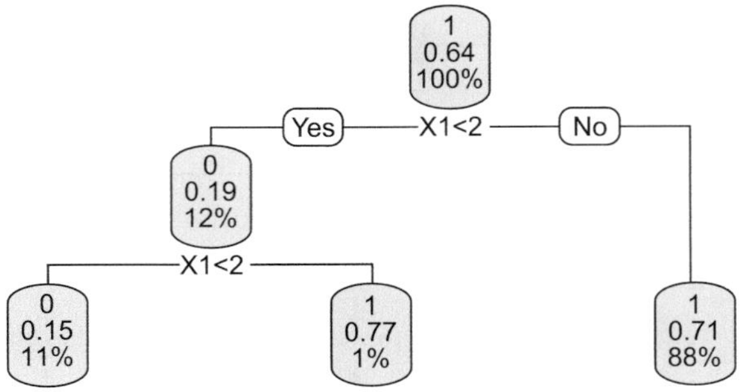

Figure 13.4: The classification and regression tree plot of the selected data set.

using cross-validation, and the optimal tree size is determined by applying the complexity parameter.

When we have a dataset with a binary response, each node has values showing the predicted class and probability, as well as the percentage of observations in the node. In Figure 13.4, if the variable wanting to adopt is not smaller than 2 (for our data it means that it is 2, representing not wanting to adopt,) it is possible to say that, for the people who do not want to adopt, the expected desire to be a foster family is 0.71, and 88% of samples will fall into that leaf. For the people who want to adopt, the expected desire to be a foster family is 0.19, and 12% of samples will fall into that leaf. Similarly, for the people who want to adopt and consider at the same time people who to be a foster family if the couple is unable to have a child naturally as appropriate, the expected desire to be a foster family is 0.15, and 11% of samples will fall into that leaf. Lastly, for the people who want to adopt and at the same time people who consider the thought to be a foster family if the couple is unable to have a child naturally as appropriate, the expected desire to be a foster family is 0.77, and 1% of samples will fall into that leaf. One should also notice that the total percentage of the last leaf nodes roughly adds up to 100%.

13.5.2 Random Forest Algorithm

In the field of machine learning, there are two main categories, namely, supervised learning, which includes classification and regression, and unsupervised learning, which covers clustering. Basically, classification is the process of assigning observations to the correct subgroups based on known elements and one of the popular unsupervised decision-tree methods is the random forest algorithm, which utilizes a technique, called bagging. Bagging constructs decision trees independently by selecting a bootstrap sample of the data set in such a way that each tree makes a prediction with independently trained models, and the final output is based on the majority of votes. Hereby, the random forest selects

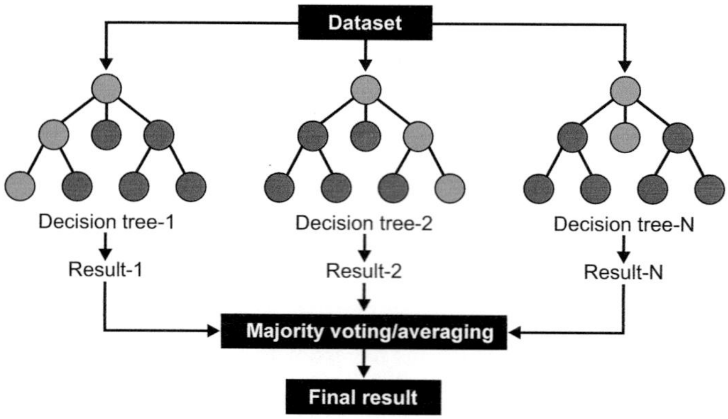

Figure 13.5: An illustration of the random forest scheme.

the most voted class from multiple classes with multiple trees [3]. But, unlike standard decision trees, the predictors in the random forest are distributed among small subsets, and the best one is used to split each node randomly. The splitting rule implemented in the calculation is based on the type of created tree. For classification trees, which include binary classes in the response variable, the most preferred rule is the Gini rule, which is efficient for obtaining classification trees quickly and is also used in the CART approach as well. A basic mathematical explanation of this process is presented below similar to Equation 13.33.

$$i(t) = 1 - \sum_{k=1}^{K} p^2(k|t), \tag{13.34}$$

where k is the class in the interval $[1, K]$ and $p(k|t)$ refers to the conditional probability of k of node t.

Hereby, the concept of a random forest involves developing a group of distinct decision trees and combining their predictions through averaging or voting to generate a final prediction. Different from CART, this approach is less likely to overfit and can estimate the importance of variables for feature selection. Additionally, it can capture complex interactions and non-linear relationships in the data successfully and is robust to outliers similar to CART, resulting in wide application in distinct data types.

Example: Application of random forest algorithm

Random forest can be used as an ensemble learning method for both classification and regression tasks. In our analyses, we perform it for regression by taking the response variable as a continuous numeric variable. Therefore, we set the number of trees in the random forest to 500. Indeed, we have applied different numbers of trees too, whereas, since the findings have not changed significantly in higher numbers, we have equated it to 500 throughout the analyses.

Moreover, the algorithm generates multiple decision trees and combines their predictions to make more accurate predictions. So, at each split in a decision tree, the algorithm randomly selects a subset of predictor variables to consider for the split. Our model tries 8 variables at each split and the mean of the squared residuals of the model's predictions is found as 0.192.

In addition, residuals are defined as the differences between the actual values and the predicted values. In the regression via the random forest, we use the squared residuals to measure the model's goodness of fit. Accordingly, the percentage of variance in the response variable that is explained by our model is computed as 16.25%. Here, a higher percentage is interpreted as the indicator of a better fit and the calculated low percentage of variance implies that this model may not be the best model for the foster family data set.

13.5.3 Artificial Neural Network

A neural network is an intricate algorithm that simulates the functioning of the human brain while generating a structure to mimic the way neurons connect and transmit signals when activated as simply seen in Figure 13.6. Here, the layers function as axons and dendrites.

Shortly, the neural network is comprised of several node layers, including the input layer, one or more hidden layers, and the output layer [8]. The inner layers are referred to *hidden* layers because of the fact that they do not interact directly with the input or output data. Indeed, each node, also known as an artificial neuron, has a weight and threshold associated with it and when a node is activated, it will send data to the next layer if the output surpasses the threshold value. Once the data has passed through all the nodes, the output is generated. Furthermore, the learning process proceeds through the backpropagation algorithm, and the weights are adjusted accordingly.

All neural network models have similar components whose mathematical representation can be summarized into the following two equations for the neuron k.

$$u_k = \sum_{j=1}^{m} w_{kj}x_j, \tag{13.35}$$

$$y_k = f(u_k + b_k), \tag{13.36}$$

where x indicates an input signal and w describes its weight. Accordingly, the summation function in the above expression is the weighted sum of all input signals. b_k denotes the bias and y_k shows the neuron output signal which is the fired weighted sum with bias.

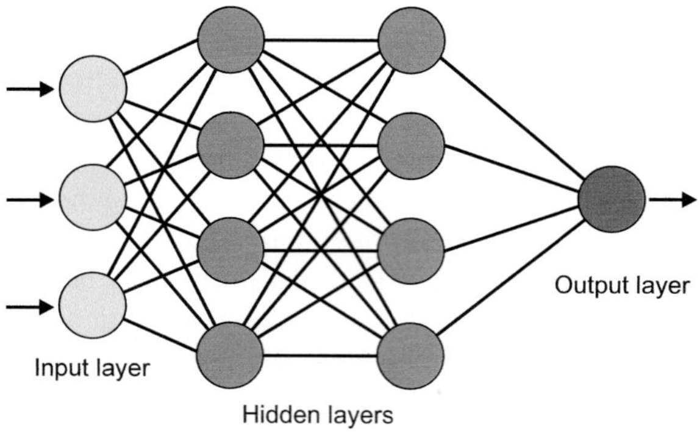

Figure 13.6: A simple architecture of the artificial neural network.

In an artificial neural network (ANN) model, the activation function plays an important role in determining whether the output neuron should be activated or not. This is done by setting conditions based on the weighted sum of the inputs with a bias term. It is possible to say that, basically, it acts as a *filter* that allows for the learning and understanding of complex, nonlinear relationships between input and target variables. When the weighted sum with the bias exceeds a certain threshold, the neuron will fire, indicating a positive output. This allows for the ANN model to make accurate predictions and decisions based on the inputs provided.

The sigmoid function, also known as the binary sigmoid or logistic sigmoid function, is frequently used as an activation function in creating ANN models. The mathematical equation for the sigmoid function can be represented below.

$$f\left(\sum_{j=1}^{m} w_{kj}x_j\right) = \frac{1}{1+e^{-x}}. \tag{13.37}$$

On the other hand, if we delve into the world of neural networks, we come across two fundamental types that exhibit differences in how neurons interconnect and how data is transmitted. These two types are commonly known as feed-forward and recurrent networks. In the feed-forward neural networks, the network begins with the input and progresses towards the output. When data is transmitted through networks, it follows a one-way flow of direction. The inputs do not receive any feedback from the outputs, thus, creating an indirect data transfer. So, the fundamental difference between the feed-forward neural network and the recurrent neural network lies in the latter's inclusion of a *feedback loop*. This means that each neuron's output is used as input for all the other neurons, allowing the output to circulate within the network.

Example: Application of neural network model

In this part, we perform neural network regression to the reduced dimensional data and check the most influential variables with respect to their weights. From the analyses, we observe that age (x_2), animal (x_9), house (x_7), and car (x_8) ownership, whether a family who cannot have a child naturally could be foster parents (x_{13}), or adopt a relative's child as their own (x_{14}), or adopt a child from a child welfare agency (x_{12}), having (x_{10}) or wanting (x_{11}) an adopted child, whether having the thought of having children giving the mother social credit (x_{19}) or strengthening the bonds between spouses (x_{21}) are the variables having the biggest absolute weights. Here, we define the value 10 as the threshold of weights for the selection of the most important variables. The ones with the minus sign (as observed for the variables *having an adopted child, whether a family who cannot have a child naturally could adopt a relative's child as their own, or adopt a child from a child welfare agency, thinking that having children giving the mother social credit*) negatively affect the response (Y) while the others affect positively.

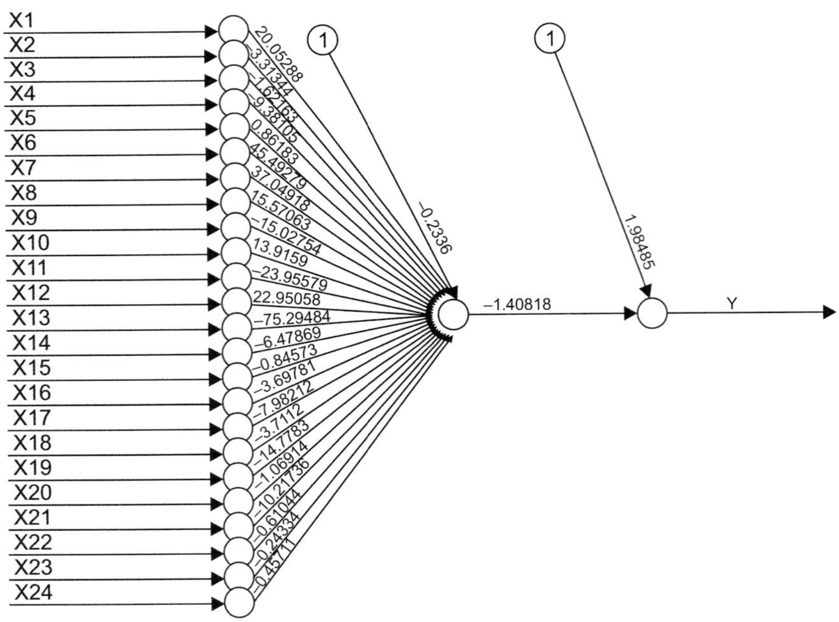

Figure 13.7: The graphical view of neural network plot of the selected foster family data set.

Table 13.1: Significant questions, their options, and calculated weights found from the analyses of being foster family as a neural network model for the full dataset.

Questions	Weights
The age you finished.	20.053
Do you own cattle/ovine animals in the last year?	15.571
Do you have adopted children?	-15.028
Do you want to adopt a child?	13.916
Please state if you agree or disagree with the following statement about children. A woman who has children is more respected than one who does not.	-14.778
Please state if you agree or disagree with the following statement about children. Having a child brings the spouses closer together.	10.217
Do you have any residence(s), other than the residence of your own household in the last year?	45.493
Do you have any automobile, jeep, etc. of your own (transport vehicles for private use) in the last year?	37.049
Do you think a couple who are unable to have children naturally, can/should be foster parents?	22.951
Do you think a couple who is unable to have children naturally, can/should adopt?	-23.956
Do you think a couple who is unable to have children naturally, can/should pass on the child of one of their relatives?	-75.295

13.6 Comparison of Models

13.6.1 *Model Comparison Criteria*

13.6.1.1 Akaike's Information Criteria (AIC)

AIC is one of the most common goodness-of-fit indices. It is an accuracy-based model selection technique and uses the log-likelihood of a model, in other words, the maximum likelihood estimation of it, as a measure of fit. When evaluating statistical models based on a specific set of data, AIC can be used to estimate the prediction errors and compare their quality. It's worth noting that a lower AIC indicates a higher level of accuracy in predicting outcomes using the given data. As a higher number of parameters increases the chance for a model to overfit the training data, AIC adds a penalty term for models with higher parameter complexity. The mathematical expression of AIC is shown as below.

$$\text{AIC} = -2\ln(L) + 2k \tag{13.38}$$

where L is the likelihood and k denotes the number of parameters.

13.6.1.2 Bayesian Information Criteria (BIC)

Both AIC and BIC penalize the model by adding parameters and they have a different method for this. Comparing the two criteria, it is safe to say that BIC penalizes more strongly than the AIC. It includes additional variables to the model.

Table 13.2: AIC and BIC values of the parametric additive models for summary dataset.

Model Name	AIC	BIC
Logistic Regression	14034.00	14220.46
Poisson Regression	22675.00	22861.48
Probit Model	14036.00	14222.12
Logit Model	14034.00	14220.46
Tobit Model	23630.00	23829.80
Generalized Linear Mixed Model	14036.20	14229.90

Like AIC, models with lower BIC values are preferred as presented in Equation 13.39.

$$\text{BIC} = \ln(n)k - 2\ln(\hat{L}) \quad (13.39)$$

where $\hat{L}$ is the maximized value of the likelihood function, n is the number of data points, and k denotes the number of free parameters.

13.6.1.3 Mean Squared Error (MSE)

Statistical model error is measured by MSE which calculates the average squared difference between predicted and observed values. The square is taken solely because it removes any negative values. It is aimed a lower MSE value for a better fit as given below.

$$\text{MSE} = \frac{\sum(y_i - \hat{y}_i)^2}{n} \quad (13.40)$$

where y_i is the ith observed value, $\hat{y}_i$ presents the corresponding predicted value, and n refers to the number of observations.

13.6.2 Comparing the Models

In this section, the models that were constructed for the summary dataset will be compared using known model comparison methods.

According to Table 13.2, for both AIC and BIC, the best models are seen as the logistic regression and the logit model. For the summary data, these two models indicate the same significance with the same regression coefficients. The reason is that the logistic regression is the generalized version of the categorical data, whereas, the logit model is used for specifically binary data sets. Moreover, both models assume the normality of the variables. Hereby, in our summary data, since we implement merely the binary variables, two models share the same outputs. On the other hand, as seen from Table 13.2, the probit model is the second best model based on our selection criteria. As previously mentioned, the major difference between logit and probit models comes from their distributional differences, where the former uses the normal density, whereas, the latter applies

Table 13.3: MSE values of spline-based and machine learning methods for summary dataset.

Model Name	MSE
Multivariate Adaptive Regression Splines	0.185
Conic Multivariate Adaptive Regression Splines	0.186
Classification and Regression Trees	0.301
Random Forest Algorithm	0.209
Artificial Neural Network	0.229

the logistic distribution. The logistic distribution can be defined as similar to normal density in shape apart from its heavier tails, causing high kurtosis.

Lastly, the other two models, poisson regression and tobit model are not very good at modeling our dataset because of the fact that these are specific models generated for count and censored observations.

According to Table 13.3, the best model to explain our data set is found as the multivariate adaptive regression splines, closely followed by its extended version, conic multivariate adaptive regression splines. Based on the MSE values in this category, it is possible to say that spline-based methods can successfully describe our data set. Furthermore, machine learning methods, except classification and regression trees, are also successful in modeling of the summary set. Overall, it is possible to say that these models are suitable to use as modeling tools for these types of data.

13.7 Conclusion

Looking at all the models, we have seen that some variables used in TAYA foster family data set are repeated in the majority of models. Gender, age, desire to adopt a child, whether the thought of adopting a child, being a foster family, divorce and remarriage with other people, and applying to an egg bank for a couple who are unable to have a child naturally is considered appropriate, whether the idea of continuation of generation (lineage) can only be provided by the son is agreed, whether the idea that when the child grows up, she/he should provide financial support to her/his parent is agreed is observed as the common significant factors to request a foster family.

Furthermore, we have found that the generalized additive models are suitable models for these kinds of categorical variable-heavy data sets because of the fact that they can give robust and consistent results among alternative models. Moreover, as it is possible to apply machine learning models for classification and regression tasks, we have implemented well-known approaches in this group for our analyses. Additionally, it has been found that the spline-based models give surprisingly good results as they are mostly used for modeling complex nonlinear curves and detecting the optimal solutions.

On conclusion, we have presented various modeling and dimension reduction approaches in comprehensive analyses of social data. Although we have used them in a specific TAYA 2016 data set, we consider that they can be adapted to other social studies as their data structures are similar. Finally, we think that such types of detailed analyses can be helpful for policymakers too while developing social strategies once effective factors are determined via the underlying approaches.

Acknowledgement

Both authors would like to thank to Turkish Statistical Institute for their permission to use the TAYA survey. Moreover, Vilda Purutçuoğlu would like to thank Huawei Turkey R & D Center for their support for modeling big data sets.

References

[1] Coxe, S., West, S.G. and Aiken, L.S. 2009. The analysis of count data: A gentle introduction to poisson regression and its alternatives. *Journal of Personality Assessment*, 91(2): 121–136, February 2009.

[2] Friedman, J.H. 1991. Multivariate adaptive regression splines. *The Annals of Statistics*, 19(1): 1–67.

[3] Genuer, R., Poggi, J.-M., Tuleau-Malot, C. and Villa-Vialaneix, N. 2017. Random forests for big data. *Big Data Research*, 9: 28–46, September 2017.

[4] Lynch, S.M. 2013. *Using Statistics in Social Research*. Springer New York.

[5] Ma, X. 2018. Using Classification and Regression Trees: A Practical Primer. *Information Age Publishing Incorporated*, 2018.

[6] Pan, Z. and Lin, D.Y. 2005. Goodness-of-fit methods for generalized linear mixed models. *Biometrics*, 61(4): 1000–1009, May 2005.

[7] Park, H.-A. 2013. An introduction to logistic regression: From basic concepts to interpretation with particular attention to nursing domain. *Journal of Korean Academy of Nursing*, 43(2): 154–164.

[8] Shahid, N., Rappon, T. and Berta, W. 2019. Applications of artificial neural networks in health care organizational decision-making: A scoping review. *PLOS ONE*, 14(2): e0212356, February 2019.

[9] Weber, G.-W., Batmaz, I., Köksal, G., Taylan, P. and Yerlikaya-Özkurt, F. 2011. CMARS: A new contribution to nonparametric regression with multivariate adaptive regression splines supported by continuous optimization. *Inverse Problems in Science and Engineering*, 20(3): 371–400, October 2011.

[10] Güneri, O.I. and Durmuş, B. 2020. Dependent dummy variable models: An application of logit, probit and tobit models on survey data. *International Journal of Computational and Experimental Science and Engineering*, 6(1): 63–74.

Index